T0295274

Biogenic Amines in Food: Analysis, Occurrence and Toxicity

Biogenic Amines in Food: Analysis, Occurrence and Toxicity

Editor: Freddy Lawrence

www.callistoreference.com

Callisto Reference,
118-35 Queens Blvd., Suite 400,
Forest Hills, NY 11375, USA

Visit us on the World Wide Web at:
www.callistoreference.com

ISBN: 978-1-64116-838-0 (Hardback)

Cataloging-in-Publication Data

Biogenic amines in food : analysis, occurrence and toxicity / edited by Freddy Lawrence.
    p. cm.
Includes bibliographical references and index.
ISBN 978-1-64116-838-0
1. Biogenic amines. 2. Food--Safety measures. 3. Food--Analysis. 4. Biogenic amines--Health aspects.
I. Lawrence, Freddy.
QP801.B66 B56 2023
547.042--dc23

Table of Contents

Preface

This book aims to highlight the current researches and provides a platform to further the scope of innovations in this area. This book is a product of the combined efforts of many researchers and scientists, after going through thorough studies and analysis from different parts of the world. The objective of this book is to provide the readers with the latest information of the field.

Biogenic amines (BAs) are organic and nitrogenous compounds found in certain fermented foods, such as sausage, wine, cheese, fish and fermented vegetables. They are involved in controlling various functions of the brain such as emotions, blood pressure, movements, endocrine secretion, behavior and temperature. Bacterial decarboxylation of the amino acids in food leads to the production of BAs. The consumption of large amounts of BAs in food can have toxic effects and may also lead to gastrointestinal, neurological, respiratory and cardiovascular issues, as well as hypertensive, psychotropic and vasoactive effects. The regulation of decarboxylase activity of amino acids can help in controlling the concentrations of BAs in food items. Their levels can be decreased through various techniques including hydrostatic pressure, smoking, temperature, packing, irradiation, starter culture, oxidizing formed biogenic amine, additives, and pasteurization. This book explores all the important aspects of biogenic amines, including their occurrence and toxic effects in the present day scenario. It will also provide interesting topics for research, which interested readers can take up. This book is a resource guide for experts as well as students.

I would like to express my sincere thanks to the authors for their dedicated efforts in the completion of this book. I acknowledge the efforts of the publisher for providing constant support. Lastly, I would like to thank my family for their support in all academic endeavors.

Editor

Amines in Fish and Fish Products

Pierina Visciano ⓘ, Maria Schirone *ⓘ and Antonello Paparella ⓘ

Faculty of Bioscience and Technology for Food, Agriculture and Environment, University of Teramo,
Via R. Balzarini, 1, 64100 Teramo, Italy; pvisciano@unite.it (P.V.); apaparella@unite.it (A.P.)
* Correspondence: mschirone@unite.it

Abstract: The occurrence of biogenic amines in fish is directly associated with microorganisms with decarboxylase activity. These compounds are generally detoxified by oxidases in the intestinal tract of humans, but some conditions, such as alcohol consumption, enzyme deficiency, or monoamino-oxidase antidepressant use, can make their intake by food dangerous. Due to its toxicity, histamine is the unique biogenic amine with regulatory limits for fishery products. This review focuses on biogenic amines in fish, with a detailed picture of the number of alert notifications or intoxication events reported in the last years. The favoring conditions for their formation, as well as the main preventive and control measures to ensure public health, are also reviewed.

Keywords: biogenic amines; histamine food poisoning; outbreaks; seafood safety

1. Introduction

Biogenic amines (BAs) are nitrogenous compounds resulting from the free amino acid decarboxylation or the amination of carbonyl-containing organic compounds through the metabolism of different microorganisms. Thus, their accumulation in food can be considered a good indicator of spoilage [1]. BAs are distinguished based on chemical structure into heterocyclic, aliphatic, or aromatic compounds (Table 1), or based on the number of amine groups into monoamines (tyramine and phenylethylamine), diamines (histamine, putrescine, and cadaverine), or polyamines (spermidine and spermine) [2].

Table 1. Classification of biogenic amines based on chemical structure.

Precursors	Biogenic Amines		
	Aliphatic	Aromatic	Heterocyclic
Arginine	Agmatine		
Lysine	Cadaverine		
Ornithine	Putrescine		
Phenylalanine		Phenylethylamine	
Tyrosine		Tyramine	
Histidine			Histamine
Tryptophan			Tryptamine

Fish products constitute an important part of the human diet because they are an excellent source of nutrients, including proteins, vitamins, salt minerals, and polyunsaturated fatty acids [3]. Nevertheless, they are very perishable due to postmortem modifications followed by the formation of spoilage compounds, such as organic acids, aldehydes and ketones, alcohols, sulfides, and BAs [4].

The inappropriate storage of fish and/or temperature abuse can lead to BA formation due to microbial enzymatic activities.

Gram-positive and -negative bacteria associated with fish spoilage can produce BAs. They are generally located on skin, gills, or in the gastrointestinal tract [5–11] (Table 2), and can spread to muscle tissue during butchering or gutting through rupture or spillage of gastric contents [12]. The most frequent species belong to *Enterobacteriaceae* and include mesophilic and psychrotolerant bacteria, such as *Morganella*, *Enterobacter*, *Hafnia*, *Proteus*, and *Photobacterium* [13]. Also, *Pseudomonas* spp. and lactic acid bacteria belonging to *Lactobacillus* and *Enterococcus* genera can cause BA formation [14].

Table 2. Microorganisms producing biogenic amines in seafood.

Biogenic Amine	Microorganisms	References
Histamine	*Morganella morganii, Morganella psychrotolerans, Hafnia alvei, Photobacterium phosphoreum, Photobacterium psychrotolerans, Klebsiella pneumoniae, Clostridium* spp., *Pseudomonas fluorescens, Pseudomonas putida, Pseudomonas cepaciae, Aeromonas* spp., *Aeromonas hydrophila, Acinetobacter lowffi, Plesiomonas shigelloides, Proteus vulgaris, Proteus mirabilis, Serratia fonticola, Serratia liquefaciens, Enterobacter cloacae, Enterobacter aerogenes, Klebsiella oxytoca, Citrobacter freundii, Raoultella planticola, Staphylococcus xylosus, Staphylococcus epidermidis, Bacillus* spp., *Vibrio alginolyticus, Vibrio* spp., *Escherichia* spp.	Hungerforf, 2010 [5] Biji et al., 2016 [2] Doeun et al., 2017 [6] Barbieri et al., 2019 [7] Xu et al., 2020 [8]
Tiramine	Lactic acid bacteria (including *lactobacilli, lactococci, enterococci* and *carnobacteria*)	Marcobal et al., 2012 [9]
Putrescine	*Enterobacter* spp., *Hafnia alvei, Pantoea agglomerans, Serratia liquefaciens, Photobacterium phosphoreum, Aeromonas* spp., *Lactobacillus curvatus, Lactobacillus sakei, Carnobacterium divergens*	Wunderlichová et al., 2014 [10]
Cadaverine	Pseudomonads, *Enterobacteriaceae*	Paleologos et al., 2004 [11] Kuley et al., 2017 [4]

The amounts of BAs ingested by food are regulated in the human organism by a detoxification system formed mainly by mono and diamino oxidases. High intake of BAs or the presence of factors such as the use of alcohol or medication that reduce the effectiveness of such detoxifying enzymes, rather than a genetic deficiency, can lead to intoxication with different symptoms depending on the type of BA. Histamine is the most toxic, as it can act as neurotransmitter and vasodilator, causing headache, hypotension, heart palpitations, asthma attacks, and cutaneous (edema and flushing of the face, neck, and upper arms) or gastrointestinal (difficulties in swallowing, vomiting, and diarrhea) effects [7]. Bronchospasm, respiratory distress, and vasodilatory shock are also described [15]. A detailed summary of histamine intoxication is reported in Table 3. Instead, tyramine may increase the cardiac frequency or cause respiratory disorders, but also nausea and vomiting, while phenylethylamine can be a migraine inductor. Further symptoms associated with tyramine and phenylethylamine are hypertension and cerebral hemorrhage. Even if putrescine and cadaverine are not toxic, they can potentiate the adverse effects of the above cited BAs, as they favor their adsorption or interfere with the detoxification system [16,17].

In the literature, several reviews are reported on BAs in fish. Some authors [2,6,18] described in detail the content of histidine in some fish species, the decarboxylation reactions, some indexes of quality, and toxicological effects of many BAs, not only histamine, alongside the presence of these compounds in various foods. Instead, the focus of this review is the statement of outbreaks reported in the Member States of European Union (EU) and the notifications of the Rapid Alert System of Food and Feed (RASFF). In addition, factors affecting BA formation are also discussed.

Table 3. Symptoms of scombroid poisoning.

Apparatus	Symptoms
Integumentary	Face, neck, and upper arm flushing, itchy rash, hives, localized swelling, redness, urticaria, pruritus
Cardiovascular	Hypotension with distributive shock, cardiac arrhythmias, myocardial disfunction, acute pulmonary edema, oral numbness, tingling
Gastrointestinal	Abdominal pain, stomach cramps, nausea, vomiting, diarrhea
Neurological	Throbbing headache, migraines, loss of sight, dizziness, faintness, anxiety, tremor
Respiratory	Asthma attacks, respiratory distress, rhinitis, bronchoconstriction, dyspnea
Other	Metallic or peppery taste, oral numbness, difficulties in swallowing and thirst, feeling of warmth around the mouth

2. Biogenic Amines in Seafood: Focus on Scombroid Poisoning Outbreaks

Histamine, tyramine, putrescine, and cadaverine are the most common BAs found in fish and derive from decarboxylation of corresponding free amino acids by microorganisms [18]. Their accumulation depends on the presence of the precursor amino acids, the growth or activity of decarboxylating bacteria, and a favorable environment. The main factors that influence the microbiota and its enzymatic activity are temperature, pH, water activity, oxygen availability, concentration of NaCl, some additives, and competition among microorganisms. The combination of these factors can be responsible of the variability of BA content within the same batch, and also within individual fish [12,19].

In the flesh of fish, BAs can be considered an indicator of good handling and storage procedures, and some authors reported different chemical indices for fish quality as combinations of amines [20]. However, the levels of different BAs are associated with the predominant muscle type in fish. Fish with dark muscles have more histidine content compared to those with white muscles, and therefore they accumulate more histamine when kept under elevated temperatures. On the contrary, fish belonging to the second category (white muscles) can show high cadaverine and putrescine concentrations due to poor handling, but also temperature abuse. For cephalopods, agmatine is used as a quality indicator, while in the case of crustaceans, shrimp, and lobsters, putrescine and cadaverine are considered [21]. Individual BA or a combination of various amines are considered as a quality index of fish freshness. Some authors reported that a histamine content lower than a value of 10 mg/kg denotes fish of good quality, while concentrations between 30 and 50 mg/kg represent important and definite deterioration, respectively [22]. Instead, the consumption of meals with histamine concentrations of 8–40 mg can cause only slight intoxication, while values of 40–100 mg or higher than 100 mg are associated with intermediate and severe intoxication, respectively [2]. The Food and Drug Administration (FDA) established a defect action level of 50 mg/kg for histamine (according to the revised Guidance in year 2020) [23], whereas the Commission Regulation (EC) No 2073/2005, which is currently in force, fixed maximum levels of 200 and 400 mg/kg for raw fish and fishery products subjected to enzyme maturation treatment in brine, respectively. These limits apply to the fish families *Scombridae*, *Scomberosocidae*, *Engraulidae*, *Clupeidae*, *Coriphaenidae*, and *Pomatomidae*, with high concentrations of free histidine in the muscle tissue. Some outbreak reports [24] and scientific studies confirmed that fish with low levels of the precursor amino acid showed no toxic histamine content in their flesh, whereas the opposite situation was described by Visciano et al. [25] in fish experimentally subjected to temperature abuse and belonging to the above cited families with regulatory limits.

Scombroid poisoning occurs worldwide and the largest numbers of such events are described in the United States, the United Kingdom, Australia, and Japan. Up to 40% of foodborne outbreaks reported in Europe and the United States can be ascribed to histamine intoxication [26]. In Table 4, the histamine poisoning and human cases in the EU from 2010 to 2018 are shown. Even if a positive trend was observed during the years 2010–2017, a sudden decrease was described in 2018, probably due to the lack of case reports from some Member States [27–35].

Table 4. Number of histamine poisoning outbreaks reported in the European Union during the years 2010–2018 [27–35].

Year	Outbreaks (N)			Cases (N)	
	Strong-Evidence	Weak-Evidence	Total	Human Cases	Hospitalized
2018	24	56	80	488	115
2017	56	61	117	572	51
2016 *	28	78	106	489	74
2015	23	57	80	437	43
2014	35	- **	35	164	15
2013	42	-	42	231	30
2012	34	-	34	241	14
2011	58	-	58	259	31
2010	33	-	33	185	12

Legend: * Data reported as other causative agents that include chemical agents, histamine, lectin, marine biotoxins, mushroom toxins, and scombrotoxin; ** data of weak-evidence outbreaks not reported.

RASFF organized among the EU Member States for the notification of risks to human health was established by the Regulation (EC) No 178/2002, with histamine representing one of the most common causes of such notifications. The list of notifying countries and the countries of origin of notifications for histamine presence in fish products from 2015 to 2020 are shown in Figures 1 and 2, respectively. Italy is the major notifying Member State, with a total of 74 alert or information for attention notifications, followed by France, with 33. Instead, the main origin of fish and fish products contaminated by histamine is linked to Spain (56 cases). Most notifications derived from the analysis of products showed high concentrations of histamine (Table 4), whereas some cases (n = 22) were associated with foodborne outbreaks generally occurring in the notifying country. The highest percentage of both alert (30.2%) and information for attention (34.4%) notifications reported a histamine content (Table 5) that slightly exceeded the maximum limits set by the Commission Regulation (EC) No 2073/2005. Instead, the highest values, i.e., >1000 or even >2000 mg/kg, referred to scombroid poisoning caused by chilled or canned fish, particularly regarding tuna species. In Figures 3 and 4, the fish species and types of fish products (i.e., raw or processed) involved in the RASFF reports from 2015 to August 2020 are shown. Tuna is the most representative species because it presents high concentrations of histidine in muscle tissue for its high speed and long duration swimming as predator fish [36]. Moreover, harvesting practices, such as longlining and gillnetting, can contribute to histamine formation due to the long period in which the fish remains in the sea before it is brought onboard the vessel. Such conditions are particularly dangerous for tuna species, which can generate heat into their body that exceeds the environmental temperature, thereby favoring the growth of histamine-forming microorganisms [23].

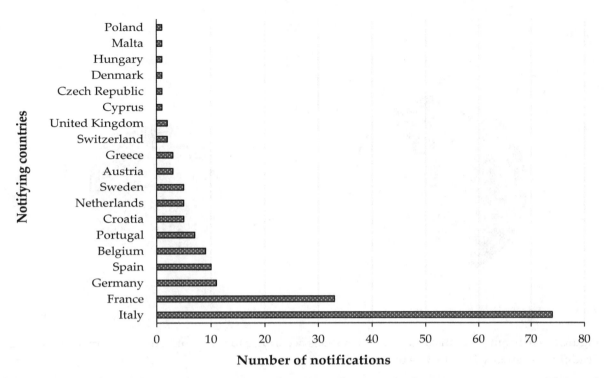

Figure 1. List of notifying countries and number of notifications for histamine presence in "fish and fish products" by the Rapid Alert System for Food and Feed from 1 January 2015 to 31 August 2020.

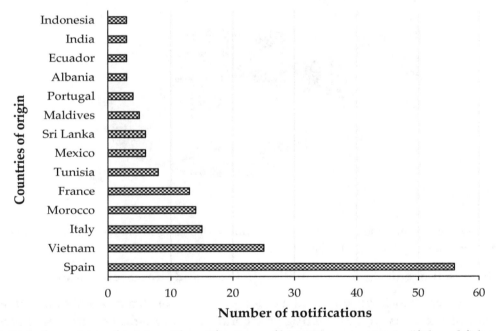

Figure 2. List of countries of origin with notifications of histamine presence in "fish and fish products" by the Rapid Alert System for Food and Feed from 1 January 2015 to 31 August 2020.

Table 5. Histamine concentrations reported in "fish and fish products" by the Rapid Alert System for Food and Feed from 1 January 2015 to 31 August 2020.

Notification	Histamine Concentrations (mg/kg)					
	≤200	>200 and ≤500	>500 and ≤1000	>1000 and ≤2000	>2000	Total
IA *	31 (20.6%)	52 (34.4%)	23 (15.2%)	29 (19.2%)	16 (10.6%)	151 (100%)
Alert	15 (12.9%)	35 (30.2%)	17 (14.7%)	26 (22.4%)	23 (19.8%)	116 (100%)
BR **	8 (22.2%)	18 (50%)	9 (25%)	-	1 (2.8%)	36 (100%)

Legend: * information for attention; ** border rejection.

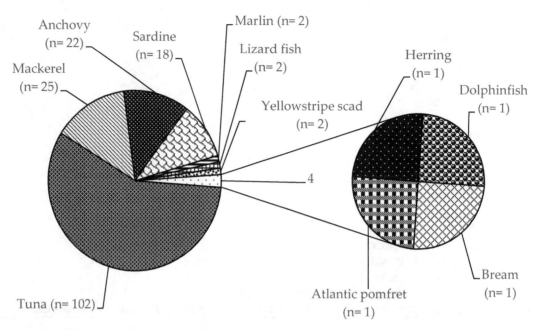

Figure 3. Distribution of histamine in fish species according to the Rapid Alert System for Food and Feed from 1 January 2015 to 31 August 2020.

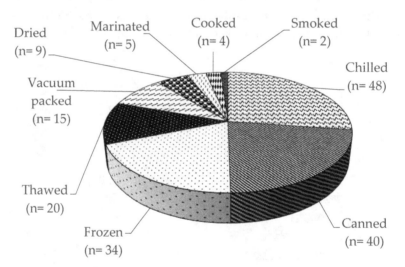

Figure 4. Presence of histamine in different fish products according to the Rapid Alert System for Food and Feed from 1 January 2015 to 31 August 2020.

3. Factors Affecting Biogenic Amine Formation in Fresh and Processed Seafood

Several parameters associated with food (i.e., temperature, NaCl, redox potential, pH, water activity (a_w), oxygen supply, etc.), as well as the hygienic conditions of manufacturing practices, can play a significant role in BA formation. The highest production by microorganisms occurs at temperatures ranging from 20 to 37 °C, so that the cold chain maintenance after harvesting of fish may prevent BA accumulation by reducing both bacterial growth and enzyme activity [37]. However, some decarboxylases continue their functions even if the microbial cells are not active. This phenomenon was demonstrated for histidine decarboxylase in Gram-negative bacteria such as *Morganella morganii*, *Photobacterium damselae*, *Photobacterium phosphoreum*, and *Raoultella planticola* [38]. Moreover, the specific metabolism of the microorganisms, the variability of strains belonging to the same species, as well as the complex matrix of analysis can influence decarboxylase responses to the environmental factors [39].

Even if it is important to ice fish as quickly as possible after catching, this practice cannot prevent/inhibit enzyme activities or microbial spoilage. Superchilling is a low temperature-based technique that consists in the decrease of temperature to 1–2 °C below freezing point (i.e., 0 °C) so that only a minor part of the water content of fish is frozen [40]. It reduces most autolytic and microbial reactions compared with normal chilling, and therefore its application can extend the shelf-life of many fish products [41]. Also, freezing at temperatures ranging from −18 to −30 °C inhibits microbial growth, but some enzymatic and nonenzymatic reactions can persist at lower rates and the formation of large ice crystals during such process may increase the risk of texture damage, loss of water holding capacity, and oxidation [42].

Besides temperature, pH and redox potential of the medium can influence amino acid decarboxylase activity. At low pH, microorganisms are more induced to generate decarboxylases as a protective mechanism from acidity, whereas conditions bringing about a diminished redox potential, enhance histamine formation [37]. *Photobacterium* spp., enterobacteria, and pseudomonads produce low quantities of BAs when NaCl concentrations correspond to 4–5%, even if the decarboxylation reactions are still operating [39].

Some other hurdles can be useful to preserve the characteristics and shelf-life of fish products, i.e., application of osmotic dehydration process, preservatives, and competitive microorganisms, such as lactic acid bacteria (LAB). The use of LAB and their metabolites as biopreservation techniques received much attention over the last two decades [43]. They are generally used for their ability to generate bacteriocins, organic acids, and hydrogen peroxide as inhibitory compounds [44]. Lee et al. [45] described the combined supplement of salt with fermentation by a starter culture (*Bacillus polymyxa*), decreasing histamine and other BA formation in fish.

The combination of two or more preservation methods (hurdle technology) often shows a greater inhibitory effect against the targeted microorganisms than any single treatment [46]. The application of modified atmospheres with low a_w and the addition of nisin extended the shelf-life of chilled fillets of gilthead seabream stored at 0–15 °C [47]. In particular, the modified atmosphere packaging (MAP) technique is based on the use of the three principal gases (i.e., $\%CO_2$, $\%O_2$, and $\%N_2$) inside the package and provides optimal conditions for the effective retardation of both microbiological and chemical processes [48]. The use of MAP and additives containing quercetin reduced the risk of BA production in Pacific white shrimp at 4 °C [49], whereas the shelf-life of striped red mullet was extended by MAP with ozone treatment [50]. The application of MAP together with ultra-violet (UV) radiation caused the reduction of putrescine concentrations during storage at 4 °C for 22 days in fillets of rainbow trout [51]. Also, Yew et al. [52] demonstrated strong reductions in histamine, cadaverine, and putrescine contents in Indian mackerel packaged with MAP (100% CO_2) after 12 days of storage at 5 °C. Indeed, other authors [53] reported increases of some BAs when the association of vacuum package and UV treatment was applied in fillets of tambacu during storage at 4 °C for six days. Two different doses of gamma radiation were investigated in samples of sea bream stored in ice, obtaining different results according to BAs. When increases in agmatine, tryptamine, and spermine were observed, cadaverine and putrescine levels decreased [54].

High pressure processing (HPP) is another technique able to inactivate microorganisms and autolytic enzymes at low temperature, thus extending the shelf-life of fish products [55]. The effect of HPP on BA formation was studied by Doeun et al. [56] in half-dried fish at different temperatures for 28 days. The authors observed a decrease of cadaverine and spermidine, while tyramine and spermine increased in concentration.

Many bioactive compounds deriving from plants were investigated for their use against pathogens and spoilage microorganisms [57]. Essential oils (EOs) are produced by different part of plants and consist of complex mixtures of hundreds of individual aromatic volatile oily compounds [58], even if only 300 are used in the food industry [59]. They are distinguished into several groups (i.e., terpenes, terpenoids, aromatics, and other compounds) according to their chemical structure [60]. According to the literature, EOs from oregano, rosemary, thyme, laurel, sage, cinnamon, clove, and basil are the most described antimicrobial and antioxidant agents in fish and fishery products. They can be applied to inhibit bacterial growth or for their bactericidal actions at high concentrations [61]. With regard to their effect on BAs, Özogul et al. [62] reported that rosemary and sage tea extracts could reduce histamine, putrescine, and cadaverine content in fillets of sardine during storage at 3 °C, 100 times smaller than the control group. Similarly, Cai et al. [63] found lower histamine levels in fillets of red drum stored at 4 °C treated with cumin, clove, and spearmint as essential oils. The authors supposed that such treatment inhibited the growth of microorganisms with histidine decarboxylase activity. Vacuum-packed fillets of sardine were stored after the addition of ethanolic extracts from mint and artemisia at 3 °C for 21 days. The contents of histamine, tyramine, and cadaverine were lower in treated than in control samples, and extracts of mint were more efficient than artemisia [64]. Kuley et al. [65] evaluated the inhibitory effects of safflower and bitter lemon extracts on both fish spoilage and growth of pathogenic bacteria. Such effects varied depending on the bacterial strains and specific amines. A general decrease in BA accumulation was observed and histamine production by *P. phosphoreum* was considerably suppressed. The effects of a microemulsion containing 0.3% or 1% lemon EO on the quality of salted sardines during 150 days of ripening were reported by Alfonzo et al. [66]. The results showed a reduction in *Enterobacteriaceae*, staphylococci, and rod LAB counts and a lower accumulation of histamine in the treated sardines compared to the control. However, as some EOs can have a negative impact on the sensory characteristics of seafood, even at low doses, some authors suggested the use of edible coating films enriched with EOs as an alternative and interesting option in order to reduce the required doses [67,68]. A more recent approach is active food packaging, i.e., the incorporation of EOs into the food package with a controlled release in order to maintain the organoleptic properties and microbiological integrity of food [69,70].

Also, other natural compounds, such as tea polyphenols and sage extracts, are used for food preservation [71–73]. The application of chitosan is becoming more frequent in the seafood industry due to its antibacterial and antioxidant characteristics [74,75].

4. Preventive and Control Measures

Fish products are subjected to spoilage because the chemical composition of meat rather than the microbial load on the skin or gastrointestinal tract can affect shelf-life, therefore, they must be chilled as soon as possible after catching. The formation of BAs in seafood is mostly dependent on time/temperature conditions from harvest through consumption. The best practices to control the growth of microorganisms producing BAs, histamine included, are chilling and freezing, even if it was reported that some microorganisms can grow at low temperatures and produce decarboxylases [76].

Many preservation processes, such as freezing, drying, margination, or salting can control the development of spoilage bacteria in fish [44]. The heating treatment before canning can eliminate both histamine-producing bacteria and their enzymes, but as histamine is heat stable, it can be found in the final product because it was present before the technological process started. Moreover, histamine can be still produced after thermal treatment when temperature abuse or recontamination appear [12]. Therefore, a dual approach for the control of BA increase in seafood is based on the quality of raw material, as well as the implementation of specific conditions able to inhibit or eliminate microorganisms with potential BA formation activity [77].

The prevention of BA formation in raw fish is primarily based on the rapid cooling after catching and its subsequent storage at ice temperature, as well as correct handling and hygiene practices on-board vessels. Ice, slurry ice, or mechanically refrigerated sea water can be used to chill fish after harvesting. The amount of ice used, as well as the size and temperature of fish brought on board and the air temperature on the deck and in storage hold, can affect the rate at which the internal temperature of fish refrigerate. For instance, it is well known that large fish chill more slowly than small fish. Moreover, the practice of placing ice into the gut cavity of fish could allow faster refrigeration, but if it is inappropriate, it can cause contamination from bacteria from the visceral cavity to the flesh fish and consequently the process of histamine production is accelerated. According to FDA recommendations (revised guidance, year 2020), if the water temperature is above 28.3 °C, the caught fish must be put into ice or in refrigerated seawater at 4.4 °C as soon as possible; instead, if the water temperature is below 28.3 °C the fish must be chilled within a maximum of 9 h from death. This period corresponds to 12 h if the fish are gilled and gutted before chilling [23].

According to the EU legislation (Regulation EC No 853/2004, currently in force) fish and fishery products must be kept at temperature of the melting ice as soon as possible after catching and during steps such as production, transport, storage, and distribution. Also, the operations of heading and gutting must be carried out quickly, and then whole and gutted fresh fishery products must be stored at ice temperature, while frozen fishery products must be kept at a temperature of no more than −18 °C.

All these requirements are necessary not only to ensure fish freshness and quality, but also to avoid potential histamine formation in fish families with a high histidine levels in muscle tissue. Indeed, fishermen and all food business operators involved in the fish food chain must ensure that the regulatory limits of histamine are not exceeded.

With regard to processed fishery products, it is known that some histamine-forming microorganisms are halotolerant or halophilic and can produce histamine at low pH levels. Moreover, technological processes such as salting, drying, fermenting, smoking, and pickling can contribute to histamine formation. For fermented seafood, the addition of negative amine-producing starter cultures could prevent BA formation during processing and storage [21].

Many regulatory organizations around the world adopted maximum limits for histamine in fish and fishery products. The EU, the Codex Alimentarius, Australia, and New Zealand established histamine values above 200 mg/kg as a maximum limit in raw fish, whereas the FDA considered 50 mg/kg for United States and Canada [78]. The last value was derived from the different distribution of amines in large fish, i.e., if in one section of a fish the histamine concentration is 50 ppm, in other parts of the same fish 500 ppm can be found [23]. Instead, in the Russian Federation the histamine maximum content is 100 mg/kg for salmon, herring, tuna, and mackerel [79], according to "The Sanitary and Epidemiological Rules and Regulations, SanPin 2.3.2.1078-2001".

Sampling and analysis also constitute important components of the control strategy. In Table 6, the sampling plan for histamine determination according to the Commission Regulation (EC) No 2073/2005 and FDA Guidance is well described. The main difference regards the number of samples to be analyzed. The collection and analysis of 18 samples described in the FDA scheme can allow the detection of nonconforming lots with a higher probability than the EU sampling plan [80]. Instead, the last scheme consists of nine samples, except for fish collected at the retail level, where only a single sample can be analyzed; if the result is higher than the maximum limit (200 mg/kg), the whole batch is considered unsafe. Also, for fish sauce produced by fermentation of fishery products, one sample can be considered representative, because histamine is uniformly distributed in these products.

Table 6. Sampling plan for histamine determination according to official control rules.

Country	Limit	Reference
EU	• Fishery products from fish species associated with a high amount of histidine n = 9, c = 2, m = 100 mg/kg, M = 200 mg/kg • Fishery products undergoing enzyme maturation treatment in brine, manufactured from fish species associated with a high amount of histidine n = 9 *, c = 2, m = 200 mg/kg, M = 400 mg/kg	Commission Regulation (EC) No 2073/2005 and further amendments
	Fish sauce produced by fermentation of fishery products n = 1, m = 400 mg/kg	Commission Regulation (EU) No 1019/2013
US FDA	n = 18, c = 1, m = 50 mg/kg, M = 500 mg/kg n = 18; c = 0, m = 500 mg/kg	FAO/WHO, 2012

Legend: * Single samples may be taken at retail level. If one of nine samples analyzed is found to be above M, the whole batch shall be deemed unsafe (Commission Regulation EU No 1019/2013).

The Commission Regulation (EC) No 2073/2005 also establishes the analytical reference method (EN ISO 19343:2017) for histamine detection in fish by high-performance liquid chromatography (HPLC).

5. Trend and Challenges of Detection Technologies

The detection of BA in fish is generally difficult, requiring long time of analysis and expert technicians. Moreover, a phase of preconcentration is often necessary for complex matrices and the separation of histamine from interference compounds, such as histidine or carnosine, needs careful and long pretreatment of the sample with precolumn or postcolumn derivatization, which is time-consuming and prolongs the entire analytical process [81]. Some other hurdles are described for BA determination, such as strong polar characteristics, the simultaneous occurrence of several BAs, and variable concentration ranges [82].

Many simple and rapid techniques are described in the literature for monitoring histamine levels in fish and fishery products, such as biosensors, immune-enzymatic assays (ELISA), and colorimetric methods. Alonso-Lomillo et al. [83] reported the use of amine oxidase-based electrodes to catalyze the oxidative deamination of BAs producing other compounds, such as ammonia and hydrogen peroxide. The commercial ELISA kits are functional as screening methods due to their quickness and simplicity, even if they give only semiquantitative results [84]. Also, colorimetric methods can be applied for routine analysis [85], as well as capillary electrophoresis, which is less sensitive than other methods but is rapid and cheap, allowing the analysis of many samples in a short time [82].

In the fishery industry and on an own-check basis, such simple detection technologies used as screening tests must be reinforced by confirmatory methods if positive results are achieved. Among them, thin-layer chromatography [86], gas chromatography [87], and electrochemical assays [88] are known to be sensitive and specific but require trained technicians and sophisticated, expensive instruments [89]. In particular, thin-layer chromatography is a technique that allows the fast determination of small quantities of compounds, as well as the simultaneous analysis of many samples. With regard to gas chromatography, some analytical problems were reported and it is better to use it in combination with derivative technologies. Finally, the electrochemical detectors are rapid and easy tests with high sensitivity [90].

In recent years, new technologies based on nanomaterials, such as carbon nanotubes [91], graphene [92], and metal nanoparticles [93,94], significantly improved the speed and cost of analysis, low sample volume requirement and field deployability. The use of gold nanoparticles for the detection of two important volatile biogenic markers, i.e., dimethyl sulfide and histamine, was described by Chow et al. [95], showing excellent selectivity in the presence of other volatiles commonly produced during fish spoilage and a low limit of detection of 0.5 and 0.035 µg/mL, respectively. A dual detection approach based on colorimetric sensor and laser desorption–ionization mass spectrometry was also reported by Siripongpreda et al. [96] for screening and quantitative determination of BAs.

6. Conclusions

In fish and fishery products, BAs can be considered as indicators of both quality and safety, and their formation depends on the harvest method, the handling and other operations onboard vessels, post catching contamination, inadequate chilling, or temperature abuse. In fish exposed to high temperatures even for a short time, large populations of microorganisms can grow and produce decarboxylases. Even if during the subsequent refrigeration bacterial growth is reduced, residual enzyme activity can continue and, therefore, BA levels increase. The application of appropriate preventive strategies and control procedures represent the most efficient tool for both consumers and the fish industry.

Author Contributions: Conceptualization, P.V. and M.S.; writing—original draft preparation, P.V. and M.S.; writing—review and editing, P.V., M.S., and A.P.; supervision, A.P. All authors have read and agreed to the published version of the manuscript.

References

1. Zhong, X.; Huo, D.; Fa, H.; Luo, X.; Wang, Y.; Zhao, Y.; Hou, C. Rapid and ultrasensitive detection of biogenic amines with colorimetric sensor array. *Sens. Actuators B Chem.* **2018**, *274*, 464–471. [CrossRef]
2. Biji, K.B.; Ravishankar, C.N.; Venkateswarlu, R.; Mohan, C.O.; Srinivasa Gopal, T.K. Biogenic amines in seafood: A review. *J. Food Sci. Technol.* **2016**, *53*, 2210–2218. [CrossRef]
3. Visciano, P. Chemicals and Safety of Chemical Contaminants in Seafood. In *Food Safety Chemistry. Toxicant Occurrence, Analysis and Mitigation*; Liangli, L.Y., Shuo, W., Bao-Guo, S., Eds.; CRC Press; Taylor & Francis Group: Boca Raton, FL, USA, 2015; pp. 215–236.
4. Kuley, E.; Durmus, M.; Balikci, E.; Ucar, Y.; Regenstein, J.M.; Özoğul, F. Fish spoilage bacterial growth and their biogenic amine accumulation: Inhibitory effect of olive by-products. *Int. J. Food Prop.* **2017**, *20*, 1029–1043. [CrossRef]
5. Hungerford, J.M. Scombroid poisoning: A review. *Toxicon* **2010**, *56*, 231–243. [CrossRef] [PubMed]
6. Doeun, D.; Davaatseren, M.; Chung, M.S. Biogenic amines in foods. *Food Sci. Biotechnol.* **2017**, *26*, 1463–1474. [CrossRef] [PubMed]
7. Barbieri, F.; Montanari, C.; Gardini, F.; Tabanelli, G. Biogenic amine production by Lactic Acid Bacteria: A review. *Foods* **2019**, *8*, 17. [CrossRef] [PubMed]
8. Xu, Y.; Zang, J.; Regenstein, J.M.; Xia, W. Technological roles of microorganisms in fish fermentation: A review. *Crit. Rev. Food Sci. Nutr.* **2020**, 1–13. [CrossRef]
9. Marcobal, A.; De Las Rivas, B.; Landete, J.M.; Tabera, L.; Muñoz, R. Tyramine and phenylethylamine biosynthesis by food bacteria. *Crit. Rev. Food Sci. Nutr.* **2012**, *52*, 448–467. [CrossRef]
10. Wunderlichová, L.; Buňková, L.; Koutný, M.; Jančová, P.; Buňka, F. Formation, degradation, and detoxification of putrescine by foodborne bacteria: A review. *Compr. Rev. Food Sci. Food Saf.* **2014**, *13*, 1012–1030. [CrossRef]
11. Paleologos, E.K.; Savvaidis, I.N.; Kontominas, M.G. Biogenic amines formation and its relation to microbiological and sensory attributes in ice-stored whole, gutted and filleted Mediterranean Sea bass (*Dicentrarchus labrax*). *Food Microbiol.* **2004**, *21*, 549–557. [CrossRef]
12. FAO/WHO (Food and Agriculture Organization of the United Nations/World Health Organization). *Public Health Risks of Histamine and Other Biogenic Amines from Fish and Fishery Products. Meeting Report*; FAO Headquarters: Rome, Italy, 2012.
13. Comas-Basté, O.; Latorre-Moratalla, M.L.; Sánchez-Pérez, S.; Veciana-Nogués, M.T.; Vidal-Carou, M.C. Histamine and other biogenic amines in food. From Scombroid Poisoning to histamine intolerance. In *Biogenic Amines*; Proestos, C., Ed.; IntechOpen: London, UK, 2019.
14. Fusek, M.; Michálek, J.; Buňková, L.; Buňka, F. Modelling biogenic amines in fish meat in Central Europe using censored distributions. *Chemosphere* **2020**, *251*, 126390. [CrossRef] [PubMed]
15. Stratta, P.; Badino, G. Scombroid poisoning. *CMAJ* **2012**, *184*, 674. [CrossRef] [PubMed]
16. Palomino-Vasco, M.; Acedo-Valenzuela, M.I.; Rodríguez-Cáceres, M.I.; Mora-Diez, N. Automated chromatographic method with fluorescent detection to determine biogenic amines and amino acids. Application to craft beer brewing process. *J. Chromatogr. A* **2019**, *1601*, 155–163. [CrossRef] [PubMed]
17. Silva, I.P.; Dias, L.G.; da Silva, M.O.; Machado, C.S.; Paula, V.M.B.P.; Evangelista-Barreto, N.S.; de Carvalho, C.A.L.; Estevinho, L.M. Detection of biogenic amines in mead of social bee. *LWT Food Sci. Technol.* **2020**, *121*, 108969. [CrossRef]
18. Prester, L. Biogenic amines in fish, fish products and shellfish: A review. *Food Addit. Contam.* **2011**, *28*, 1547–1560. [CrossRef] [PubMed]
19. Visciano, P.; Schirone, M.; Tofalo, R.; Suzzi, G. Biogenic amines in raw and processed seafood. *Front. Microbiol.* **2012**, *3*, 188. [CrossRef]
20. Mattsson, L.; Xu, J.; Preininger, C.; Bui, B.T.S.; Haupt, K. Competitive fluorescent pseudo-immunoassay exploiting molecularly imprinted polymers for the detection of biogenic amines in fish matrix. *Talanta* **2018**, *181*, 190–196. [CrossRef]
21. Sedaghati, M.; Mooraki, N. Biogenic amines in sea products. *J. Surv. Fish. Sci.* **2019**, *6*, 1–8. [CrossRef]
22. Muscarella, M.; Lo Magro, S.; Campaniello, M.; Armentano, A.; Stacchini, P. Survey of histamine levels in fresh fish and fish products collected in Puglia (Italy) by ELISA and HPLC with fluorometric detection. *Food Control* **2013**, *31*, 211–217. [CrossRef]

23. FDA (Food and Drug Administration). *Fish and Fishery Products Hazards and Controls Guidance*, 4th ed.; Chapter 7—Scombrotoxin (Histamine) Formation; FDA: Silver Spring, MD, USA, 2020; pp. 113–152.

24. Colombo, F.M.; Cattaneo, P.; Confalonieri, E.; Bernardi, C. Histamine food poisonings: A systematic review and meta-analysis. *Crit. Rev. Food Sci.* **2018**, *58*, 1131–1151. [CrossRef]

25. Visciano, P.; Campana, G.; Annunziata, L.; Vergara, A.; Ianieri, A. Effect of storage temperature on histamine formation in *Sardina pilchardus* and *Engraulis encrasicolus* after catch. *J. Food Biochem.* **2007**, *31*, 577–588. [CrossRef]

26. Guergué-Díaz de Cerio, O.; Barrutia-Borque, A.; Gardeazabal-García, J. Scombroid poisoning: A practical approach. *Actas Dermodifiliogr.* **2016**, *107*, 567–571. [CrossRef] [PubMed]

27. EFSA and ECDC (European Food Safety Authority and European Centre for Disease Prevention and Control). The European Union One Health 2018 Zoonoses Report. *EFSA J.* **2019**, *17*, 5926.

28. EFSA and ECDC (European Food Safety Authority and European Centre for Disease Prevention and Control). The European Union summary report on trends and sources of zoonoses, zoonotic agents and food-borne outbreaks in 2017. *EFSA J.* **2018**, *16*, 5500.

29. EFSA and ECDC (European Food Safety Authority and European Centre for Disease Prevention and Control). The European Union summary report on trends and sources of zoonoses, zoonotic agents and food-borne outbreaks in 2016. *EFSA J.* **2017**, *15*, 55077.

30. EFSA and ECDC (European Food Safety Authority and European Centre for Disease Prevention and Control). The European Union summary report on trends and sources of zoonoses, zoonotic agents and food-borne outbreaks in 2015. *EFSA J.* **2016**, *14*, 4634.

31. EFSA and ECDC (European Food Safety Authority and European Centre for Disease Prevention and Control). The European Union summary report on trends and sources of zoonoses, zoonotic agents and food-borne outbreaks in 2014. *EFSA J.* **2015**, *13*, 4329.

32. EFSA and ECDC (European Food Safety Authority and European Centre for Disease Prevention and Control). The European Union summary report on trends and sources of zoonoses, zoonotic agents and food-borne outbreaks in 2013. *EFSA J.* **2015**, *13*, 3991. [CrossRef]

33. EFSA and ECDC (European Food Safety Authority and European Centre for Disease Prevention and Control). The European Union summary report on trends and sources of zoonoses, zoonotic agents and food-borne outbreaks in 2012. *EFSA J.* **2014**, *12*, 3547. [CrossRef]

34. EFSA and ECDC (European Food Safety Authority and European Centre for Disease Prevention and Control). The European Union summary report on trends and sources of zoonoses, zoonotic agents and food-borne outbreaks in 2011. *EFSA J.* **2013**, *11*, 3129. [CrossRef]

35. EFSA and ECDC (European Food Safety Authority and European Centre for Disease Prevention and Control). The European Union summary report on trends and sources of zoonoses, zoonotic agents and food-borne outbreaks in 2010. *EFSA J.* **2012**, *10*, 2597.

36. Mercogliano, R.; Santonicola, S. Scombroid fish poisoning: Factors influencing the production of histamine in tuna supply chain. A review. *LWT Food Sci. Technol.* **2014**, *114*, 108374. [CrossRef]

37. Ekici, K.; Omer, A.K. Biogenic amines formation and their importance in fermented foods. In *BIO Web of Conferences*; EDP Sciences: Ulis, France, 2020; Volume 17, p. 00232.

38. Gardini, F.; Özogul, Y.; Suzzi, G.; Tabanelli, G.; Özogul, F. Technological factors affecting biogenic amines content in foods: A review. *Front. Microbiol.* **2016**, *7*, 1218. [CrossRef] [PubMed]

39. Tabanelli, G.; Montanari, C.; Gardini, F. Biogenic amines in food: A review of factors affecting their formation. *Ref. Modul. Food Sci. Encycl. Food Chem.* **2019**, *1*, 337–343.

40. Stonehouse, G.G.; Evans, J.A. The use of supercooling for fresh foods: A review. *J. Food Eng.* **2015**, *148*, 74–79. [CrossRef]

41. Sampels, S. The effects of storage and preservation technologies on the quality of fish products: A review. *J. Food Process. Preserv.* **2015**, *39*, 1206–1215. [CrossRef]

42. Karoui, R.; Hassoun, A.; Ethuin, P. Front face fluorescence spectroscopy enables rapid differentiation of fresh and frozen- thawed sea bass (*Dicentrarchus labrax*) fillets. *J. Food Eng.* **2017**, *202*, 89–98. [CrossRef]

43. Ghanbari, M.; Jami, M.; Domig, K.J.; Kneifel, W. Seafood biopreservation by lactic acid bacteria—A review. *LWT Food Sci. Technol.* **2013**, *54*, 315–324. [CrossRef]

44. Tsironi, T.; Houhoula, D.; Taokis, P. Hurdle technology for fish preservation. *Aquac. Fish.* **2020**, *5*, 65–71. [CrossRef]

45. Lee, Y.C.; Kung, H.F.; Huang, C.Y.; Huang, T.C.; Tsai, Y.H. Reduction of histamine and biogenic amines during salted fish fermentation by *Bacillus polymyxa* as a starter culture. *J. Food Drug Anal.* **2016**, *24*, 157–163. [CrossRef]

46. Khan, I.; Tango, C.N.; Miskeen, S.; Lee, B.H.; Oh, D.H. Hurdle technology: A novel approach for enhanced food quality and safety-a review. *Food Control* **2017**, *73*, 1426–1444. [CrossRef]

47. Tsironi, T.N.; Taokis, P.S. Modeling microbial spoilage and quality of gilthead seabream fillets: Combined effect of osmotic pre-treatment, modified atmosphere packaging and nisin on shelf life. *J. Food Sci.* **2010**, *75*, 243–251. [CrossRef] [PubMed]

48. Hassoun, A.; Çoban, O.E. Essential oils for antimicrobial and antioxidant applications in fish and other seafood products. *Trends Food Sci. Technol.* **2017**, *68*, 26–36. [CrossRef]

49. Qian, Y.; Yang, S.; Ye, J.X.; Xie, J. Effect of quercetin-containing preservatives and modified atmospheric packaging on the production of biogenic amines in Pacific white shrimp (*Litopenaeus vannamei*). *Aquac. Fish.* **2018**, *3*, 254–259. [CrossRef]

50. Bono, G.; Badalucco, C. Combining ozone and modified atmosphere packaging (MAP) to maximize shelf-life and quality of striped red mullet (*Mullus surmuletus*). *LWT-Food Sci. Technol.* **2012**, *47*, 500–504. [CrossRef]

51. Rodrigues, B.L.; Alvares, T.D.S.; Sampaio, G.S.L.; Cabral, C.C.; Araujo, J.V.A.; Franco, R.M.; Mano, S.B.; Conte-Junior, C.A. Influence of vacuum and modified atmosphere packaging in combination with UV-C radiation on the shelf life of rainbow trout (*Oncorhynchus mykiss*) fillets. *Food Control* **2016**, *60*, 596–605. [CrossRef]

52. Yew, C.C.; Bakar, F.A.; Rahman, R.A.; Bakar, J.; Zaman, M.Z.; Velu, S.; Shariat, M. Effects of modified atmosphere packaging with various carbon dioxide composition on biogenic amines formation in Indian mackerel (*Rastrelliger kanagurta*) stored at 5 ± 1 °C. *Packag. Technol. Sci.* **2014**, *27*, 249–254. [CrossRef]

53. Bottino, F.D.O.; Rodrigues, B.L.; De Nunes Ribeiro, J.D.; Lázaro, C.A.; Conte-Junior, C.A. Influence of UV-C radiation on shelf life of vacuum package tambacu (*Colossoma macropomum* × *Piaractus mesopotamicus*) fillets. *J. Food Process. Preserv.* **2016**, *41*, e13003. [CrossRef]

54. Özogul, F.; Özden, Ö. The effects of gamma irradiation on the biogenic amine formation in sea bream (*Sparus aurata*) stored in ice. *Food Bioprocess. Technol.* **2013**, *6*, 1343–1349. [CrossRef]

55. Truong, B.Q.; Buckow, R.; Stathopoulos, C.E.; Nguyen, M.H. Advances in high-pressure processing of fish muscles. *Food Eng. Rev.* **2014**, *7*, 109–129. [CrossRef]

56. Doeun, D.; Shin, H.-S.; Chung, M.-S. Effects of storage temperatures, vacuum packaging, and high hydrostatic pressure treatment on the formation of biogenic amines in Gwamegi. *Appl. Biol. Chem.* **2016**, *59*, 51–58. [CrossRef]

57. Baptista, R.C.; Horita, C.N.; Sant'Ana, A.S. Natural products with preservative properties for enhancing the microbiological safety and extending the shelf-life of seafood: A review. *Food Res. Int.* **2020**, *127*, 108762. [CrossRef] [PubMed]

58. Calo, J.R.; Crandall, P.G.; O'Bryan, C.A.; Ricke, S.C. Essential oils as antimicrobials in food systems—A review. *Food Control* **2015**, *54*, 111–119. [CrossRef]

59. Bakkali, F.; Averbeck, S.; Averbeck, D.; Idaomar, M. Biological effects of essential oils—A review. *Food Chem. Toxicol.* **2008**, *46*, 446–475. [CrossRef]

60. Hyldgaard, M.; Mygind, T.; Meyer, R.L. Essential oils in food preservation: Mode of action, synergies, and interactions with food matrix components. *Front. Microbiol.* **2012**, *3*, 12. [CrossRef] [PubMed]

61. Swamy, M.K.; Akhtar, M.S.; Sinniah, U.R. Antimicrobial properties of plant essential oils against human pathogens and their mode of actions: An updated review. *Evid. Based Complement. Altern. Med.* **2016**, *21*, 3012462. [CrossRef]

62. Özogul, F.; Kuley, E.; Kenar, M. Effects of rosemary and sage tea extract on biogenic amines formation of sardine (*Sardina pilchardus*) fillets. *Int. J. Food Sci. Technol.* **2011**, *46*, 761–766. [CrossRef]

63. Cai, L.; Cao, A.; Li, Y.; Song, Z.; Leng, L.; Li, Y. The effects of essential oil treatment on the biogenic amines inhibition and quality preservation of red drum (*Sciaenops ocellatus*) fillets. *Food Control* **2015**, *56*, 1–8. [CrossRef]

64. Houicher, A.; Kuley, E.; Özogul, F.; Bendeddouche, B. Effect of natural extracts (*Mentha spicata* L. and *Artemisia campestris*) on biogenic amine formation of sardine vacuum-packed and refrigerated (*Sardina pilchardus*) fillets. *J. Food Process. Pres.* **2015**, *39*, 2393–2403. [CrossRef]

65. Kuley, E.; Yavuzer, M.N.; Yavuzer, E.; Durmuş, M.; Yazgan, H.; Gezginç, Y.; Özogul, F. Inhibitory effects of safflower and bitter melon extracts on biogenic amine formation by fish spoilage bacteria and food borne pathogens. *Food Biosci.* **2019**, *32*, 100478. [CrossRef]

66. Alfonzo, A.; Martorana, A.; Guarrasi, V.; Barbera, M.; Gaglio, R.; Santulli, A.; Settanni, L.; Galati, A.; Moschetti, G.; Francesca, N. Effect of the lemon essential oils on the safety and sensory quality of salted sardines (*Sardina pilchardus* Walbaum 1972). *Food Control* **2017**, *73*, 1265–1274. [CrossRef]

67. Doğan, G.; İzci, L. Effects on quality properties of smoked rainbow trout (*Oncorhynchus mykiss*) fillets of chitosan films enriched with essential oils. *J. Food Process. Preserv.* **2017**, *41*, e12757. [CrossRef]

68. Yuan, G.; Lv, H.; Tang, W.; Zhang, X.; Sun, H. Effect of chitosan coating combined with pomegranate peel extract on the quality of Pacific white shrimp during iced storage. *Food Control* **2016**, *59*, 818–823. [CrossRef]

69. Atarés, L.; Chiralt, A. Essential oils as additives in biodegradable films and coatings for active food packaging. *Trends Food Sci. Technol.* **2016**, *48*, 51–62. [CrossRef]

70. Ribeiro-Santos, R.; Andrade, M.; Ramos de Melo, N.; Sanches-Silva, A. Use of essential oils in active food packaging: Recent advances and future trends. *Trends Food Sci. Technol.* **2017**, *61*, 132–140. [CrossRef]

71. Emir Çoban, O.; Özpolat, E. The effects of different concentrations of rosemary (*rosmarinus officinalis*) extract on the shelf life of hot-smoked and vacuum-packed luciobarbus esocinus fillets. *J. Food Process. Preserv.* **2013**, *37*, 269–274. [CrossRef]

72. Li, T.; Li, J.; Hu, W.; Zhang, X.; Li, X.; Zhao, J. Shelf-life extension of crucian carp (*Carassius auratus*) using natural preservatives during chilled storage. *Food Chem.* **2012**, *135*, 140–145. [CrossRef]

73. Pezeshk, S.; Ojagh, S.M.; Alishahi, A. Effect of plant antioxidant and antimicrobial compounds on the shelf-life of seafood—A review. *Czech J. Food Sci.* **2015**, *33*, 195–203. [CrossRef]

74. Alishahi, A.; Äider, M. Applications of chitosan in the seafood industry and aquaculture: A review. *Food Bioproc. Technol.* **2012**, *5*, 817–830. [CrossRef]

75. Yuan, G.; Chen, X.; Li, D. Chitosan films and coatings containing essential oils: The antioxidant and antimicrobial activity, and application in food systems. *Food Res. Int.* **2016**, *89*, 117–128. [CrossRef]

76. Naila, A.; Flint, S.; Fletcher, G.; Bremer, P.; Meerdink, G. Control of biogenic amines in food—Existing and emerging approaches. *J. Food Sci.* **2010**, *75*, R139–R150. [CrossRef] [PubMed]

77. EFSA Panel on Biological Hazards (BIOHAZ). Scientific Opinion on Scientific Opinion on risk based control of biogenic amine formation in fermented foods. *EFSA J.* **2011**, *9*, 2393. [CrossRef]

78. Lázaro, C.A.; Conte-Junior, C.A. Detection of biogenic amines: Quality and toxicity indicators in food of animal origin. In *Food Control and Biosecurity*; Holban, A.M., Grumezescu, A.M., Eds.; Elsevier: Amsterdam, The Netherlands, 2018; Volume 16, pp. 225–257.

79. Verkhivker, Y.; Altman, E. Influence parameters of storage on process of formation the histamine in fish and fish products. *J. Water Res. Ocean. Sci.* **2018**, *7*, 10–14. [CrossRef]

80. Visciano, P.; Schirone, M.; Tofalo, R.; Suzzi, G. Histamine poisoning and control measures in fish and fishery products. *Front. Microbiol.* **2014**, *5*, 500. [CrossRef] [PubMed]

81. Surya, T.; Sivaraman, B.; Alamelu, V.; Priyatharshini, A.; Arisekar, U.; Sundhar, S. Rapid methods for histamine detection in fishery products. *Int. J. Curr. Microbiol. Appl. Sci.* **2019**, *8*, 2035–2046. [CrossRef]

82. Papageorgiou, M.; Lambropoulou, D.; Morrison, C.; Kłodzińska, E.; Namieśnik, J.; Płotka-Wasylka, J. Literature update of analytical methods for biogenic amines determination in food and beverages. *Trends Analyt. Chem.* **2018**, *98*, 128–142. [CrossRef]

83. Alonso-Lomillo, M.A.; Domínguez-Renedo, O.; Matos, P.; Arcos-Martínez, M.J. Disposable biosensors for determination of biogenic amines. *Anal. Chim. Acta* **2010**, *665*, 26–31. [CrossRef]

84. Önal, A. A Review: Current analytical methods for the determination of biogenic amines in foods. *Food Chem* **2007**, *103*, 1475–1486. [CrossRef]

85. Patange, S.B.; Mukundan, M.K.; Ashok Kumar, K. A simple and rapid method for colorimetric determination of histamine in fish flesh. *Food Control* **2005**, *16*, 465–472. [CrossRef]

86. Tao, Z.; Sato, M.; Han, Y.; Tan, Z.; Yamaguchi, T.; Nakano, T. A simple and rapid method for histamine analysis in fish and fishery products by TLC determination. *Food Control* **2011**, *22*, 1154–1157. [CrossRef]

87. Machiels, D.; Van Ruth, S.M.; Posthumus, M.A.; Istasse, L. Gas chromatography-olfactometry analysis of the volatile compounds of two commercial Irish beef meat. *Talanta* **2003**, *60*, 755–764. [CrossRef]

88. Young, J.A.; Jiang, X.; Kirchhoff, J.R. Amperometric detection of histamine with a pyrroloquinoline-quinone modified electrode. *Electroanalysis* **2013**, *25*, 1589–1593. [CrossRef]

89.   Yadav, S.; Nair, S.S.; Sai, V.V.R.; Satija, J. Nanomaterials based optical and electrochemical sensing of histamine: Progress and perspectives. *Food Res. Int.* **2019**, *119*, 99–109. [CrossRef] [PubMed]

90.   Zhang, Y.-J.; Zhang, Y.; Zhou, Y.; Li, G.-H.; Yang, W.-Z.; Feng, X.-S. A review of pretreatment and analytical methods of biogenic amines in food and biological samples since 2010. *J. Chromatogr. A* **2019**, *1605*, 360361. [CrossRef]

91.   Zang, X.; Zhou, Q.; Chang, J.; Liu, Y.; Lin, L. Graphene and carbon nanotube (CNT) in MEMS/NEMS applications. *Microelectron. Eng.* **2015**, *132*, 192–206. [CrossRef]

92.   Liao, M.; Koide, Y. Carbon-based materials: Growth, properties, MEMS/NEMS technologies, and MEM/NEM switches. *Crit. Rev. Solid State* **2011**, *36*, 66–101. [CrossRef]

93.   Jian, R.S.; Huang, R.X.; Lu, C.J. A micro GC detector array based on chemiresistors employing various surface functionalized monolayer-protected gold nanoparticles. *Talanta* **2012**, *88*, 160–167. [CrossRef]

94.   Rajdi, N.N.Z.M.; Salleh, S.M.; Bakir, A.A.; Yuen, A.; Wicaksono, D.H.B.; Harun, F.K.C. Silver nanoparticles stamping for the production of fabrics-based bio-MEMS. In Proceedings of the International Conference on Robotics, Biomimetics, Intelligent Computational Systems, Jogjakarta, Indonesia, 25–27 November 2013; pp. 10–14.

95.   Chow, C.F. Biogenic amines- and sulfides-responsive gold nanoparticles for real-time visual detection of raw meat, fish, crustaceans, and preserved meat. *Food Chem.* **2020**, *311*, 125908. [CrossRef]

96.   Siripongpreda, T.; Siralertmukul, K.; Rodthongkum, N. Colorimetric sensor and LDI-MS detection of biogenic amines in food spoilage based on porous PLA and graphene oxide. *Food Chem.* **2020**, *329*, 127165. [CrossRef]

2

Formation of Biogenic Amines in *Pa* (Green Onion) Kimchi and *Gat* (Mustard Leaf) Kimchi

Jun-Hee Lee, Young Hun Jin, Young Kyoung Park, Se Jin Yun and Jae-Hyung Mah *⬤

Department of Food and Biotechnology, Korea University, 2511 Sejong-ro, Sejong 30019, Korea;
bory92@korea.ac.kr (J.-H.L.); younghoonjin3090@korea.ac.kr (Y.H.J.); eskimo@korea.ac.kr (Y.K.P.);
ysj529@korea.ac.kr (S.J.Y.)
* Correspondence: nextbio@korea.ac.kr

Abstract: In this study, biogenic amine content in *Pa* (green onion) kimchi and *Gat* (mustard leaf) kimchi, Korean specialty kimchi types, was determined by high-performance liquid chromatography (HPLC). Many kimchi samples contained low levels of biogenic amines, but some samples had histamine and tyramine content over the safe levels. Based on the comparative analysis between the ingredient information on food labels and biogenic amine content of kimchi samples, *Myeolchi-aekjeot* appeared to be an important source of biogenic amines in both kimchi. Besides, through the 16s rRNA sequence analysis, *Lactobacillus brevis* appeared to be responsible for the formation of biogenic amines (tyramine, β-phenylethylamine, putrescine, and cadaverine) in both kimchi, in a strain-dependent manner. During fermentation, a higher accumulation of tyramine, β-phenylethylamine, and putrescine was observed in both or one (for putrescine) of kimchi types when *L. brevis* strains served as inocula. The addition of *Myeolchi-aekjeot* affected the initial concentrations of most biogenic amines (except for spermidine in *Gat* kimchi) in both kimchi. Therefore, this study suggests that using appropriately salted and fermented seafood products for kimchi preparation and using biogenic amine-negative and/or biogenic amine-degrading starter cultures would be effective in reducing biogenic amine content in *Pa* kimchi and *Gat* kimchi.

Keywords: kimchi; *Pa* kimchi; *Gat* kimchi; Korean specialty kimchi; biogenic amines; lactic acid bacteria; *Lactobacillus brevis*

1. Introduction

Biogenic amines (BA), including vasoactive amines (tryptamine, β-phenylethylamine, histamine, and tyramine), putrefactive amines (putrescine and cadaverine), and polyamines (spermidine and spermine), are present in a wide range of food products. Particularly, various types of fermented foods have been reported to contain BA, because these nitrogenous compounds are produced by microbial decarboxylation of amino acids [1]. While amine oxidases are able to detoxify most BA, excessive consumption of BA and/or amine oxidase inhibition by drugs, alcohol, and gastrointestinal disease have been reported to reduce the efficiency of the enzymes [1,2]. Consequently, unoxidized BA may cause symptoms such as diarrhea, hyperhidrosis, urticaria, hypotension, hypertension, headache, burning mouth, nausea, hot flush, respiratory distress, and cardiac arrest [2,3]. According to Silla Santos [3], the scombroid fish poisoning and cheese reaction are representative examples of foodborne illness associated with the intolerance to dietary histamine and tyramine, respectively. Thus, several studies have suggested the limits for consumption of BA as follows: β-phenylethylamine, 30 mg/kg; histamine, 100 mg/kg; tyramine, 100–800 mg/kg [4]; total BA, 1000 mg/kg [3].

According to the Codex standard [4], kimchi (in reality, *Baechu* kimchi) is defined as a salted and fermented food product made of Chinese cabbage (the main ingredient; *Baechu* in Korean) with a seasoning mixture consisting of red pepper powder, garlic, ginger, green onion, and radish. Other optional ingredients include fruits, glutinous rice paste, nuts, sugar, and salted and fermented seafood products [4]. Among various salted and fermented seafood products, such as *Jeotgal* (salted and fermented seafood) and *Aekjeot* (liquid part of *Jeotgal*), *Myeolchi-jeotgal* (salted and fermented anchovy), *Myeolchi-aekjeot* (salted and fermented anchovy sauce), and *Saeu-jeot* (salted and fermented shrimp) are the most popular condiments for the preparation of kimchi [5]. Meanwhile, kimchi has been used as a term standing for various types of lactic acid fermented vegetables in Korea. Nearly 190 types of kimchi have been developed based on regional styles and ingredients used [5,6]. Hence, people who live in eight provinces in Korea have differences in preferences for the types of kimchi, which include a number of kimchi varieties made of different main ingredients as well as *Baechu* kimchi [5]. For instance, people in North and South Jeolla provinces prefer *Pa* (green onion) kimchi and *Gat* (mustard leaf) kimchi, respectively, to *Baechu* kimchi. The main ingredients of *Pa* kimchi and *Gat* kimchi (the most popular regional specialty kimchi types) are green onion (*Allium wakegi* Araki) and mustard leaf (*Brassica juncea*), respectively, and have been reported to possess various health benefits, including antimicrobial and antioxidative effects [7–10]. Thus, the two types of regional specialty kimchi have gained increasing attention in Korea. In the meantime, Chang et al. [11] reported that kimchi varieties are sources for functional lactic acid bacteria (LAB) with probiotic activities; for example, *Lactobacillus acidophilus* with anti-neoplastic activity. Consequently, it has been expected (and scientifically proven) that kimchi varieties, including *Pa* kimchi and *Gat* kimchi, provide health benefits such as antimicrobial, antioxidative, antimutagenic, anticarcinogenic, anti-obesity, and anti-diabetic effects [6,12].

Differently from health benefits of LAB in kimchi, some species of LAB isolated from *Baechu* kimchi, *Kkakdugi*, and *Chonggak* kimchi have displayed tyramine production [13,14]. Since tyramine may lead to hypertensive crisis, headache, cerebral hemorrhage, and heart attack [15,16], the levels of BA, particularly tyramine, in kimchi varieties need to be investigated. While several studies have reported BA content in *Baechu* kimchi [17–19], only a single study has described BA content in *Pa* kimchi and *Gat* kimchi [17]. Therefore, the present study investigated bacterial production of BA as well as BA content in *Pa* kimchi and *Gat* kimchi. Fermentation of the two types of kimchi was also carried out to determine the predominant LAB species capable of producing BA throughout the fermentation period.

2. Materials and Methods

2.1. Samples Used

Fifteen samples of each type of kimchi (*Pa* kimchi and *Gat* kimchi) were purchased from retail markets, and stored at 4 °C until use. The broth of kimchi samples was subjected to analyses of BA as well as physicochemical and microbial properties.

2.2. Measurements of Physicochemical Properties

Several physicochemical properties of *Pa* kimchi and *Gat* kimchi samples were measured as following. The pH was measured with a pH meter (Orion 3-star Benchtop, Thermo Scientific, Waltham, MA, USA). The salinity and titratable acidity were measured according to the Association of Official Analytical Chemists (AOAC) methods [20]. The water activity was measured by a water activity meter (AquaLab Pre, Meter Group, Inc., Pullman, WA, USA).

2.3. Measurements of Microbial Properties

Lactic acid bacterial count and total aerobic bacterial count were determined on de Man, Rogosa, and Sharpe (MRS) agar (Laboratorios Conda, Madrid, Spain) supplemented with Bromo-cresol purple (BCP; Samchun chemical, Ltd., Pyeongtaek, Korea) and Plate Count Agar (PCA; Difco, Becton Dickinson, Sparks, MD, USA), respectively. The incubation conditions were as follows: LAB, 37 °C for 48–72 h in an anaerobic chamber (Coy laboratory products Inc., Grass Lake, MI, USA) containing an atmosphere of 95% nitrogen and 5% hydrogen; total aerobic bacteria, 37 °C for 24–48 h in an incubator. After incubation, enumeration was carried out on plates with 30–300 colonies.

2.4. Isolation and Identification of LAB Strains

For the isolation of LAB strains, individual yellow colonies on MRS-BCP agar used for enumeration were picked and streaked on MRS agar, and incubated at 37 °C for 48–72 h. Single colonies were then streaked again on MRS agar, and incubated under the same condition to obtain pure cultures. The strains were then grown in MRS broth (Laboratorios Conda) under the same condition, and stored in the presence of 20% glycerol (v/v) at −80 °C.

For the identification of LAB strains, individual strains grown on MRS agar were subjected to sequence analysis of 16s rRNA gene amplified with the universal bacterial primer pair 518F and 805R (Solgent Co., Daejeon, Korea). The identities of sequences were determined using the Basic Local Alignment Search Tool (BLAST) of the National Center for Biotechnology Information (NCBI; http://www.ncbi.nlm.nih.gov/BLAST/).

2.5. Preparation and Fermentation of Pa Kimchi and Gat Kimchi

Pa kimchi and Gat kimchi were prepared according to the recipes described in previous studies [21,22] with minor modifications. Green onions or mustard leaves (the main ingredients of respective kimchi) were soaked in 10% (w/v) salt brine for 3 h. The salted main ingredients were rinsed with tap water three times, drained for 2 h, and mixed with a seasoning mixture consisting of red pepper powder, garlic, glutinous rice paste, and sugar. The desired values of salinity of Pa kimchi and Gat kimchi were 2.5%, respectively. The Pa kimchi and Gat kimchi samples were divided into five experimental groups, respectively, based on the presence or absence of Myeolchi-aekjeot (a condiment for kimchi) and LAB inoculum, as shown in Table 1. For this, three types of LAB strains for each kimchi, including a reference strain (L. brevis JCM1170 for both types of kimchi), two isolated strains capable of producing tyramine (L. brevis, PK3M04 for Pa kimchi, GK2M08 for Gat kimchi), and two isolated strains incapable of producing BA (L. plantarum, PK2M30 for Pa kimchi, GK2M12 for Gat kimchi), were inoculated onto the kimchi samples (all prepared with Myeolchi-aekjeot) at final concentrations of about 10^7 CFU/g, if required. The isolated strains of L. brevis and L. plantarum served as inocula were originated from samples of the corresponding kimchi types, respectively. Two groups of non-inoculated kimchi were prepared with and without Myeolchi-aekjeot, respectively. Finally, the samples of five groups belonging to each type of kimchi were placed in polypropylene containers (25 × 15 × 20 cm), and Pa kimchi and Gat kimchi samples were fermented at 25 °C for 5 days and 25 °C for 3 days, respectively. The fermentation was carried out in duplicate.

Table 1. Ingredients used for preparation of Pa (green onion) kimchi and Gat (mustard leaf) kimchi.

Ingredients (g)	Experimental Groups				
	B (Blank group)	C (Control group)	R (Reference group)	LB (L. brevis group)	LP (L. plantarum group)
Main Ingredient [1]	1500	1500	1500	1500	1500
Red Pepper Powder	120	120	120	120	120
Garlic	30	30	30	30	30
Ginger	30	30	30	30	30
Glutinous Rice Paste [2]	225	225	225	225	225
Sugar	30	30	30	30	30
Myeolchi-aekjeot	-	150	150	150	150
L. brevis JCM1170	-	-	10^7 CFU/g	-	-
L. brevis [3]	-	-	-	10^7 CFU/g	-
L. plantarum [4]	-	-	-	-	10^7 CFU/g

[1] Salted green onion or salted mustard leaf; [2] glutinous rice flour:water = 1:10; [3] L. brevis PK3M04 for Pa kimchi, L. brevis GK2M08 for Gat kimchi; [4] L. plantarum PK2M30 for Pa kimchi, GK2M12 for Gat kimchi.

2.6. BA Extraction from Kimchi Samples for HPLC Analysis

Analysis of BA in *Pa* kimchi and *Gat* kimchi samples was carried out based on the procedures developed by Eerola et al. [23] and modified by Ben-Gigirey et al. [24]. For BA extraction, 20 mL of 0.4 M perchloric acid (Sigma-Aldrich Chemical Co., St. Louis, MO, USA) were added to 5 g of kimchi broth, homogenized using a vortex mixer (Vortex-Genie, Scientific industries, Inc., Bohemia, NY, USA), and reacted in a cold chamber at 4 °C for 2 h. The mixture was centrifuged at $3000 \times g$ at 4 °C for 10 min, and the supernatant was collected. Subsequently, the residue was extracted again with an equal volume of 0.4 M perchloric acid under the same condition. Both supernatants were combined, and the final volume was adjusted to 50 mL with 0.4 M perchloric acid. The extract was filtered through Whatman paper No. 1 (Whatman International Ltd., Maidstone, UK).

2.7. BA Extraction from Lactic Acid Bacterial Cultures for HPLC Analysis

BA in lactic acid bacterial cultures were measured according to the procedures developed by Eerola et al. [23] and modified by Ben-Gigirey et al. [24,25]; the only exception was the culture medium used in this study. LAB strains were inoculated in 5 mL of MRS broth (pH 5.8) supplemented with 0.5% of L-histidine monohydrochloride monohydrate, L-tyrosine disodium salt hydrate, L-ornithine monohydrochloride, and L-lysine monohydrochloride, as well as with 0.0005% of pyridoxal-HCl (all from Sigma-Aldrich). After incubation at 37 °C for 48 h, 100 μL of the lactic acid bacterial broth culture was transferred to the same broth, and incubated under the same condition. Subsequently, the broth culture was filtered through a 0.2 μm membrane (Millipore Co., Bedford, MA, USA). Nine milliliters of 0.4 M perchloric acid were then added to 1 mL of the filtered broth culture, and mixed using a vortex mixer. The mixture was reacted in a cold chamber at 4 °C for 2 h, and centrifuged at $3000 \times g$ at 4 °C for 10 min. The supernatant (viz., the BA extract from LAB culture) was filtered through Whatman paper No. 1.

2.8. Preparation of Standard Solutions for HPLC Analysis

Stock standard solutions of BA, including tryptamine, β-phenylethylamine hydrochloride, putrescine dihydrochloride, cadaverine dihydrochloride, histamine dihydrochloride, tyramine hydrochloride, spermidine trihydrochloride, and spermine tetrahydrochloride (all from Sigma-Aldrich), were separately prepared at 10,000 mg/L concentration in deionized water. Working solutions at 1000 mg/L concentration were prepared by diluting 1 mL of each stock solution in deionized water to bring to a final volume of 10 mL. The concentrations of all standard solutions used were 0, 10, 50, 100, and 1000 mg/L. In addition, 1,7-diaminoheptane (Sigma-Aldrich) served as an internal standard.

2.9. Derivatization of Extracts and Standards

Derivatization of BA was carried out according to the procedures developed by Ben-Gigirey et al. [24]. One milliliter of extract (or standard solution) prepared above was mixed with 200 μL of 2 M sodium hydroxide and 300 μL of saturated sodium bicarbonate (all from Sigma-Aldrich). Two milliliters of a dansyl chloride (Sigma-Aldrich) solution (10 mg/mL) prepared in acetone were added to the mixture, and reacted at 40 °C for 45 min. Residual dansyl chloride was removed by adding 100 μL of 25% ammonium hydroxide (Sigma-Aldrich). After reaction at 25 °C for 30 min, the final volume was adjusted to 5 mL with acetonitrile. Finally, the mixture was centrifuged at $3000 \times g$ for 5 min, and the supernatant was filtered through a 0.2 μm-pore-size filter (Millipore). The filtered supernatant was kept at −25 °C until assayed by HPLC.

2.10. Chromatographic Separations

Chromatographic separation of BA was carried out according to the procedures developed by Ben-Gigirey et al. [24] with minor modifications. An HPLC unit (YL9100, YL Instruments Co., Ltd.,

Anyang, Korea) equipped with a UV-vis detector (YL Instruments) and with Autochro-3000 data system (YL Instruments) was used. A Nova-Pak C_{18} 4 μm column (150 mm × 4.6 mm, Waters, Milford, MA, USA) held at 40 °C was used with 0.1 M ammonium acetate (solvent A; Sigma-Aldrich) and acetonitrile (solvent B; SK chemicals, Ulsan, Korea) as the mobile phases adjusted to the flow rate of 1 mL/min. The program was set for a linear gradient starting from 50% of solvent B to reach 90% of the solvent at 19 min. The injection volume was 10 μL, and monitored at 254 nm. The limits of detection were approximately 0.1 μg/mL for all BA in standard solutions and in lactic acid bacterial cultures, and about 0.1 mg/kg for BA in food matrices. The procedure for BA analysis, from extraction to HPLC analysis, was illustrated in our previous article [14].

2.11. Statistical Analyses

The data were presented as means and standard deviations of triplicate experiments. Statistical outliers in data were eliminated according to the Grubbs outlier test (α = 0.05). The significance of differences was determined by one-way analysis of variance (ANOVA) with Fisher's multiple comparison module of the Minitab statistical software, version 17.1 (Minitab Inc., State College, PA, USA). Differences with p values of <0.05 were considered statistically significant. Differences with p values of <0.05 were considered statistically significant.

3. Results and Discussion

3.1. BA Content in Pa Kimchi and Gat Kimchi

As shown in Table 2, concentrations of BA in commercial products of *Pa* kimchi and *Gat* kimchi were measured to determine whether the amounts of BA in both kimchi are within the safe levels for consumption based on the suggestions of Ten Brink et al. [1] and Silla Santos [3]. Among the data obtained from 15 samples of each type of kimchi, statistical outliers in datasets generated from two samples of each kimchi type were respectively excluded from the results, and thereby data from 13 samples of each kimchi type were used for calculation and interpretation of the experimental results. The ranges (minimum to maximum) of BA content measured in *Pa* kimchi were as follows: Tryptamine, not detected (ND)–15.95 mg/kg; β-phenylethylamine, ND–5.97 mg/kg; putrescine, ND–254.47 mg/kg; cadaverine, ND–123.29 mg/kg; histamine, 8.67–386.03 mg/kg; tyramine, ND–181.10 mg/kg; spermidine, 2.32–18.74 mg/kg; spermine, ND–33.84 mg/kg (the upper part of Table 2). In *Gat* kimchi samples, the ranges of BA in the same order as above were ND–26.74 mg/kg, ND–15.75 mg/kg, 1.89–720.82 mg/kg, 2.12–52.43 mg/kg, 3.30–232.10 mg/kg, 1.28–142.06 mg/kg, 12.26–32.62 mg/kg, and ND–61.94 mg/kg, respectively (the lower part of Table 2). In agreement with the results, a wide range of BA content in *Baechu* kimchi has been reported previously [18,19]. On the other hand, Mah et al. [17] reported the low levels (below 30 mg/kg) of putrescine, cadaverine, histamine, tyramine, spermidine, and spermine in both *Pa* kimchi and *Gat* kimchi, all showing narrow ranges of BA content. Somewhat disparate observations between the present and previous studies might result from manufacturing methods, ingredients, and storage condition of kimchi samples [26].

Table 2. Biogenic amines (BA) content in *Pa* (green onion) kimchi and *Gat* (mustard leaf) kimchi samples.

Samples [1]	BA Content (mg/kg) [2]							
	TRP	PHE	PUT	CAD	HIS	TYR	SPD	SPM
PK11	14.92 ± 3.35 A,3	3.39 ± 0.01 C	254.47 ± 16.83 A	63.43 ± 5.77 C	286.04 ± 26.18 BC	150.81 ± 11.90 B	12.87 ± 1.72 C	27.90 ± 9.08 ABC
PK12	11.56 ± 1.20 BC	4.65 ± 0.94 B	158.33 ± 21.71 B	53.83 ± 7.36 CD	318.67 ± 19.21 B	181.10 ± 18.00 A	13.01 ± 1.46 BC	24.37 ± 1.72 ABCD
PK13	15.95 ± 3.91 A	5.97 ± 1.20 A	94.99 ± 5.91 D	82.73 ± 8.38 B	386.03 ± 33.13 A	179.98 ± 14.26 A	12.21 ± 1.98 C	25.87 ± 2.72 ABCD
PK14	6.33 ± 0.42 DE	2.34 ± 0.15 D	137.71 ± 12.78 C	38.56 ± 6.53 DE	249.07 ± 9.01 C	141.12 ± 2.10 BC	7.15 ± 3.63 DE	26.97 ± 7.00 ABC
PK15	10.26 ± 0.73 C	1.81 ± 0.23 D	135.55 ± 0.64 C	36.83 ± 0.14 DE	226.72 ± 0.39 C	127.59 ± 3.98 C	5.75 ± 2.05 EF	ND F4
PK16	13.74 ± 3.13 AB	3.28 ± 0.25 C	69.98 ± 1.74 E	119.68 ± 32.16 A	300.73 ± 59.36 B	ND D	16.55 ± 1.92 AB	10.40 ± 1.84 E
PK21	5.20 ± 1.37 DE	ND E	6.05 ± 0.25 G	11.02 ± 2.12 F	62.60 ± 2.55 D	16.31 ± 2.62 D	7.79 ± 1.49 DE	24.88 ± 1.74 ABCD
PK31	ND G	ND E	18.65 ± 1.81 FG	123.29 ± 8.97 A	45.72 ± 4.51 DEF	5.53 ± 0.86 D	12.40 ± 1.36 C	22.02 ± 4.24 BCD
PK41	3.30 ± 0.59 EF	ND E	7.56 ± 0.22 G	ND F	23.45 ± 0.18 EFG	15.99 ± 2.12 D	4.72 ± 1.15 EF	17.81 ± 2.68 CDE
PK42	1.12 ± 1.28 FG	ND E	ND G	ND F	8.67 ± 9.91 G	7.50 ± 8.72 D	2.32 ± 2.95 F	16.07 ± 14.20 DE
PK43	1.96 ± 0.53 FG	ND E	6.43 ± 0.86 G	2.52 ± 0.92 F	14.72 ± 0.05 FG	17.65 ± 3.20 D	5.34 ± 3.10 EF	33.84 ± 6.70 A
PK51	6.48 ± 0.96 D	ND E	31.47 ± 9.56 F	10.91 ± 1.33 F	53.12 ± 2.35 DE	20.30 ± 1.49 D	18.74 ± 2.77 A	29.71 ± 8.02 AB
PK52	ND G	1.57 ± 0.64 D	24.34 ± 1.87 F	30.08 ± 3.49 E	50.57 ± 3.90 DE	5.58 ± 0.73 D	10.03 ± 0.78 CD	22.86 ± 1.10 BCD
Average	6.99 ± 5.74	1.77 ± 2.04	78.79 ± 79.00	44.07 ± 42.85	155.85 ± 139.26	66.88 ± 74.91	9.91 ± 4.89	21.75 ± 8.94
GK11	4.86 ± 0.15 DE	ND E	9.38 ± 1.19 HI	8.61 ± 0.90 CD	11.27 ± 1.13 C	9.99 ± 0.05 GH	13.01 ± 0.36 FG	24.83 ± 7.96 DE
GK21	19.53 ± 6.88 AB	ND E	25.35 ± 1.52 GH	9.90 ± 0.96 CD	30.81 ± 0.86 C	138.45 ± 3.83 B	12.26 ± 5.86 G	ND F
GK22	18.06 ± 5.31 BC	ND E	36.52 ± 1.68 FG	16.18 ± 1.47 C	39.08 ± 1.76 BC	142.06 ± 7.71 A	13.07 ± 4.65 EFG	ND F
GK23	26.74 ± 0.79 A	4.30 ± 0.08 C	720.82 ± 37.04 A	47.73 ± 0.66 AB	3.30 ± 0.25 C	139.97 ± 10.17 B	28.49 ± 3.57 AB	61.94 ± 9.34 A
GK24	2.58 ± 0.44 E	7.41 ± 0.96 B	88.50 ± 10.64 D	5.56 ± 1.16 CD	3.62 ± 0.63 C	44.43 ± 1.70 E	26.09 ± 2.95 BC	58.57 ± 5.85 AB
GK25	10.58 ± 0.39 CD	4.33 ± 0.49 C	499.94 ± 23.93 B	7.66 ± 0.92 CD	3.30 ± 0.52 C	149.77 ± 7.17 A	32.62 ± 3.71 A	53.86 ± 1.62 B
GK31	23.04 ± 2.55 AB	15.75 ± 4.22 A	1.89 ± 0.68 I	2.12 ± 1.44 D	5.77 ± 0.12 C	7.33 ± 1.98 GHI	16.09 ± 0.64 EFG	21.25 ± 0.05 E
GK32	10.97 ± 1.18 CD	2.70 ± 0.06 CDE	178.97 ± 2.80 C	52.43 ± 0.91 A	206.95 ± 1.53 A	117.60 ± 5.10 C	16.50 ± 1.07 EFG	29.54 ± 6.64 D
GK33	ND E	3.00 ± 1.60 CDE	63.20 ± 0.40 E	48.60 ± 4.60 AB	232.10 ± 11.90 A	107.00 ± 5.40 D	18.60 ± 3.20 DE	ND F
GK41	5.13 ± 0.23 DE	ND E	25.65 ± 2.24 GH	6.67 ± 0.75 CD	41.13 ± 3.92 BC	18.41 ± 0.43 F	18.25 ± 1.61 DEF	40.63 ± 3.20 C
GK42	7.14 ± 0.38 DE	3.18 ± 2.18 CD	52.86 ± 9.96 EF	35.74 ± 0.45 B	88.21 ± 2.60 B	1.28 ± 0.14 I	23.57 ± 2.04 BCD	56.22 ± 2.06 AB
GK51	10.56 ± 1.84 CD	2.69 ± 0.62 CDE	16.99 ± 1.17 GHI	9.57 ± 0.61 CD	49.00 ± 4.50 BC	13.01 ± 2.99 FG	22.58 ± 5.56 CD	31.97 ± 3.26 D
GK52	6.64 ± 0.35 DE	1.34 ± 0.46 DE	34.44 ± 1.16 FG	15.73 ± 1.00 C	45.15 ± 2.29 BC	2.75 ± 0.38 HI	22.95 ± 1.31 BCD	28.11 ± 4.09 DE
Average	11.22 ± 8.23	3.44 ± 4.30	134.96 ± 220.53	20.5 ± 18.52	58.44 ± 75.77	76.15 ± 65.91	20.31 ± 6.35	31.30 ± 22.35

[1] PK: *Pa* kimchi, GK: *Gat* kimchi; [2] TRP: tryptamine, PHE: β-phenylethylamine, PUT: putrescine, CAD: cadaverine, HIS: histamine, TYR: tyramine, SPD: spermidine, SPM: spermine; [3] mean values (±standard deviation) in the same column that are followed by different letters (A–I) are significantly different ($p < 0.05$); [4] ND: not detected (<0.1 mg/kg).

In the present study, many samples of *Pa* kimchi and *Gat* kimchi had low levels of β-phenylethylamine (below 30 mg/kg) and other BA (below 100 mg/kg), which were all within the safe levels for consumption. However, 6 of 13 (46%) *Pa* kimchi and 2 of 13 (15%) *Gat* kimchi samples contained about two to four times higher levels of histamine than the safe level of the amine (100 mg/kg) suggested by Ten Brink et al. [1]. Meanwhile, 5 of 13 (38%) *Pa* kimchi and 6 of 13 (46%) *Gat* kimchi samples had approximately one to two times higher levels of tyramine compared to the safe level of the amine (100 mg/kg) suggested by Ten Brink et al. [1]. As for the levels of putrefactive amines, 4 of 13 (31%) *Pa* kimchi and 3 of 13 (23%) *Gat* kimchi samples contained putrescine over 100 mg/kg, and 2 of 13 (15%) *Pa* kimchi samples had over 100 mg/kg of cadaverine. Particularly, one (GK23) of the *Gat* kimchi samples contained 720.82 ± 37.04 mg/kg (mean ± standard deviation) of putrescine, so that total BA content was over the safe level of 1000 mg/kg suggested by Silla Santos [3]. Taken together, although many samples of *Pa* kimchi and *Gat* kimchi seem to be safe for consumption, proper monitoring and reduction of BA in both types of kimchi are required to reduce the risk of ingesting high levels of BA due to the presence of not only high levels of histamine and tyramine in several samples, but also a significant level of total BA, exceeding the safe levels. This is in disagreement with a previous report in which low levels of BA were detected in *Pa* kimchi and *Gat* kimchi as described above [17]. Different findings between the present and previous studies imply that the BA content in both types of kimchi can be reduced.

Though the main ingredients in kimchi varieties are vegetables, salted and fermented seafood products such as *Jeotgal* and *Aekjeot* are commonly used as condiments for the preparation of kimchi. Among the salted and fermented seafood products, *Myeolchi-aekjeot* is one of the most widely used condiments for kimchi preparation and makes up approximately 8%–15% of *Pa* kimchi (on the basis of weight percent) and 5%–9% of *Gat* kimchi, respectively [7,21,22,27]. Likely, most samples used in the present study were prepared with *Myeolchi-aekjeot*, according to food labels, although the labels just provided the list of ingredients (but not the content). In previous studies, Mah et al. [17] and Cho et al. [18] reported that *Myeolchi-aekjeot* contained significantly high levels of histamine, tyramine, putrescine, and cadaverine. Thus, high levels of these four BA in several samples of *Pa* kimchi and *Gat* kimchi in this study might be, at least partially, originated from *Myeolchi-aekjeot*. This speculation is in part supported by a previous report by Kang [19] that histamine content was higher in *Baechu* kimchi prepared with *Myeolchi-aekjeot* (592.78 ± 3.43 mg/kg) than in that without *Myeolchi-aekjeot* (77.13 ± 0.39 mg/kg). Therefore, when preparing *Pa* kimchi and *Gat* kimchi, condiment adjustments (such as using *Myeolchi-aekjeot* with low BA content, reducing the amount of *Myeolchi-aekjeot* used, and/or replacing *Myeolchi-aekjeot* with other salted and fermented seafood products) would be helpful in reducing BA content in the kimchi varieties.

3.2. Physicochemical and Microbial Properties of Pa Kimchi and Gat Kimchi

The values of pH, salinity, titratable acidity, water activity, total aerobic bacterial count, and lactic acid bacterial count in *Pa* kimchi and *Gat* kimchi were measured to see whether the properties are related to BA content. The properties measured in *Pa* kimchi were as follows: pH, 4.26–5.70 (5.19 ± 0.37, mean ± standard deviation); salinity, 1.28–2.73% (2.10 ± 0.40%); titratable acidity, 0.41–1.28% (0.68 ± 0.27%); water activity, 0.980–0.990 (0.990 ± 0.010); total aerobic bacterial count, 6.41–9.04 Log CFU/mL (7.43 ± 0.78 Log CFU/mL); lactic acid bacterial count, 5.78–8.74 Log CFU/mL (7.04 ± 0.84 Log CFU/mL). In *Gat* kimchi, the properties (in the same order as above) were as follows: pH, 4.26–5.30 (4.57 ± 0.37); salinity, 1.34–2.35% (1.96 ± 0.34%); titratable acidity, 0.53–1.23% (0.81 ± 0.26%); water activity, 0.973–0.989 (0.980 ± 0.010); total aerobic bacterial count, 6.37–8.11 Log CFU/mL (7.39 ± 0.65 Log CFU/mL); lactic acid bacterial count, 6.37–8.21 Log CFU/mL (7.18 ± 0.80 Log CFU/mL). The values are in accordance with those of previous reports on *Pa* kimchi and *Gat* kimchi [7,21]. Silla Santos [3] and Lu et al. [28] suggested that certain physicochemical and microbial properties are related to BA formation in fermented foods. In this study, linear regression analyses were carried out between the values of the properties and BA content. However, the content of respective BA had weak correlations with the respective properties in this study (data not shown). Therefore, the BA content detected in *Pa* kimchi and *Gat* kimchi might be affected by complex combinations of more than one factors (or/as multiple variable), including the addition of *Myeolchi-aekjeot*.

3.3. BA Production by LAB Strains Isolated from Pa *Kimchi and* Gat *Kimchi*

A total of 202 LAB strains (99 and 103 strains from *Pa* kimchi and *Gat* kimchi, respectively) were tested for in vitro BA production. Most LAB strains produced low levels of BA (below the detection limit). However, 16 of 99 LAB strains from *Pa* kimchi displayed three different patterns of BA production: (i) Strong production of tyramine along with β-phenylethylamine (seven strains), (ii) strong production of putrescine and cadaverine (eight strains), and (iii) simultaneous production of tyramine, β-phenylethylamine, putrescine, and cadaverine (one strain). Of the 103 LAB strains from *Gat* kimchi, 11 LAB strains produced BA with two different patterns, i.e., 9 LAB strains produced BA following pattern (i), and 2 LAB strains followed pattern (ii). Considering the high levels of histamine, tyramine, putrescine, and cadaverine measured in several samples of *Pa* kimchi and *Gat* kimchi (see Section 3.1), the LAB strains capable of producing tyramine, putrescine, and cadaverine might significantly contribute to the content of, at least, these three BA in both kimchi. On the other hand, histamine production was not detected from all the LAB strains. Thus, histamine content in both kimchi appeared to be attributed to the *Myeolchi-aekjeot* rather than LAB strains as described in Section 3.1.

To determine LAB species responsible for BA production in *Pa* kimchi and *Gat* kimchi, a total of 27 BA-producing LAB strains were identified based on 16s rRNA sequence analysis, as shown in Table 3. The LAB strains were identified as *L. brevis* (21 strains), *L. sakei* (2 strains), *Enterococcus faecium* (2 strains), and *Leuconostoc mesenteroides* (2 strains). Of the *L. brevis* strains, 10 strains produced 1.27–2.39 μg/mL of β-phenylethylamine and 278.57–365.96 μg/mL of tyramine, whereas another 10 strains produced 313.43–322.21 μg/mL of putrescine and 39.24–53.06 μg/mL of cadaverine (all 20 strains did not produce other BA). The other one strain simultaneously produced β-phenylethylamine, tyramine, putrescine, and cadaverine, exhibiting a somewhat weaker production of the former two BA and a little stronger production of the latter two BA. In accordance with the results, *L. brevis* has displayed a strain-dependent (as well as species-dependent) capacity to produce BA in previous studies [29,30]. Meanwhile, BA production by the rest six LAB strains, producing only β-phenylethylamine and tyramine, were as follows: *L. sakei*, 1.00 and 3.96 μg/mL of β-phenylethylamine, and 113.98 and 131.36 μg/mL of tyramine; *E. faecium*, 3.51 and 3.88 μg/mL of β-phenylethylamine, and 259.10 and 269.57 μg/mL of tyramine; *Leu. mesenteroides*, 1.47 and 1.91 μg/mL of β-phenylethylamine, and 145.14 and 301.67 μg/mL of tyramine. Similarly, the production of β-phenylethylamine and tyramine by *L. sakei*, *Leu. mesenteroides*, and *E. faecium* has been reported in previous studies [13,29,31]. Nevertheless, *L. brevis* is most likely responsible for BA formation in *Pa* kimchi and *Gat* kimchi, because the species was not only more abundant, but also revealed a larger production of most BA (except for β-phenylethylamine production by *E. faecium*) than other species in this study. In addition, all of the tyramine-producing LAB strains simultaneously produced a trace amount of β-phenylethylamine. This might be because bacterial tyrosine decarboxylase produces not only tyramine from tyrosine, but also β-phenylethylamine from phenylalanine [30,32].

For the 175 LAB strains incapable of producing BA (below the detection limit), about 30 strains were randomly selected, and subsequently identified as *L. brevis* (3 strains), *L. plantarum* (8 strains), *Leu. mesenteroides* (18 strains), and *Pediococcus pentosaceus* (1 strain). All the species have been reported as dominant species in previous studies on *Baechu* kimchi [6,13]. Thus, all the strains seem to have potential as BA-negative starter cultures for kimchi fermentation if they fulfill the criteria of starter culture [33]. Interestingly, it appeared in this study that, while certain strains of *L. brevis* and *Leu. mesenteroides* could produce high levels of BA, other strains belonging to the same species could not, which suggests that the ability of LAB strains to produce BA is determined at strain level. Therefore, although a bacterial species has been commonly considered to produce BA, some strains of this species may be able to serve as BA-negative and/or BA-degrading starter cultures depending on their abilities [34].

Table 3. BA produced in assay media by lactic acid bacteria (LAB) strains isolated from *Pa* (green onion) kimchi and *Gat* (mustard leaf) kimchi samples.

Samples	Isolates	N [1]	BA Production (µg/mL) [2]			
			PHE	PUT	CAD	TYR
Pa kimchi	*Lactobacillus brevis*	5	1.67 ± 0.46 (1.27–2.39) [3]	ND [4]	ND	293.27 ± 9.11 (278.57–301.52)
		8	ND	317.07 ± 4.47 (314.13–322.21)	48.43 ± 4.47 (41.69–53.06)	ND
		1	0.98 [5]	362.44	54.79	190.50
	Lactobacillus sakei	2	1.92 ± 1.18 (1.00–3.96)	ND	ND	122.67 ± 12.29 (113.98–131.36)
Gat kimchi	*Enterococcus faecium*	2	3.70 ± 0.26 (3.51–3.88)	ND	ND	264.34 ± 7.40 (259.10–269.57)
	Lactobacillus brevis	5	1.55 ± 0.53 (1.47–2.34)	ND	ND	338.51 ± 25.65 (300.61–365.96)
		2	ND	318.05 ± 4.94 (313.43–320.42)	42.99 ± 6.71 (39.24–47.73)	ND
	Leuconostoc mesenteroides	2	1.69 ± 0.31 (1.47–1.91)	ND	ND	122.67 ± 12.29(145.14–301.67)

[1] N: the number of strains tested; [2] PHE: β-phenylethylamine, PUT: putrescine, CAD: cadaverine, TYR: tyramine (tryptamine, histamine, spermidine, and spermine were all not detected); [3] mean ± standard deviation (the range from minimum to maximum); [4] ND, not detected (<0.1 µg/mL); [5] mean value obtained from a single strain.

3.4. Changes in BA Content during Fermentation of Pa Kimchi and Gat Kimchi

As shown in Table 1, five experimental groups of *Pa* kimchi and *Gat* kimchi were prepared with/without *Myeolchi-aekjeot* (a condiment used for kimchi preparation) and with/without selected LAB strains (*L. brevis* and *L. plantarum* as inocula) to investigate the contributions of the condiment and inoculum to BA content during the fermentation of both types of kimchi. *L. brevis* strains of PK3M04 (for *Pa* kimchi) and GK2M08 (for *Gat* kimchi) with the highest tyramine production capabilities among the isolated LAB strains were employed as inocula to investigate whether the strains of *L. brevis* practically contribute to BA formation (particularly tyramine) in *Pa* kimchi and *Gat* kimchi (refer to Section 3.3). On the other hand, *L. plantarum* strains of PK2M30 (for *Pa* kimchi) and GK2M12 (for *Gat* kimchi) were used as counterparts of *L. brevis* strains because the strains were unable to produce BA in this study, which is also supported by previous reports that *L. plantarum* was incapable of producing BA [29,30].

As presented in Figure 1, changes in physicochemical and microbial properties of *Pa* kimchi and *Gat* kimchi were similar with those of previous studies [7,21]. In detail, *Pa* kimchi groups displayed a slower acidification than *Gat* kimchi groups. Similarly, Lee et al. [21] reported a slower fermentation progress in *Pa* kimchi than in *Baechu* kimchi. Thus, fermentation of *Pa* kimchi and *Gat* kimchi was carried out for five days and three days, respectively. Changes in both pH and titratable acidity measured in the LP group of *Pa* kimchi were slightly faster than those in the other groups of the kimchi, and the values of all the groups remained constant in the later period of fermentation. In *Gat* kimchi, the pH and titratable acidity measured in the groups inoculated with LAB strains (LB, LP, and R groups) were altered much faster than non-inoculated groups (B and C groups), and the values remained constant in the later period of fermentation. Meanwhile, all groups of *Pa* kimchi and *Gat* kimchi exhibited an increment in microbial counts (either total aerobic bacteria or lactic acid bacteria) during the fermentation for 1.5 days, and remained constant thereafter. The salinity of *Pa* kimchi and

Gat kimchi was altered from 2.37 ± 0.06 to 2.07 ± 0.04, and from 3.03 ± 0.16 to 2.79 ± 0.06, respectively, showing a slight decline in the value (data not shown). The steady reduction of salinity might be due to osmosis between the main ingredient (either green onion or mustard leaf) and kimchi broth (containing seasoning mixture), as suggested by Shin et al. [35]. The water activity of Pa kimchi and Gat kimchi remained constant at 0.970 ± 0.002 and 0.959 ± 0.004, respectively, throughout the fermentation period (data not shown).

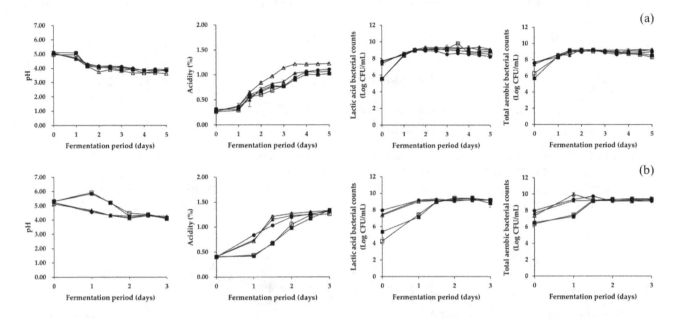

Figure 1. Changes in physicochemical and microbial properties during fermentation of (a) Pa kimchi and (b) Gat kimchi. □: B (no addition of Myeolchi-aekjeot, no inoculum), ■: C (addition of Myeolchi-aekjeot, no inoculum), ●: R (addition of Myeolchi-aekjeot, L. brevis JCM1170 for both kimchi), ▲: LB (addition of Myeolchi-aekjeot, L. brevis PK3M04 for Pa kimchi, L. brevis GK2M08 for Gat kimchi), △: LP (addition of Myeolchi-aekjeot, L. plantarum PK2M30 for Pa kimchi, L. plantarum GK2M12 for Gat kimchi).

As shown in Figures 2 and 3, overall changes in the content of respective BA during the fermentation of Pa kimchi and Gat kimchi were similar with those of a previous report on Kkakdugi and Chonggak kimchi [14]. In detail, tyramine content steadily increased in most groups of both kimchi, except for the LP group of Pa kimchi. The increment might be caused by either inoculated or indigenous tyramine-producing LAB strains (most likely L. brevis; refer to Section 3.3). Thus, R and LB groups of both kimchi that had been inoculated with tyramine-producing L. brevis strains revealed a larger increment of tyramine content than the other groups. In contrast, the tyramine content in LP groups either remained constant (in Pa kimchi) or increased relatively slightly (in Gat kimchi) because L. plantarum inocula were selected due to their incapability of BA production [29,30]. The levels of tyramine in B and C groups of both kimchi were much lower than those in R and LB groups, but either slightly higher than that in LP group of Pa kimchi or slightly lower than that in LP group of Gat kimchi. Thus, tyramine in the B and C groups might be formed by indigenous tyramine-producing LAB strains other than L. plantarum. Moreover, since tyramine content was higher in C group than in B group, bacterial strains derived from Myeolchi-aekjeot might also affect the content. In addition, -phenylethylamine in all groups exhibited almost the same patterns of alterations as tyramine in the groups throughout the fermentation of both kimchi, as expected based on previous reports [30,32], although the content was much lower than tyramine.

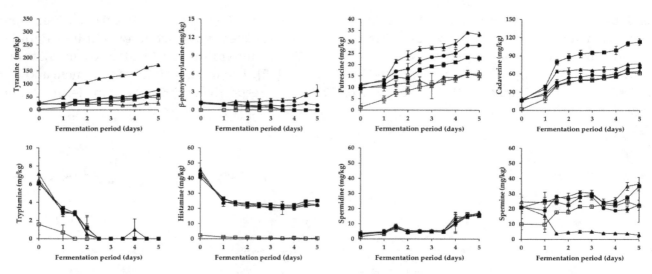

Figure 2. Changes in BA content in *Pa* kimchi during fermentation. □: B (no addition of *Myeolchi-aekjeot*, no inoculum), ■: C (addition of *Myeolchi-aekjeot*, no inoculum), •: R (addition of *Myeolchi-aekjeot*, *L. brevis* JCM1170), ▲: LB (addition of *Myeolchi-aekjeot*, *L. brevis* PK3M04), △: LP (addition of *Myeolchi-aekjeot*, *L. plantarum* PK2M30).

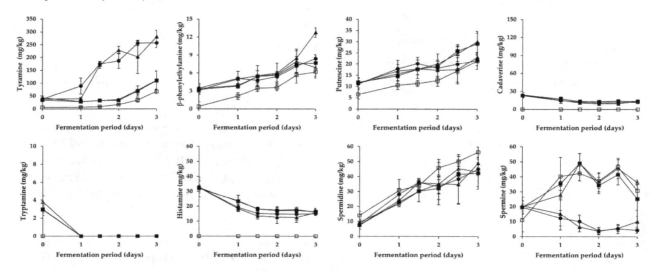

Figure 3. Changes in BA content in *Gat* kimchi during fermentation. □: B (no addition of *Myeolchi-aekjeot*, no inoculum), ■: C (addition of *Myeolchi-aekjeot*, no inoculum), •: R (addition of *Myeolchi-aekjeot*, *L. brevis* JCM1170), ▲: LB (addition of *Myeolchi-aekjeot*, *L. brevis* GK2M08), △: LP (addition of *Myeolchi-aekjeot*, *L. plantarum* GK2M12).

Differently from tyramine and β-phenylethylamine, tryptamine content in all groups of both kimchi steadily decreased during the earlier period of fermentation. Among the groups, B group (prepared without *Myeolchi-aekjeot*) had the lowest initial concentration, regardless of the types of kimchi. Likely, the lowest histamine content was detected in the B groups of both kimchi. Thus, the initial concentrations of tryptamine and histamine seem to come from *Myeolchi-aekjeot*. This is in agreement with previous studies suggesting that *Myeolchi-aekjeot* is a source of histamine in kimchi [14,19]. In addition to the tryptamine and histamine, similar differences in the content of other BA were also observed between B group and the other groups of both types of kimchi (except for spermidine in *Gat* kimchi) in this study. Therefore, it turned out that *Myeolchi-aekjeot* serves as an important source of not only histamine, but also other BA in kimchi. Supporting this assumption, as mentioned in Section 3.1, significantly high levels of BA, including tryptamine (<296.8 mg/kg), β-phenylethylamine (<54.1 mg/kg), putrescine (<182.1 mg/kg), cadaverine (<263.6 mg/kg), histamine (<1154.7 mg/kg), tyramine (<611.3 mg/kg), spermidine (<358.6 mg/kg), and spermine (<12.2 mg/kg), have been previously reported in *Myeolchi-aekjeot* [17,18]. Meanwhile, in most groups (except for B group), histamine content significantly decreased during the earlier period of fermentation, and remained constant in the later period, which is in accordance with previous reports by Kim et al. [36] and Jin et al. [14] that inoculated and indigenous LAB strains degraded histamine during the fermentation period.

As for putrefactive amines, putrescine content in all groups of *Pa* kimchi and *Gat* kimchi steadily increased throughout the fermentation period, as reported in previous studies on *Baechu* kimchi, *Kkakdugi*, and *Chonggak* kimchi [14,17]. In *Pa* kimchi, the groups inoculated with *L. brevis* (LB and R groups in order of amount) contained higher levels of putrescine than C and LP groups (in the same order), which indicates that while *L. brevis* significantly produced putrescine, *L. plantarum* likely degraded the putrescine because the lowest putrescine level was observed in LP group. In contrast, in the case of *Gat* kimchi, the levels of putrescine in C and LP groups were higher than those in LB and R groups. Thus, it seems that the *L. brevis* and *L. plantarum* work differently depending upon the types of kimchi. In the meantime, cadaverine content in *Pa* kimchi was the highest in C group, and followed by the groups inoculated with LAB strains (LB, R, and LP groups). This suggests that in *Pa* kimchi, bacterial strains originated from *Myeolchi-aekjeot* might influence the cadaverine content, and LAB strains could somehow (most likely their antimicrobial action) inhibit the cadaverine production by those bacterial strains from *Myeolchi-aekjeot*. In addition, cadaverine content in all groups constantly increased during the fermentation period of *Pa* kimchi. On the contrary, the cadaverine content in all groups of *Gat* kimchi slightly decreased throughout the fermentation period. Taken together, *Pa* kimchi and *Gat* kimchi displayed different patterns of putrefactive amine production (or degradation), which might be attributed to complex factors (and combinations thereof), including distinct indigenous strains as well as physicochemical properties resulting from different ingredients (especially main ingredients). Further research is required to clarify this issue.

Spermidine content in all groups of *Pa* kimchi and *Gat* kimchi steadily increased throughout the fermentation period, which is in agreement with observations in a previous study in which the spermidine content increased during the fermentation of *Kkakdugi* and *Chonggak* kimchi [14]. Particularly, *Pa* kimchi showed a much smaller increment of spermidine content than *Gat* kimchi. *Pa* kimchi also exhibited a statistically insignificant difference in spermidine content among all the groups on each day of the fermentation, while B group of *Gat* kimchi contained a statistically higher level of spermidine than C, R, and LP groups in the later period of fermentation. As described above, B groups of both kimchi showed lower BA levels than the other groups. However, the spermidine content in the B group of *Gat* kimchi is one exception to this rule, and the reason for this is unclear. Aside from spermidine, the content of spermine (another polyamine) in LB group of *Pa* kimchi and in LB and R groups of *Gat* kimchi revealed a decrement during the earlier period of fermentation, which indicates that *L. brevis* could degrade this polyamine. Similarly, there is a report of spermine-degrading *L. brevis* isolated from a traditional Italian cheese [37]. In the meantime, spermine content in the other groups of both kimchi in this study gradually increased during the earlier period of fermentation, and then decreased during the later period (except for C and LP groups of *Pa* kimchi showing a continuous increment). In both types of kimchi, interestingly, the levels of tyramine and β-phenylethylamine were higher in the LB and R groups inoculated with *L. brevis* than in the other groups, whereas an almost complete reversed order of the groups was found when spermine content was compared. Similar patterns were also observed in a previous study [14]. Thus, spermine content seems to be negatively related to microbial production of tyramine and β-phenylethylamine. Further research is required to make it clear. Together with the observations described in Sections 3.1 and 3.3, it turned out that *Myeolchi-aekjeot* and LAB strains influence either the formation or the degradation of BA in several ways during the fermentation of *Pa* kimchi and *Gat* kimchi.

4. Conclusions

This study indicated that the amounts of BA in many samples of *Pa* kimchi and *Gat* kimchi were within the safe levels for human consumption, but several samples contained histamine and tyramine over the safe levels of the respective BA (100 mg/kg for both BA, respectively). It was also found that while *Myeolchi-aekjeot* was an important source of BA in both *Pa* kimchi and *Gat* kimchi, *L. brevis* strains from both types of kimchi had a strain-dependent capacity to produce both β-phenylethylamine and tyramine, and/or both putrescine and cadaverine. The physicochemical and microbial properties of both types of kimchi exhibited weak correlations with BA content in the corresponding kimchi types in the present study. Through the fermentation of *Pa* kimchi and *Gat* kimchi, it turned out that *L. brevis* is responsible for the formation of BA, including β-phenylethylamine, tyramine, and putrescine, in both types of kimchi, and *Myeolchi-aekjeot* significantly affects the BA content in both *Pa* kimchi and *Gat* kimchi, except for spermidine in *Gat* kimchi.

Interestingly, the groups inoculated with *L. brevis* strains contained higher levels of tyramine and β-phenylethylamine, but much lower levels of spermine (probably due to degradation by the strains) than the other groups throughout the fermentation of *Pa* kimchi and *Gat* kimchi. Also, inoculated and indigenous LAB strains significantly degraded both tryptamine and histamine during the fermentation of both types of kimchi. The results imply that the LAB strains, including *L. brevis*, may have a potential to degrade specific BA in kimchi. In addition, differences in the production patterns of putrescine, cadaverine, and spermidine were observed between *Pa* kimchi and *Gat* kimchi. This might be because distinct complex factors (and their combinations) present in the respective types of kimchi differently influence the BA production in the kimchi varieties, which may need to be further studied.

Taken together, this study suggests two important measures for reducing BA in *Pa* kimchi and *Gat* kimchi, as follows: (i) The alteration of the ratio of *Myeolchi-aekjeot* to other ingredients used for kimchi preparation, and (ii) use of starter cultures other than BA-producing *L. brevis* strains, particularly BA-negative and/or -degrading LAB starter cultures.

Author Contributions: Conceptualization: J.-H.M.; investigation: J.-H.L., Y.H.J., Y.K.P., and S.J.Y.; writing—original draft: J.-H.L. and J.-H.M.; writing—review and editing: J.-H.L. and J.-H.M.; supervision: J.-H.M.

Acknowledgments: The authors thank Jae Hoan Lee and Junsu Lee of Department of Food and Biotechnology at Korea University for English editing and technical assistance, respectively.

References

1. Ten Brink, B.; Damink, C.; Joosten, H.M.L.J.; Huis In't Veld, J.H.J. Occurrence and formation of biologically active amines in foods. *Int. J. Food Microbiol.* **1990**, *11*, 73–84. [CrossRef]
2. Maintz, L.; Novak, N. Histamine and histamine intolerance. *Am. J. Clin. Nutr.* **2007**, *85*, 1185–1196. [CrossRef]
3. Silla Santos, M.H. Biogenic amines: Their importance in foods. *Int. J. Food Microbiol.* **1996**, *29*, 213–231. [CrossRef]
4. Codex Alimentarius Commission. *Codex Standard for Kimchi, Codex Stan 223-2001*; Food and Agriculture Organization of the United Nations: Rome, Italy, 2001.
5. Cha, Y.-J.; Lee, Y.-M.; Jung, Y.-J.; Jeong, E.-J.; Kim, S.-J.; Park, S.-Y.; Yoon, S.-S.; Kim, E.-J. A nationwide survey on the preference characteristics of minor ingredients for winter *kimchi*. *J. Korean Soc. Food Sci. Nutr.* **2003**, *32*, 555–561.
6. Lee, C.-H. Lactic acid fermented foods and their benefits in Asia. *Food Control* **1997**, *8*, 259–269. [CrossRef]
7. Kang, C.-H.; Chung, K.-O.; Ha, D.-M. Inhibitory effect on the growth of intestinal pathogenic bacteria by *kimchi* fermentation. *Korean J. Food Sci. Technol.* **2002**, *34*, 480–486.
8. Benkeblia, N. Antimicrobial activity of essential oil extracts of various onions (*Allium cepa*) and garlic (*Allium sativum*). *LWT-Food Sci. Techol.* **2004**, *37*, 263–268. [CrossRef]

9. Kim, J.-I.; Choi, J.-S.; Kim, W.-S.; Woo, K.-L.; Jeon, J.-T.; Min, B.-T.; Cheigh, H.-S. Antioxidant activity of various fractions extracted from mustard leaf (*Brassica juncea*) and their kimchi. *J. Life Sci.* **2004**, *14*, 286–290.

10. Ryu, S.-H.; Song, W.-S. Amino acid analysis and antioxidation activity in *Allium wakegi* Araki. *Korean J. Plan. Res.* **2004**, *17*, 35–40.

11. Chang, J.-H.; Shim, Y.Y.; Cha, S.-K.; Chee, K.M. Probiotic characteristics of lactic acid bacteria isolated from kimchi. *J. Appl. Microbiol.* **2010**, *109*, 220–230. [CrossRef] [PubMed]

12. Jung, J.Y.; Lee, S.H.; Jeon, C.O. Kimchi microflora: History, current status, and perspectives for industrial kimchi production. *Appl. Microbiol. Biotechnol.* **2014**, *98*, 2385–2393. [CrossRef] [PubMed]

13. Kim, M.-J.; Kim, K.-S. Tyramine production among lactic acid bacteria and other species isolated from kimchi. *LWT-Food Sci. Technol.* **2014**, *56*, 406–413. [CrossRef]

14. Jin, Y.H.; Lee, J.H.; Park, Y.K.; Lee, J.-H.; Mah, J.-H. The occurrence of biogenic amines and determination of biogenic amine-producing lactic acid bacteria in *Kkakdugi* and *Chonggak* kimchi. *Foods* **2019**, *8*, 73. [CrossRef]

15. Smith, T.A. Amines in Food. *Food Chem.* **1981**, *6*, 169–200. [CrossRef]

16. Park, Y.K.; Lee, J.H.; Mah, J.-H. Occurrence and reduction of biogenic amines in traditional Asian fermented soybean foods: A review. *Food Chem.* **2019**, *278*, 1–9. [CrossRef]

17. Mah, J.-H.; Kim, Y.J.; No, H.-K.; Hwang, H.-J. Determination of biogenic amines in *kimchi*, Korean traditional fermented vegetable products. *Food Sci. Biotechnol.* **2004**, *13*, 826–829.

18. Cho, T.-Y.; Han, G.-H.; Bahn, K.-N.; Son, Y.-W.; Jang, M.-R.; Lee, C.-H.; Kim, S.-H.; Kim, D.-B.; Kim, S.-B. Evaluation of biogenic amines in Korean commercial fermented foods. *Korean J. Food Sci. Technol.* **2006**, *38*, 730–737.

19. Kang, H.-W. Characteristics of *kimchi* added with anchovy sauce from heat and non-heat treatments. *Culin. Sci. Hosp. Res.* **2013**, *19*, 49–58.

20. AOAC. *Official Methods of Analysis of AOAC International*, 18th ed.; AOAC International: Gaithersburg, MD, USA, 2005.

21. Lee, H.-J.; Joo, Y.-J.; Park, C.-S.; Lee, J.-S.; Park, Y.-H.; Ahn, J.-S.; Mheen, T.-I. Fermentation patterns of Green Onion *Kimchi* and Chinese Cabbage *Kimchi*. *Korean J. Food Sci. Technol.* **1999**, *31*, 488–494.

22. Kim, H.-R.; Cho, K.-J.; Kim, J.-S.; Lee, I.-S. Quality changes of mustard leaf (*Dolsangat*) kimchi during low temperature storage. *Korean J. Food Sci. Technol.* **2006**, *38*, 609–614.

23. Eerola, S.; Hinkkanen, R.; Lindfors, E.; Hirvi, T. Liquid chromatographic determination of biogenic amines in dry sausages. *J. AOAC Int.* **1993**, *76*, 575–577. [PubMed]

24. Ben-Gigirey, B.; De Sousa, J.M.V.B.; Villa, T.G.; Barros-Velazquez, J. Changes in biogenic amines and microbiological analysis in albacore (*Thunnus alalunga*) muscle during frozen storage. *J. Food Prot.* **1998**, *61*, 608–615. [CrossRef] [PubMed]

25. Ben-Gigirey, B.; De Sousa, J.M.V.B.; Villa, T.G.; Barros-Velazquez, J. Histamine and cadaverine production by bacteria isolated from fresh and frozen albacore (*Thunnus alalunga*). *J. Food Prot.* **1999**, *62*, 933–939. [CrossRef] [PubMed]

26. Cheigh, H.-S.; Park, K.-Y. Biochemical, microbiological, and nutritional aspects of kimchi (Korean fermented vegetable products). *Crit. Rev. Food Sci. Nutr.* **1994**, *34*, 175–203. [CrossRef] [PubMed]

27. Kang, K.-O.; Lee, S.-H.; Cha, B.-S. A study on the material ratio Kimchi products of Seoul and Chung Cheong area and chemical properties of the fermented Kimchis. *Korean J. Soc. Food Sci.* **1995**, *11*, 487–493.

28. Lu, S.; Xu, X.; Shu, R.; Zhou, G.; Meng, Y.; Sun, Y.; Chen, Y.; Wang, P. Characterization of biogenic amines and factors influencing their formation in traditional Chinese sausages. *J. Food Sci.* **2010**, *75*, M366–M372. [CrossRef] [PubMed]

29. Bover-Cid, S.; Hugas, M.; Izquierdo-Pulido, M.; Vidal-Carou, M.C. Amino acid-decarboxylase activity of bacteria isolated from fermented pork sausages. *Int. J. Food Microbiol.* **2001**, *66*, 185–189. [CrossRef]

30. Landete, J.M.; Pardo, I.; Ferrer, S. Tyramine and phenylethylamine production among lactic acid bacteria isolated from wine. *Int. J. Food Microbiol.* **2007**, *115*, 364–368. [CrossRef]

31. González de Llano, D.; Cuesta, P.; Rodriguez, A. Biogenic amine production by wild lactococcal and leuconostoc strains. *Lett. Appl. Microbiol.* **1998**, *26*, 270–274. [CrossRef] [PubMed]

32. Pessione, E.; Pessione, A.; Lamberti, C.; Coïsson, D.J.; Riedel, K.; Mazzoli, R.; Bonetta, S.; Eberl, L.; Giunta, C. First evidence of a membrane-bound, tyramine and β-phenylethylamine producing, tyrosine decarboxylase in *Enterococcus faecalis*: A two-dimensional electrophoresis proteomic study. *Proteomics* **2009**, *9*, 2695–2710. [CrossRef] [PubMed]

33. Holzapfel, W.H. Appropriate starter culture technologies for small-scale fermentation in developing countries. *Int. J. Food Microbiol.* **2002**, *75*, 197–212. [CrossRef]

34. Mah, J.-H.; Park, Y.K.; Jin, Y.H.; Lee, J.-H.; Hwang, H.-J. Bacterial production and control of biogenic amines in Asian fermented soybean foods. *Foods* **2019**, *8*, 85. [CrossRef] [PubMed]

35. Shin, Y.-H.; Ann, G.-J.; Kim, J.-E. The changes of hardness and microstructure of Dongchimi according to different kinds of water. *Korean J. Food Cook. Sci.* **2004**, *20*, 86–94.

36. Kim, S.-H.; Kim, S.H.; Kang, K.H.; Lee, S.; Kim, S.J.; Kim, J.G.; Chung, M.J. Kimchi probiotic bacteria contribute to reduced amounts of *N*-nitrosodimethylamine in lactic acid bacteria-fortified kimchi. *LWT-Food Sci. Technol.* **2017**, *84*, 196–203. [CrossRef]

37. Guarcello, R.; De Angelis, M.; Settanni, L.; Formiglio, S.; Gaglio, R.; Minervini, F.; Moschetti, G.; Gobbetti, M. Selection of amine-oxidizing dairy lactic acid bacteria and identification of the enzyme and gene involved in the decrease of biogenic amines. *Appl. Environ. Microbiol.* **2016**, *82*, 6870–6880. [CrossRef] [PubMed]

3

Liquid Chromatographic Determination of Biogenic Amines in Fish based on Pyrene Sulfonyl Chloride Pre-Column Derivatization

Elvira S. Plakidi [†], **Niki C. Maragou** [*,‡](ID), **Marilena E. Dasenaki**(ID), **Nikolaos C. Megoulas** [§], **Michael A. Koupparis and Nikolaos S. Thomaidis**(ID)

Laboratory of Analytical Chemistry, Department of Chemistry, National and Kapodistrian University of Athens, Panepistimioupolis Zografou, 15771 Athens, Greece; eplakidi@hcmr.gr (E.S.P.); mdasenaki@chem.uoa.gr (M.E.D.); n.megkoulas@qualimetrix.com (N.C.M.); koupparis@chem.uoa.gr (M.A.K.); ntho@chem.uoa.gr (N.S.T.)

*   Correspondence: nmarag@chem.uoa.gr
†   Present affiliation for E.S.P.: Institute of Oceanography, Hellenic Centre for Marine Research (H.C.M.R.), Mavro Lithari, 19013 Anavyssos, Attiki, Greece.
‡   Present affiliation for N.C.M: Department of Pesticides Control and Phytopharmacy, Benaki Phytopathological Institute, 8 St. Delta str, 14561 Kifissia, Attica, Greece.
§   Present affiliation for N.C.M: Qualimetrix SA, 579 Mesogeion avenue, 15343 Agia Paraskevi, Attica, Greece.

Abstract: Monitoring of biogenic amines in food is important for quality control, in terms of freshness evaluation and even more for food safety. A novel and cost-effective method was developed and validated for the determination of the main biogenic amines: histamine, putrescine, cadaverine, spermidine and spermine in fish tissues. The method includes extraction of amines with perchloric acid, pre-column derivatization with Pyrene Sulfonyl Chloride (PSCl), extraction of derivatives with toluene, back-dissolution in ACN after evaporation and determination by reversed phase high performance liquid chromatography with UV and intramolecular excimer fluorescence detection. The structure of the pyrene-derivatives was confirmed by liquid chromatography–mass spectrometry with electrospray ionization. The standard addition technique was applied for the quantitation due to significant matrix effect, while the use of 1,7-diaminoheptane as internal standard offered an additional confirmation tool for the identification of the analytes. Method repeatability expressed as %RSD ranged between 7.4–14% for the different amines and recovery ranged from 67% for histamine up to 114% for spermine. The limits of detection ranged between 0.1–1.4 mg kg^{-1} and the limits of quantification between 0.3–4.2 mg kg^{-1}. The method was applied to canned fish samples and the concentrations of the individual biogenic amines were below the detection limit up to 40.1 mg kg^{-1}, while their sum was within the range 4.1–49.6 mg kg^{-1}.

Keywords: bioamines; polyamines; scombroid poisoning; seafood; dietary exposure; HPLC; pyrene probe; excimer fluorescence; intramolecular excitement

1. Introduction

Biogenic amines (BAs) have been widely associated with food quality and safety [1–4]. Although they are naturally occurring substances in animals and humans, their presence in food is mainly a result of bacterial growth and spoilage through the decarboxylation path of free amino acids [3,4]. Histamine, putrescine, cadaverine, spermidine and spermine (Figure 1), are considered among the most important biogenic amines occurring in food, and they have been used for the generation of a chemical quality index of canned tuna [5] and Mediterranean hake stored in ice [1].

Apart from the effect on the quality of food as regards aspects like the freshness, the presence of BAs in food can endanger food safety, since the dietary exposure of humans to BAs, through consumption of food and beverages with high levels of these compounds, can have serious toxicological effects on human health [4]. The most well-known biogenic amine is histamine, which is present in great abundance in fish and fishery products, and is the main component in "scombroid poisoning" or "histamine poisoning" [2]. Putrescine and cadaverine, are also associated with this illness and their presence has been reported to enhance the toxicity of histamine [2,6]. Moreover, it has been reported that biogenic amines, especially putrescine and cadaverine, may also be considered as carcinogens because of their ability to react with nitrites to form potentially carcinogenic nitrosamines [6,7]. Foods likely to contain high levels of biogenic amines include fish, fish products and fermented foodstuffs (meat, dairy, vegetables, beers and wines) [4].

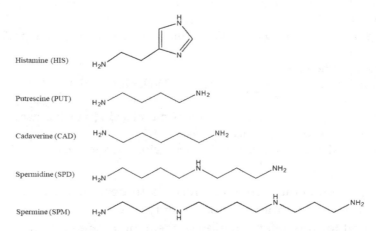

Figure 1. Molecular structures of the biogenic polyamines.

In order to protect human health, the European Union has established safety criteria for histamine levels in fishery products that range between 100 and 400 mg kg^{-1} [8]. On the other hand, in the United States of America, according to FDA and EPA safety levels, the decomposition criteria limit of histamine for scombrotoxin-forming fish, like tuna, mahi-mahi, and related fish is 50 mg kg^{-1}, while the toxic limit of histamine is 500 mg kg^{-1} [9].

According to the scientific opinion on risk-based control of biogenic amine formation in fermented foods published by the European Food Safety Authority (EFSA), no adverse health effects were observed after exposure to 50 mg of histamine in food per person per meal for healthy individuals, but below detectable limits for those with histamine intolerance, while for putrescine and cadaverine, the information was insufficient in that respect [4]. As pointed out in the EFSA scientific opinion, monitoring of BAs' concentrations in fermented foods during the production process, and along the food chain, would be beneficial for controls and further knowledge. The opinion concludes that additional research is required on several aspects of BAs including validation of the methods of analysis.

Based on the above, it becomes evident that accurate and of high detectability analytical methods for the determination of BAs in food are necessary in order to generate reliable data for a safe estimation of human dietary exposure to biogenic amines. As a result, a number of analytical methods has been developed for the determination of BAs in foodstuff, including capillary electrophoresis (CE) [10–12], gas chromatography (GC) with mass spectrometric detection [13,14], ion chromatography with conductimetric and amperometric detection [15] and thin layer chromatography with densitometric scanning [16]. Amongst them, undoubtedly, the most widely applied analytical technique is liquid chromatography (LC) with ultraviolet absorbance (UV) [17–25], or fluorescence detector (FLD) [26–36]. The use of mass spectrometric detection is still relatively limited [37–40].

Due to the lack of chromophore or fluorophore groups in the most of the BAs structures, a pre-column [17–28,30,33–36] or post-column [29,31,32] derivatization step is required.

The derivatization reagents tested for the liquid chromatographic determination of BAs include dialdehydes such as o-phthalaldehyde (OPA) [27,29–32] and naphthalene-2,3-dicarboxaldehyde (NDA) [33], sulfonyl chlorides such as dansyl chloride (Dns-Cl) [17,19,25], dabsyl chloride (Dabs-Cl) [19,25] and 10-ethyl-acridine-3-sulfonyl chloride [19,26], benzoyl chloride [18,19,22,23], succinimidyl reagents like 6-aminoquinolyl-N-hydroxy-succinimidyl carbamate [21], N-(9-fluorenylmethoxycarbonyloxy) succinimide (Fmoc-OSu) [36] and 1-pyrenebutanoic acid succinimidyl ester (PSE) [35], as well as diethyl ethoxymethylenemalonate [20,24].

Among the aforementioned derivatization agents, 1-pyrenebutanoic acid succinimidyl ester (PSE) presents special interest because pyrene derivatives of some biogenic polyamines exhibit intramolecular excimer fluorescence [35]. In case of pyrene derivatives with two or more pyrene moieties and favorable structural conformation, excimer fluorescence is generated by the energy transfer from one pyrene moiety to the other, which results in intramolecular excitement [41]. Excimer fluorescence induces a significant shifting of the emission to higher wavelengths and, therefore, selectivity is increased compared to mono-pyrene derivatives of biogenic monoamines. This derivatization agent has been used for the fluorescence determination of polyamines, putrescine, cadaverine, spermidine and spermine as indicators of food decomposition without chromatographic separation [41], and for their determination in different solvents after liquid chromatography [42]. However, the chromatographic determination of histamine in solvents after derivatization with PSE was not possible [42]. Moreover, when PSE was used for the chromatographic determination of the same polyamines in fish samples, only putrescine and cadaverine were satisfactorily determined [35]. The chromatographic determination of histamine using pyrene reagents for the generation of the excimer fluorescence has been reported to be successful in urine samples with PSE where histamine was the sole analyte [43] and in soy sauce using 2-chloro-4-methoxy-6-(4-(pyren-4-yl)butoxy)-1,3,5-triazine, a derivatization reagent which was laboratory synthesized [44].

The present work proposes 1-pyrenesulfonyl chloride (PSCl), as a new derivatization reagent for the simultaneous determination of histamine (HIS), putrescine (PUT), cadaverine (CAD), spermidine (SPD) and spermine (SPM) in fish samples based on fluorescence excimer as presented in Figure 2, and UV detection in series, after high performance liquid chromatography. Pyrene sulphonyl chloride has been used in the past as a reagent for quantitation of estrogens in human serum after liquid chromatography and fluorescence detection [45]. The conditions of the derivatization reaction, including: pH, amount of PSCl, temperature and time, were optimized and the structures of the produced derivatives of the tested polyamines were confirmed by liquid chromatography coupled with electrospray mass spectrometry (LC-ESI-MS). Liquid chromatographic separation and detection wavelengths were also optimized in order to obtain the maximum selectivity and sensitivity. The proposed LC-UV-FLD method was validated and the stability of the PSCl derivatives of BAs was investigated.

To the best of our knowledge this is the first study that presents the simultaneous determination of five significant biogenic amines, including histamine, for the quality and safety control of fish using pyrene sulfonyl chloride in order to induce intramolecular excimer fluorescence. In addition, the present study enlightens some aspects of histamine chromatographic behavior after derivatization with a pyrene reagent, reporting for the first time the existence of three chromatographic peaks which were identified with absorbance and emission spectra, as well as with mass spectrometry, addressing to some extend the reported difficulties of the chromatographic determination of histamine after derivatization with pyrene reagents [35,42].

Comparing to other derivatization agents, like Dns-Cl and OPA, the most significant advantage of the use of PSCl is the increased selectivity towards the biogenic polyamines investigated in the present study, which are derivatized to compounds that exhibit intramolecular excimer fluorescence at high wavelengths, where there is no interference from other endogenous biogenic monoamines or the excess of the derivatization reagent. Moreover, comparing to the derivatization of amines with Dns-Cl [17], the derivatization with PSCl requires much shorter reaction time, 15 min versus 20 h, and offers the possibility of the use of both detectors UV and fluorescence, increasing the degree of analyte confirmation, whereas Dns-Cl derivatized amines are detected only with UV. Finally, the smaller amount of sample and organic solvents used for the sample preparation of the present study (2 g sample, 300 μL toluene and 300 μL acetonitrile) comparing to studies using Dns-Cl (40 g sample, 3 mL heptane and 1.5 mL acetonitrile) [17] and OPA (50 g sample, 6 mL ethyl acetate and 1 mL acetonitrile) [27] for the derivatization of BAs in fish, as well as the use of a low-cost readily available HPLC-UV/FLD instrumentation comparing to mass spectrometry instruments [38,39], render the present method rather cost-effective and eco-friendly.

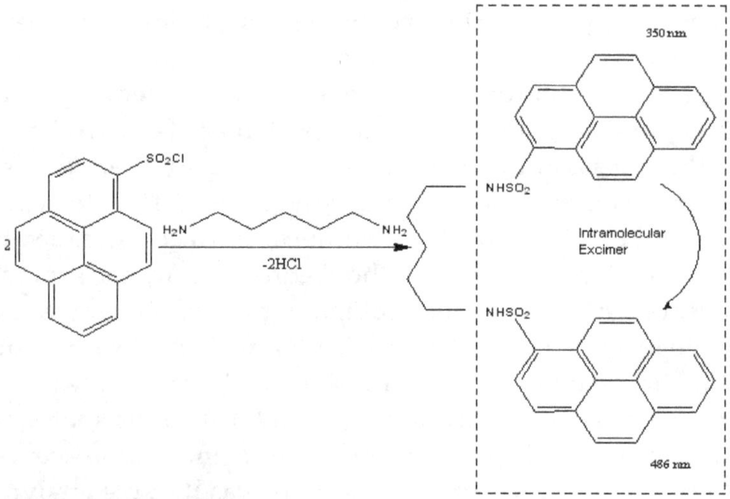

Figure 2. Intramolecular excimer-forming fluorescence derivatization of cadaverine with PSCl.

2. Materials and Methods

2.1. Standards and Reagents

HPLC-grade water (specific resistance > 17.8 MΩ cm) was obtained by a Milli-Q water purification system (Millipore, Billerica, MA, USA), acetonitrile (ACN) of HPLC grade was purchased from LAB SCAN (Dublin, Ireland) and toluene from Mallinckrodt (Surrey, UK). The derivatization reagent Pyrene Sulfonyl Chloride (PSCl, $C_{16}H_9ClO_2S$, MW: 308 g mol^{-1}) was obtained from Molecular Probes (Eugene, Oregon, OR, USA). A 3.0 mM PSCl solution was prepared by dissolution of 45 mg of the reagent in 50 mL ACN and was stored at −20 °C for 6 months. For the extraction procedure, an aqueous solution of 0.2 M perchloric acid ($HClO_4$, 70–72% reagent grade, Merck, Whitehouse Station, New Jersey, NJ, USA) and a saturated sodium carbonate (Na_2CO_3, purity ≥ 99.5%, Merck) solution in water were prepared.

The hydrochloric salts of the biogenic amines: (a) histamine dihydrochloride (99%, Sigma-H7250), (b) putrescine dihydrochloride (98%, Sigma-P7505), (c) cadaverine dihydrochloride (98%, Sigma-C8561), (d) spermidine trihydrochloride (98%, Sigma-S2501) and (e) spermine tetrahydrochloride (Sigma-S2876), as well as 1,7-diaminoheptane dihydrochloride (98%, Aldrich-D 17408), used as internal standard, and L-proline (99.0%, Fluka-81710) were purchased from Sigma-Aldrich (St. Louis, Missouri, USA). Individual standard stock solutions were prepared by dissolution of the appropriate hydrochloric salt equivalent to 10.0 mg of each amine in 100.0 mL of water and were stored at 4 °C for one month. Working standards solutions of mixtures of the BAs were prepared by appropriate dilutions of the stock solutions with 0.2 M $HClO_4$. A 2% *w/v* L-proline aqueous solution was also prepared, used as quenching agent of the derivatization.

2.2. Instrumentation

A centrifuge equipment (Rotofix 32-Hettich, Merck, Darmstadt, Germany), a pH meter (Metrohm, Herisau, Switzerland) and a thermostated evaporation aluminum block under nitrogen flow were used for the sample treatment.

HPLC-UV-FLD analysis was carried out on an Agilent 1100 LC modular system (Agilent Technologies, Wilmington, DE, USA) consisting of a G1311A quant pump, a G1379A degasser, a G1321A fluorescence detector (FLD), a G1314A ultra-violet detector (UV) and a 7725i Rheodyne (California-USA) manual sample injector equipped with a 20 µL loop.

The structural elucidation of PSCl derivatives was performed with a Thermo Finnigan LC–MS system (San Jose-CA) consisted of a Spectra System P 4000 pump, a Spectra System AS 3000 autosampler with the volume injection set to 20 µL and a MSQ quadrupole mass spectrometer equipped with an electrospray ionization interface (ESI).

2.3. Fish Samples

Six kinds of canned fish including (a) sea bass, (b) anchovy, (c) anchovy marinated, (d) mackerel smoked, (e) sardines in oil and (f) tuna in water and salt, were obtained from local markets in Athens. The whole content of each can was homogenized with a laboratory homogenizer and stored at −15 °C. Aliquots of the homogenized samples were used for method optimization and validation. All the samples in total were measured according to the optimum protocol for the determination of biogenic amines content.

2.4. Method Development

2.4.1. Extraction of Biogenic Amines

Aliquots of 2 g of fish homogenate, accurately weighted, were placed in a 15 mL plastic centrifuge tube and spiked with 2.4 µg of internal standard (24 µL of 100 µg mL^{-1} standard solution). Eight mL of 0.2 M perchloric solution were added, the tube was capped and vortexed for 5 min. Afterwards, the tube was centrifuged at 4000 rpm for 15 min and the supernatant layer was isolated for the derivatization of biogenic amines.

2.4.2. Optimization of Derivatization of Biogenic Amines–Final Protocol

The effect of pH, temperature, reaction time and amount of derivatization reagent on the derivatization yield was investigated using a standard mixture containing 1 mg L^{-1} of the tested biogenic amines and 0.3 mg L^{-1} of the internal standard and also with anchovy sample spiked with biogenic amines at a measured concentration of 1 mg L^{-1}. The optimization was performed according to the univariant technique and the effect of the tested parameters was evaluated based on the peak area of the FLD chromatograms.

Study of pH: Since the pyrene sulfonyl chloride derivatization reaction with the biogenic amines is a nucleophilic substitution, a moderate alkaline medium is required so that the biogenic amines become deprotonated. However, a highly alkaline medium must be avoided, since a successive quick decomposition by alkaline hydrolysis of the produced pyrene derivatives may be observed. The pH values tested ranged between 7 and 11, adjusted with the addition of different volumes of a saturated sodium carbonate solution. A volume of 400 µL of 3 mM PSCl solution was used and the derivatization took place at 60 °C for 15 min.

Study of temperature and time: The tested temperatures for the derivatization were 22, 30, 40, 50, 60, 70 and 80 °C with a reaction time of 15 min. The optimization of the derivatization temperature was conducted at pH of 9.6 with the addition of 30 µL of saturated sodium carbonate solution and 400 µL of 3 mM PSCl solution. Afterwards, the derivatization time at the optimum temperature was examined for a duration between 15 and 60 min.

Study of derivatization reagent amount: For the determination of the optimum amount of the derivatization reagent, different volumes between 200–1200 µL of 3 mM PSCl solution in ACN were tested. The pH was set at 9.6 with the addition of 30 µL of saturated sodium carbonate solution and the derivatization was left to take place at 60 °C for 15 min.

Final protocol of derivatization: First, 300 µL of the supernatant layer, which was isolated during the extraction step (Section 2.4.1), was transferred into a 10 mL glass tube, wrapped with aluminium foil to protect the derivatives from light. Subsequently, 30 µL of saturated sodium carbonate solution was added in order to set the pH at 9.6. Then, 600 µL of 3.0 mM solution of the derivatization reagent PSCl was added, the tube was capped, vortexed and heated in a water bath at 60 °C temperature for 15 min. After being cooled down at room temperature, 100 µL of a 2% *w/v* L-proline aqueous solution was added to stop the derivatization reaction. The tube was left for 15 min at ambient temperature. Derivatives were extracted from the aqueous phase by liquid-liquid extraction with 300 µL of toluene, followed by centrifugation at 3000 rpm for 10 min. The supernatant organic phase was collected into an Eppendorf tube and toluene was evaporated to dryness under a gentle nitrogen stream. The dry residue was reconstituted in 300 µL ACN and then filtered through a 0.2 µm syringe filter prior to LC injection.

2.4.3. HPLC-UV-FLD Analysis

Chromatography was performed on an Agilent Zorbax Eclipse XDB-C18 (150 mm × 4.6 mm, 5 µm) analytical column. Gradient elution of mobile phase consisting of water and acetonitrile was optimized in order to achieve satisfactory chromatographic separation, within the minimum possible analysis time, taking into consideration the interferences of the complex matrix of fish samples. The optimum and finally selected linear gradient elution program started with 40% ACN which increased linearly to 80% up to 25th min and further increased to 100% up to 31st min, afterwards the percentage of ACN decreased to the initial value of 40% up to 42nd min and remained constant for five more min for equilibration of the column, reaching a total analysis time of 47 min. The mobile phase flow rate was 1.2 mL min^{-1} and the column temperature was set at 30 °C.

For the selection of the optimum wavelengths, the emission spectra with constant the excitation wavelength (λ_{exc}) and the excitation spectra with constant emission wavelength (λ_{em}) for all the pyrene-derivatized amines were recorded. The final selected FLD excitation wavelength (λ_{exc}) was 350 nm and the final selected emission wavelengths were 489 nm for PUT (until 23.5 min of the chromatographic analysis), 486 nm for CAD, HIS and internal standard (until 29 min), 484 nm for SPD (until 31 min) and 486 nm SPM (until the end of chromatographic analysis). The UV absorption wavelength was set at 350 nm.

2.4.4. LC-MS Analysis

For the LC-MS analysis the chromatographic conditions were as follows: LiChrospher 100-RP18 analytical column (250 × 4.0 mm, 5 µm particle size) and isocratic elution with a mobile phase consisting of ACN-H$_2$O 80:20 *v/v*, at 1.0 mL min^{-1} flow rate was applied for the analysis of the derivatives of all the tested BAs, except for the analysis of the derivative of SPM, where a mobile phase of 100% ACN was used. ESI was applied in the positive ionization mode and the capillary was held at a potential of 3.5 kV. The cone voltage was 20 V and the ionization source was set at a temperature of 350 °C. A standard solution of 5.0 µg mL^{-1} was prepared for each BA which was derivatized according to the proposed protocol. The full scan spectrum (*m/z* 150–1500) and the total ion chromatogram for each pyrene derivative were acquired. The reconstructed ion chromatograms (RICs) of the most abundant *m/z* ions of the spectrum at the retention time of the eluted peak, which could be attributed to specific species of the tested analyte were generated.

2.5. Method Validation

The optimized method was evaluated using standard solutions of derivatized BAs and spiked samples of sea bass, anchovy, anchovy marinated, mackerel, sardines and tuna.

The linearity of the derivatization and the response of the HPLC–UV–FLD system was examined with a calibration curve, obtained by triplicate analysis of seven standard solutions in the range of 0.1 to 10.0 µg mL^{-1}, which were treated according to the proposed derivatization procedure described in Sections 2.4.1 and 2.4.2, using 300 μL of the BAs standard solutions of the different concentrations instead of fish sample. The standard solutions contained 0.3 μg mL^{-1} of the internal standard. Linear regression analysis was performed using: (a) the area and (b) the ratio *analyte peak area/internal standard peak area* against analyte concentration and the effect of the use of the internal standard was investigated.

Matrix matched calibration curves were prepared with different fish tissues (sea bass, anchovy, anchovy marinated, mackerel, sardines and tuna). Each sample was spiked at seven fortification levels with the BAs, within the range of 0.2–32 mg kg^{-1}. Spiked samples and non-spiked samples of the same batch were analyzed according to the protocol described in Sections 2.4.1 and 2.4.2. The final linear equation of the matrix matched calibration curve resulted after the subtraction of the signal of the unfortified sample from the signal of the fortified samples.

The instrumental limit of detection (LOD) and quantification (LOQ) were defined as $(3.3 \times S_a)/b$ and as $(10 \times S_a)/b$, respectively, where S_a stands for the standard deviation of the intercept of a low level calibration curve (0.025–0.8 μg mL^{-1}) and b for the slope of the same calibration curve. The overall LOD and LOQ of the method were determined applying the same mathematical equations to the matrix matched calibration curves.

The instrumental repeatability of the HPLC-UV-FLD analysis was estimated with five replicates of a standard solution containing 0.5 μg mL^{-1} of the derivatized BAs and 0.3 μg mL^{-1} of the internal standard. For the assessment of the overall precision, the method was applied to an anchovy sample which contained the BAs at a concentration range of 2–10 mg kg^{-1}. Sample preparation and HPLC-UV-FLD measurement of six replicates during one day were conducted for the determination of the repeatability ($n = 6$, intra-day precision), and six replicates in two different days, were conducted to test for the intra-laboratory reproducibility of the method ($n = 6$ $k = 2$, inter-day precision).

For the assessment of the accuracy, the method was applied to a sea bass sample that was spiked with BAs at six fortification levels (1.6, 3.2, 8.0, 12.8, 16.0 and 32.0 mg kg^{-1}) and analyzed in triplicate. The absolute recovery ($\%R_{abs}$) of the method was calculated by dividing the slope of the sea bass matrix matched curve (b_{MM}) with the slope of the standard curve (b_{STD}), according to Equation (1). The relative recovery ($\%R_{rel}$) was calculated by subtracting the concentration measured in the non-spiked sample from that measured in the spiked sample and then dividing with the spiked concentration (C_{ADDED}) according to Equation (2). The concentrations were calculated from the matrix matched calibration curve of sea bass.

$$\%R_{abs} = \frac{b_{MM}}{b_{STD}} \times 100 \tag{1}$$

$$\%R_{rel} = \frac{C_{SPIKED\,SAMPLE} - C_{NONSPIKED\,SAMPLE}}{C_{ADDED}} \times 100 \tag{2}$$

3. Results and Discussion

3.1. Liquid Chromatographic Separation

Representative HPLC-UV and HPLC-FLD chromatograms of PSCl derivatized BAs standard solution of 0.5 μg mL^{-1} and of spiked sea bass blank sample are depicted in Figures 3 and 4, respectively. It was noted that derivatized histamine solution presented three chromatographic peaks (HIS a, HIS b, HIS c) at three retention times at approximately 16.5, 19 and 24.6 min, respectively, in the fluorescence chromatogram and two of them, HIS b and HIS c, in the UV chromatogram. The three eluted peaks of histamine exhibited the same excitation and emission spectra and the same mass spectrometric spectra which correspond to the dipyrene derivatives of histamine (Sections 3.2 and 3.5). Based on these data, it was hypothesized that the three chromatographic peaks correspond to three isomers of the pyrene derivatized histamine which are generated because of the tautomerism of the imidazole ring. The sum of the peak areas of the three chromatographic peaks were considered for all the calculations for histamine. Representative HPLC-UV and HPLC-FLD chromatograms of an unfortified sea bass sample is presented in Supplementary Material (Figure S1).

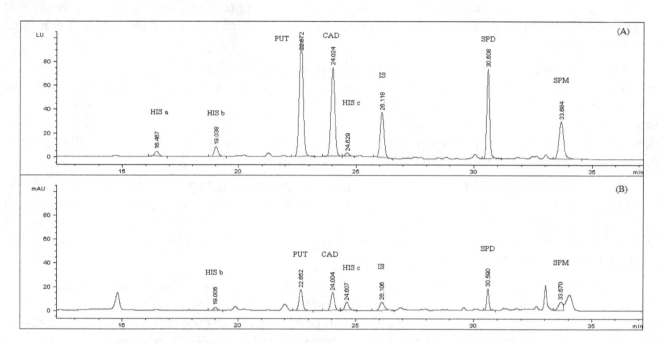

Figure 3. (**A**) HPLC-FLD; (**B**) HPLC-UV chromatogram of PSCl derivatized BAs standard solution of 0.5 μg mL^{-1} and 1,7 diaminoheptane at 0.3 μg mL^{-1}. (HIS a, HIS b, HIS c: Histamine, PUT: Putrescine, CAD: Cadaverine, SPD: Spermidine, SPM: Spermine, IS: Internal standard).

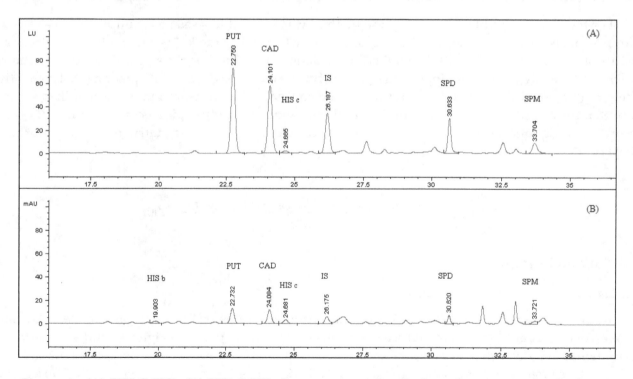

Figure 4. (**A**) HPLC-FLD; (**B**) HPLC-UV chromatogram of spiked sea bass tissue, fortified with putrescine (PUT), cadaverine (CAD), histamine (HIS), spermidine (SPD), spermine (SPM) at 8.0 mg kg^{-1} and 1,7 diaminoheptane (IS) at 4.8 mg kg^{-1}.

The chromatographic parameters, retention time, peak width at 50% height, peak shape in terms of asymmetry factor, chromatographic efficiency in terms of theoretical plates and resolution of all the amines' pyrene derivatives peaks of a spiked sea bass sample with fluorescence detection are presented in Table 1. Asymmetry factor is in the acceptable range of 0.8–1.2 and resolution is higher than 1.5 in all cases.

Table 1. Chromatographic parameters with fluorescence detection of spiked sea bass sample.

	PUT	CAD	HIS c	SPD	SPM
Retention Time (min)	22.7	24.1	24.6	30.6	33.7
Width at 50% of Peak Height (min)	0.25	0.23	0.19	0.15	0.21
Asymmetry Factor	0.92	0.95	0.98	0.87	0.88
Theoretical Plates (N) ×103	45	61	94	231	143
Resolution	1.7	1.7	2.0	2.5	2.4

3.2. Selection of Excitation and Emission Wavelengths

Figure 5 illustrates the excitation spectra of a standard solution of cadaverine derivative at the respective retention time. Similar excitation spectra were obtained for all the pyrene derivatives of the five tested biogenic amines and the internal standard with three peaks at approximately 245, 275 and 350 nm. The excitation wavelength 350 nm was selected for all the analytes.

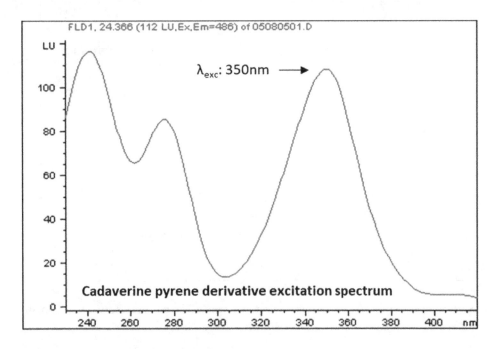

Figure 5. Excitation spectrum of cadaverine pyrene derivative chromatographic peak. Retention time approximately 24 min.

Figure 6 illustrates the emission spectra of a standard solution of cadaverine derivative (A) and spermine derivative (B) at the respective retention times of their eluted peaks. It is observed that the emission spectra of the cadaverine derivative consist of two peaks at approximately 390 nm and 486 nm, whereas in the spermine derivative emission spectrum only the second peak is present. The peak at 390 nm is attributed to the direct fluorescence of pyrene moiety, whereas the second peak which appears at significantly higher wavelengths (486 nm), is considered to be a result of the excimer fluorescence formation. Putrescine, histamine and the internal standard derivatives produced similar emission spectra with cadaverine derivatives with slight variations at the wavelength of maximum excitation, whereas the spermidine derivative emission spectra resembled that of the spermine derivative. The final selected excitation wavelengths are summarized in Section 2.4.3. The emission spectra of the PSCl-derivatized BAs of the present study, with excimer fluorescence peak at 484–489 nm are comparable with the corresponding emission spectra of the PSE-derivatized BAs with excimer fluorescence peak at 475 nm [42].

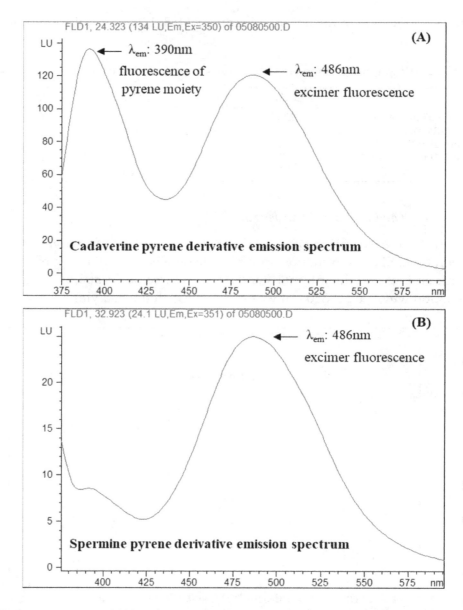

Figure 6. Emission spectra of (**A**) cadaverine and (**B**) spermine pyrene derivative chromatographic peaks. Retention times approximately 24 and 33 min, respectively.

The excimer fluorescence process is based on the neighboring of the pyrene moieties in a compound, which enables the efficient intramolecular energy transfer from one pyrene moiety to the neighboring one (intramolecular excitement). The energy transfer results in the enhancement of the fluorescence intensity and in the shifting at higher wavelengths. The most critical parameters of the excimer fluorescence are the distance and the conformation between the pyrene moieties, which can favor the intramolecular energy transfer, as for example in the case of cadaverine which is presented in Figure 2. Based on the obtained excitation spectra, the pyrene derivatives of all the tested biogenic amines presented excimer fluorescence, but among them, cadaverine and putrescine derivatives produced the higher fluorescence response factor, which is attributed to their molecular structure (Figure 1).

It is noted that the monitoring emission wavelength was selected to be the wavelength at the maximum emission of the excimer (second peak of emission spectrum) because at this wavelength no interference of the excess of the pyrene derivatization reagent is present, since it does not form fluorescence excimer, and moreover, considerably higher sensitivity for spermidine and spermine at excimer wavelength (484–486 nm) was obtained in comparison to the wavelength of the maximum emission of the pyrene moiety at 390 nm, as can be observed in the spectrum of spermine in Figure 6B.

3.3. Optimization of the Derivatization Procedure

3.3.1. Effect of pH

The effect of the pH on the completion of the derivatization reaction was studied in standard solution of biogenic amines and in spiked anchovy sample extract. Figure 7 presents the effect of pH on the peak area of the different pyrene derivatives of the biogenic amines of a standard solution. It is observed that the optimum pH was 9.6, which was achieved with the addition of 30 µL of saturated sodium carbonate solution to sample extract. At higher pH values the peak area of all the derivatized amines was reduced. It is noted that the optimum pH (9.6) is ≤ than the pKa of the tested amines (HIS-pKa_1:9.7 [46], CAD-pKa_1:10.2, pKa_2:9.1 [47], PUT-pKa:10.8 [48], SPD-pKa:10.9 [49], SPM-pKa:11.1 [50]), except for the imidazole amino group of histamine (pKa_2:6.0) [46]. As already mentioned, the pyrene sulfonyl chloride derivatization reaction with the biogenic amines is a nucleophilic substitution and amines need to be in their basic form in order to act as nucleophiles. However, at pH ≥ 10 the signal of the produced pyrene derivatives is significantly decreased potentially because of their subsequent alkaline hydrolysis. The same pattern of the peak area change of the derivatized amines in relation to pH was observed in the spiked anchovy sample as well. It is noted that the optimum pH for the derivatization of the BAs under investigation with PSCl is inside the reported pH target range for a robust reaction of putrescine and cadaverine with another pyrene reagent, PSE, which was quite wide, between pH 8 and 11 [35].

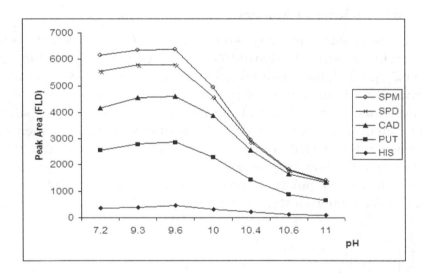

Figure 7. Effect of pH on the efficiency of the derivatization reaction of putrescine (PUT), cadaverine (CAD), histamine (HIS), spermidine (SPD) and spermine (SPM) in standard solution of 1 µg mL^{-1} with 400 µL of 3 mM PSCl.

3.3.2. Effect of Temperature and Time

Figure 8 presents the effect of reaction temperature on the obtained FLD signal of the derivatized amines. The optimum temperature for histamine, spermidine and spermine derivatization appears to be 60 °C, while for putrescine and cadaverine an optimum plateau is observed between 50–70 °C. Temperatures below 50 °C resulted in lower fluorescence signal, probably due to incomplete derivatization of all the tested amines, and at temperatures above 70 °C partial decomposition of the derivatization product is possibly taking place. The study of the derivatization time showed that incubation for a time period longer than 15 min does not increase the efficiency of the reaction. The optimum derivatization temperature and time required for the derivatization of these polyamines with PSCl (60 °C, 15 min) are very close to the optimum conditions applied for the derivatization of the same amines with other pyrene reagents like PSE (55 °C, 15 min) [35] and 2-chloro-4-methoxy-6-(4-(pyren-4-yl)butoxy)-1,3,5-triazine (50 °C, 20 min) [44]. However, more extreme conditions have also been reported for the same derivatization reaction with PSE (100 °C, 20 min) [42] and (100 °C, 90 min) [43].

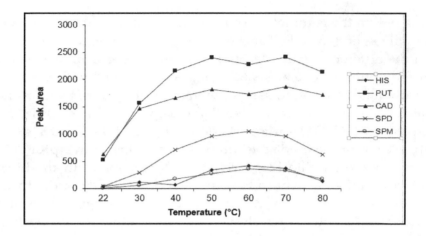

Figure 8. Effect of temperature on the efficiency of the derivatization reaction of putrescine (PUT), cadaverine (CAD), histamine (HIS), spermidine (SPD) and spermine (SPM) in standard solution of 1 μg mL^{-1} with 400 μL of 3 mM PSCl.

3.3.3. Effect of Derivatization Reagent Amount

Based on the data presented in Figure 9, a volume of 600 μL of 3.0 mM PSCl solution was selected as the optimum amount for the derivatization of 300 μL of 1 μg mL^{-1} of the five biogenic amines plus the internal standard at 0.3 μg mL^{-1}, taking into consideration that histamine, spermidine and spermine, exhibited the maximum sensitivity at this volume and at the same time sensitivity of putrescine and cadaverine was satisfactory. The addition of higher volumes, resulted in a slight decrease of the fluorescence signal for all the analytes which could be attributed to possible quenching of the excimer fluorescence. It is noted that the tested volumes of 3 mM PSCl solution, 200–1200 μL, correspond to 600–3600 nmol of the derivatization reagent which is stoichiometrically in excess, comparing to the required moles, approximately 5 nmol, for the complete derivatization of all the amino groups of the five biogenic amines plus the internal standard, which are contained in the 300 μL of the 1 μg mL^{-1} standard solution.

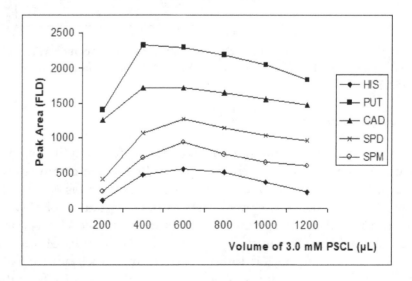

Figure 9. Effect of the volume of the 3.0 mM PSCl solution on the efficiency of the derivatization reaction with putrescine (PUT), cadaverine (CAD), histamine (HIS), spermidine (SPD) and spermine (SPM) in standard solution of 1 μg mL^{-1}.

3.4. Stability of the Derivatives

A comprehensive stability study was conducted for a two-month period. A standard solution containing 0.5 µg mL^{-1} of each amine was treated according to the optimum derivatization procedure and stored at 4 °C. Fourteen determinations were performed through the storage time. The variations expressed as %RSD ($n = 14$) of the derivatives' peak areas through this period ranged from 3.0% for cadaverine up to 15% for spermine. In addition, for all the pyrene derivatives, diagrams of the peak area against time were generated and the slope of these curves was found to be statistically equal to zero (t-test, at 95% confidence level), indicating that none of the derivatives presented significant decomposition through the two-month period.

3.5. LC-MS Identification of Biogenic Amine Derivatives

Table 2 summarizes the identified m/z ions of the tested solutions of the biogenic amines after derivatization with pyrene sulfonyl chloride. The derivatized solution of histamine presented three chromatographic peaks during the LC-MS analysis as observed in Figure S2 (Supplementary Material), similarly to the HPLC-UV-FLD analysis (Section 3.1), in contrast to all the other tested biogenic amines, whose derivatives presented one eluted peak in all the chromatographic analysis. The dipyrene-derivatives of histamine were identified with two m/z ions, the protonated molecular ion [M + H]$^+$ and the sodium adduct of the molecular ion [M + Na]$^+$.

Table 2. Identified m/z ions of the pyrene derivatives of BAs and internal standard (I.S.) by LC-ESI-MS.

Pyrene Derivatives	MW (g/mol)	[M + H]$^+$	[M+Na]$^+$	[2M+Na]$^+$	No of Pyrene Moieties
Putrescine	615.8	617.8	639.8	1256.5	2
Cadaverine	630.9	631.9	653.9	1284.6	2
Histamine	639.8	640.8	662.8	-	2
1,7-diaminoheptane (I.S)	658.9	659.8	681.8	1340.6	2
Spermidine	938.2	-	961.2	-	3
Spermine	1259.5	1260.5	-	-	4

The dipyrene-derivatives of putrescine, cadaverine and internal standard 1,7-diaminoheptane were identified with three m/z ions, the protonated molecular ion [M + H]$^+$, the sodium adduct of the molecular ion [M + Na]$^+$ and the sodium adduct of the dimeric form of the dipyrene-derivatives [2M + Na]$^+$. The monopyrene-derivatives were not detected. A representative full scan spectrum of the derivatized cadaverine is depicted in Figure S3, and the corresponding reconstructed ion chromatograms of the three m/z ions used for identification are presented in Figure S4 in Supplementary Material.

Accordingly, the tripyrene-derivative of spermidine was identified by the sodium adduct of the molecular ion, whereas no characteristic ions were identified for its mono- or dipyrene-derivatives. Finally, the tetrapyrene-derivative of spermine was identified by the protonated molecular ion. The sodium adduct was not observed in the LC-MS analysis of spermine derivative, since the mobile phase was purely organic (100% ACN). No characteristic m/z ions could be identified for the mono-, di- or tripyrene-derivatives of spermine. Based on the above, it can be concluded that all the primary and secondary amino groups of the investigated amines were completely derivatized under the optimum conditions.

3.6. Method Performance

Table 3 shows the equations of the calibration curves prepared from the analysis of standard solutions of the investigated derivatized amines using as analytical parameter the peak area. The standard deviation of the slope and the intercept, and the correlation coefficient of each equation are also given along with the instrumental limit of detection (LOD$_{instr}$) and quantification (LOQ$_{instr}$).

Table 3. Standard calibration curves, instrumental limit of detection (LOD$_{instr}$) and quantification (LOQ$_{instr}$) of the derivatized BAs with UV and Fluorescence detectors C: µg mL^{-1}. y: analyte peak area. Concentration range: 0.1–10 µg mL^{-1}.

		Calibration Curve	R	LOD$_{instr}$ (µg mL^{-1})	LOQ$_{instr}$ (µg mL^{-1})
PUT	UV	y = (352.6 ± 6.9) × C + (32 ± 30)	0.998	0.03	0.09
	FLD	y = (2148 ± 111) × C + (111 ± 114)	0.992	0.03	0.09
CAD	UV	y = (315.8 ± 3.0) × C + (9 ± 13)	0.9995	0.02	0.07
	FLD	y = (1350 ± 55) × C + (240 ± 125)	0.993	0.02	0.06
SPD	UV	y = (221.0 ± 8.4) × C + (38 ± 36)	0.993	0.04	0.10
	FLD	y = (1178 ± 45) × C + (35 ± 46)	0.996	0.03	0.10
SPM	UV	y = (168.7 ± 4.8) × C + (29 ± 21)	0.996	0.07	0.20
	FLD	y = (959 ± 14) × C + (-21 ± 31)	0.9992	0.05	0.15
HIS	UV	y = (180.1 ± 3.6) × C + (30 ± 16)	0.998	0.08	0.20
	FLD	y = (352 ± 16) × C + (37 ± 35)	0.992	0.10	0.30

Table 4 presents the corresponding information of the calibration curves prepared from the analysis of spiked sea bass sample, as well as the method limit of detection (LOD) and quantification (LOQ) for both detectors. It is demonstrated that the overall method, and therefore the derivatization procedure, presented linearity for the tested concentration range with correlation coefficients always exceeding 0.992 for both detectors. Representative matrix matched calibration graphs of sea bass are presented in Supplementary Material (Figure S5).

Table 4. Matrix matched calibration curves of sea bass sample, limit of detection (LOD) and quantification (LOQ) of the BAs with UV and Fluorescence detectors C: µg mL^{-1}. y: analyte peak area. BAs concentration range: 0.2–32 mg kg^{-1} sample, which corresponds to 0.05–8.0 µg mL^{-1} measured solution.

		Calibration Curve	R	LOD (mg kg^{-1})	LOQ (mg kg^{-1})
PUT	UV	y = (239.7 ± 3.3) × C + (-9.6 ± 3.3)	0.9992	0.2	0.6
	FLD	y = (1458 ± 22) × C + (-67 ± 22)	0.9991	0.2	0.6
CAD	UV	y = (237.4 ± 2.0) × C + (1.7 ± 2.0)	0.9997	0.1	0.3
	FLD	y = (1286 ± 16) × C + (-11 ± 16)	0.9994	0.2	0.5
SPD	UV	y = (58.6 ± 2.4) × C + (-0.8 ± 2.4)	0.993	0.5	1.6
	FLD	y = (301 ± 10) × C + (-18 ± 10)	0.995	0.4	1.3
SPM	UV	y = (20.51 ± 0.82) × C + (1.99 ± 0.81)	0.994	0.5	1.6
	FLD	y = (113.4 ± 4.6) × C + (16.3 ± 4.5)	0.994	0.5	1.6
HIS	UV	y = (118.1 ± 5.2) × C + (-20.0 ± 6.3)	0.996	0.7	2.1
	FLD	y = (134.3 ± 8.4) × C + (-46.8 ± 10.1)	0.992	1.0	3.1

The data of Tables 3 and 4 show that the FLD slopes of the calibration curves are five to six times higher than the corresponding slopes of the UV detector for the standard and sample solutions for all the derivatized BAs, except for histamine which exhibited similar sensitivity for both detectors. Among the tested BAs, the higher sensitivity was obtained for putrescine and cadaverine, probably due to their stereochemical conformation, which favored the fluorescence excimer process.

Significant matrix effect was observed for all the fish samples (sea bass, anchovy, anchovy marinated, mackerel, sardines and tuna), since t-test analysis at 95% confidence level showed that the slopes of the standard calibration curves differ significantly from the corresponding slopes of the matrix matched curves. In order to address the matrix effect, internal standardization was investigated. Compound 1,7-diaminoheptane was selected as a suitable internal standard, since its structure is similar to the linear polyamines tested, in particular to putrescine and cadaverine. Linear regression analysis was performed on the data obtained from the standard solutions and the spiked samples using the ratio analyte peak area/internal standard peak area against analyte concentration. Despite the satisfactory correlation coefficients (r > 0.99), the slopes of the matrix matched curves were still statistically different from the corresponding slopes of the standard calibration curves, indicating that the use of the internal standard could not compensate for the matrix interferences. Based on these findings it was concluded that for the quantification of the tested BAs in fish samples standard addition is necessary.

The LODs of the method for all the tested matrices for the five BAs are presented in Table 5 and they range between 0.1 to 1.4 mg kg^{-1} for both detectors, UV and FLD. The corresponding method LOQs range between 0.3–4.2 mg kg^{-1}. It is noted that similar limits of detection and quantification were obtained for both detectors, UV and FLD, with small variation among the different biogenic amines (Tables 3–5). The LODs and LOQs of the proposed method are adequate for the safety control of food products, since they are much lower than the existing regulated limits of histamine, which range between 100–500 mg kg^{-1} [8,9]. In addition, they are comparable to the LODs and LOQs of other reported analytical methods of the same scope [17,26,27,33–35].

Table 5. Limit of detection (LOD) for the different matrices.

LODs (mg kg^{-1})	PUT		CAD		SPD		SPM		HIS	
	UV	FLD	UV	FLD	UV	FLD	UV	FLD	UV	FLD
Tuna	0.3	0.3	0.4	0.3	0.7	0.5	0.6	1.0	0.7	0.6
Mackerel	1.0	1.4	0.5	0.3	0.6	0.6	1.2	0.5	0.2	0.7
Anchovy	0.6	0.8	0.5	1.0	0.5	0.5	0.5	0.6	0.5	0.1
Anchovy Marinated	0.6	0.7	0.8	0.7	1.0	0.5	1.1	0.5	0.8	0.5
Sea bass	0.2	0.2	0.1	0.2	0.5	0.4	0.5	0.5	0.7	1.0
Sardines	0.6	0.7	0.6	0.8	0.3	0.2	0.4	1.1	0.2	0.4

The instrumental repeatability, the intra- and inter-day method precision data are given in Table 6. For the evaluation of the precision of the proposed method, external standardization technique was compared to internal standardization with the use of 1,7-diaminoheptane. The repeatability of liquid chromatographic peak areas (PA) of the linear biogenic amines (putrescine, cadaverine, spermidine and spermine) PSCl derivatives ranges between 1.8–2.3% for UV detection and between 2.6–2.8% for fluorescence detection. Chromatographic repeatability is considerably improved by internal standardization and the use of ratio F, decreasing to 0.1–0.7% for both detectors. As already mentioned, standard solution of PSCl derivatized histamine exhibits a different chromatographic behaviour with three chromatographic peaks, attributed to the PSCl derivatives of three isomeric structures of the molecule. The higher %RSD values of 4.2% for UV and FLD of the peak area of derivatized histamine are potentially related to the presence of these isomers. A slight improvement by the use of the internal standard and the ratio F is observed in case of histamine.

Table 6. Instrumental and method precision data under intra and inter day conditions with UV and FLD detectors, with analytical parameter the Peak Area (PA) and the ratio *analyte peak area/internal standard peak area* (F). BAs concentrations in tested sea bass sample for the method intra and inter-day assays: 2–10 mg kg^{-1}.

	Instrumental Precision 0.5 μg mL^{-1}, %RSD, $n = 5$				Method Intra-Day Assay %RSD, $n = 6$				Method Inter-Day Assay %RSD, $n = 6$, k = 2			
	UV	UV	FLD	FLD	UV	UV	FLD	FLD	UV	UV	FLD	FLD
	PA	F	PA	F	PA	F	PA	F	PA	F	PA	F
PUT	2.3	0.1	2.6	0.4	8.2	9.3	10	8.2	8.9	12	9.6	15
CAD	2.3	0.3	2.7	0.5	8.2	9.1	7.4	6.1	9.5	12	11	16
SPD	2.4	0.3	2.7	0.4	12	8.7	12	8.8	14	13	21	15
SPM	1.8	0.7	2.8	0.5	13	15	14	11	12	19	14	23
HIS	4.2	3.9	4.2	2.3	10	9.4	11	12	11	9.4	10	14

According to Table 6 overall method repeatability (intra-day assay) ranges between 8.2–13% for UV and 7.4–14 for FLD and overall method reproducibility (inter-day assay) ranges between 8.9–14% for UV and 9.6–21 for FLD using the peak areas of the analytes as analytical parameters, while no significant improvement is gained with the use of the internal standard, ratio F. Based on these data, it can be concluded that the UV signal is slightly more robust compared to FLD and that the use of internal standard cannot compensate for the variation of the signal attributed to the complex matrix.

However, internal standard can serve as an additional confirmation tool for the identification of the presence of the biogenic amines in an unknown sample. Two-month stability data ($n = 14$) on the retention time of the derivatized amines and their relative retention time ($RT_{BA}/RT_{I.S.}$) revealed that the %RSD of the retention times ranged between 0.16–0.37% while the relative retention times ranged between 0.04–0.22.

The accuracy of the developed method expressed as absolute and relative recovery % of the biogenic amines is presented in Table 7. It is noted that for the first three eluted peaks of the derivatized putrescine, cadaverine and histamine detected by UV the absolute recovery ranges between 66% and 75% whereas it is remarkably decreased for the two last eluted peaks of the derivatized spermidine (27%) and spermine (12%). Similar results are obtained with fluorescence detector. The effect of the standard addition on the quantification of the biogenic amines in fish samples and the subsequent improvement on the accuracy of the method is described by the increased values of the relative recoveries for UV (72–112%) and FLD (67–114%).

Table 7. Absolute and relative recoveries of each compound from spiked sea bass sample at six different levels (1.6, 3.2, 8.0, 12.8, 16.0 and 32 mg kg^{-1}), $n = 6$. Effect of standard addition.

Compound	UV Absolute Recovery (%)	UV Relative Recovery % (±SD)		FLD Absolute Recovery (%)	FLD Relative Recovery % (±SD)	
PUT	68	88	(±12)	68	86	(±15)
CAD	75	102	(±2.6)	95	98	(±2.1)
SPD	27	98	(±12)	26	88	(±6.6)
SPM	12	112	(±8.0)	12	114	(±6.5)
HIS	66	72	(±13)	38	67	(±11)

3.7. Concentration Levels of Biogenic Amines in Fish Samples

Table 8 presents the mean concentration of the biogenic amines, putrescine, cadaverine, spermidine, spermine and histamine, that was determined in commercial canned fish samples and their sum. All the five biogenic amines tested were detected in mackerel, anchovy and sardine samples. In tuna sample no putrescine and histamine was detected and in sea bass sample no cadaverine and histamine was detected. The highest concentration is observed for putrescine in mackerel at 40.1 mg kg^{-1} and the highest sum of biogenic amines is observed for mackerel and sardines at 49.6 and 34.9 mg kg^{-1}, respectively. It is noted that all results are below the strictest EU limit for histamine of 100 mg kg^{-1}.

Table 8. Biogenic amines' concentrations (mg kg^{-1}) in canned fish samples.

Content of BAs (mg kg^{-1})	PUT	CAD	SPD	SPM	HIS	Sum of BAs
Tuna	ND	<0.9	2.4	7.6	ND	10.9
Mackerel	40.1	0.9	3.6	2.5	2.5	49.6
Anchovy	2.4	3.0	<1.5	<1.8	5.8	14.5
Anchovy Marinated	3.8	5.6	<1.5	<1.5	<1.5	13.9
Sea bass	<0.6	ND	1.4	2.1	ND	4.1
Sardines	2.7	10.8	8.1	9.6	3.7	34.9

ND: Not Detected (below the limit of detection). "<": Detected but below the limit of quantification.

These findings are within the range of other reported surveys on these biogenic amines' levels in similar fish products. In particular, according to a European survey the reported mean concentrations of histamine in fish and fish products range between 29.3–33.6 mg kg^{-1}, with 'dried anchovies' to be the subgroup of fish products with highest mean values of histamine, at 348 mg kg^{-1} [4]. According to a recent study in Cambodian fermented foods, histamine in fishery products ranged from <2 mg kg^{-1} up to 840 mg kg^{-1} [51]. Similarly, putrescine concentrations have been found to range between 0.11–830 mg kg^{-1} and cadaverine concentrations between 0.21–2035 mg kg^{-1} in fish and fishery products [4,31,34,50]. Limited data are available for spermidine and spermine levels in fish products, which are reported to range between 1.38–19.83 mg kg^{-1} for spermidine and 0.85–11.90 mg kg^{-1} for spermine [31,34].

4. Conclusions

The present work reports the successful use of 1-pyrenesulfonyl chloride (PSCl), a UV excitable amino-reactive fluorophore, as derivatization agent for the HPLC-UV-FLD determination of the biogenic polyamines, putrescine, cadaverine, histamine, spermidine and spermine in fish products. The optimized derivatization procedure resulted in the complete reaction of all the amino-groups, as was confirmed by the LC-ESI-MS spectra. The existence of two or more pyrene moieties in the structure of the derivatives and their favorable conformation induced intramolecular excitement (excimer fluorescence), and, therefore, the shifting of the emission to a longer wavelength. Biogenic amines were detected in all the tested samples of canned fish food, sea bass, anchovy, mackerel, sardines and tuna, collected from the Greek market but there were in all cases below the regulated limit of histamine 100 mg kg^{-1}.

Supplementary Materials: The following are available online at http://www.mdpi.com/2304-8158/9/5/609/s1, Figure S1: (A) HPLC-FLD; (B) HPLC-UV chromatogram of sea bass tissue, fortified only with the internal standard 1,7 diaminoheptane (IS) at 4.8 mg kg^{-1}, Figure S2: Reconstructed Ion Chromatograms of a 5 µg mL^{-1} histamine pyrene derivative at m/z 640.3–641.3 [M+H]+ and 662.3–653.3 [M+Na]+, Figure S3: Full scan spectrum (m/z 500–1500) at the retention time of cadaverine pyrene derivative (8.4 min) of a 5 µg mL^{-1} standard solution, Figure S4: Reconstructed Ion Chromatograms of a 5 µg mL^{-1} cadaverine pyrene derivative at m/z 631.3–632.3 [M+H]+, 652.3–653.3 [M+Na]+ and 1283.4–1284.4 [2M+Na]+, Figure S5: Matrix matched calibration curves of sea bass sample with Fluorescence detector.

Author Contributions: Conceptualization, N.S.T.; Data curation, E.S.P.; Formal analysis, E.S.P. and N.C.M. (Niki C. Maragou); Investigation, E.S.P., N.C.M. (Niki C. Maragou) and N.C.M. (Nikolaos C. Megoulas); Methodology, N.S.T.; Supervision, N.S.T.; Visualization, N.C.M (Niki C. Maragou). and N.S.T.; Writing—original draft, E.S.P. and N.C.M (Niki C. Maragou).; Writing—review and editing, N.C.M. (Niki C. Maragou), M.E.D., N.C.M. (Nikolaos C. Megoulas) and M.A.K. All authors have read and agreed to the published version of the manuscript.

Acknowledgments: The technical support of Marios Kostakis is acknowledged.

References

1.   Baixas-Nogueras, S.; Bover-Cid, S.; Veciana-Nogués, M.T.; Mariné-Font, A.; Vidal-Carou, M.C. Biogenic Amine Index for Freshness Evaluation in Iced Mediterranean Hake (Merluccius merluccius). *J. Food Prot.* **2005**, *68*, 2433–2438. [CrossRef]

2.   Ruiz-Capillas, C.; Herrero, A.M. Impact of Biogenic Amines on Food Quality and Safety. *Foods* **2019**, *8*, 62. [CrossRef] [PubMed]

3.   Doeun, D.; Davaatseren, M.; Chung, M. Biogenic amines in foods. *Food Sci. Biotechnol.* **2017**, *26*, 1463–1474. [CrossRef] [PubMed]

4.   European Food Safety Authority. Scientific Opinion on risk based control of biogenic amine formation in fermented foods. *EFSA J.* **2011**, *9*, 2393. [CrossRef]

5.   Mietz, J.L. Chemical quality index of canned tuna as determined by high-pressure liquid chromatography. *J. Food Sci.* **1977**, *42*, 155–158. [CrossRef]

6.   Shalaby, A.R. Significance of biogenic amines to food safety and human health. *Food Res. Int.* **1996**, *29*, 7–675. [CrossRef]

7.   Bulushi, I.A.; Poole, S.; Deeth, H.C.; Dykes, G.A. Biogenic Amines in Fish: Roles in Intoxication, Spoilage, and Nitrosamine Formation—A Review. *Crit. Rev. Food Sci. Nutr.* **2009**, *49*, 369–377. [CrossRef]

8.   European Commission. Commission Regulation (EC) No 2073/2005 of 15 November 2005 on microbiological criteria for foodstuffs. *Off. J. Eur. Union* **2005**, *338*, 1–26.

9.   Fish and Fishery Products Hazards and Controls Guidance. Available online: https://www.fda.gov/Food/GuidanceRegulation/GuidanceDocumentsRegulatoryInformation/Seafood/ucm2018426.htm (accessed on 17 March 2020).

10. Zhang, N.; Wang, H.; Zhang, Z.X.; Deng, Y.H.; Zhang, H.S. Sensitive determination of biogenic amines by capillary electrophoresis with a new fluorogenic reagent 3-(4-fluorobenzoyl)-2-quinolinecarboxaldehyde. *Talanta* **2008**, *76*, 791–797. [CrossRef]

11. An, D.; Chen, Z.; Zheng, J.; Chen, S.; Wang, L.; Huang, Z.; Weng, L. Determination of biogenic amines in oysters by capillary electrophoresis coupled with electrochemiluminescence. *Food Chem.* **2015**, *168*, 1–6. [CrossRef]

12. Daniel, D.; Dos Santos, V.B.; Vidal, D.T.; do Lago, C.L. Determination of biogenic amines in beer and wine by capillary electrophoresis-tandem mass spectrometry. *J. Chromatogr. A* **2015**, *1416*, 121–128. [CrossRef] [PubMed]

13. Plotka-Wasylka, J.; Simeonov, V.; Namiesnik, J. An in situ derivatization - dispersive liquid-liquid microextraction combined with gas-chromatography - mass spectrometry for determining biogenic amines in home-made fermented alcoholic drinks. *J. Chromatogr. A* **2016**, *1453*, 10–18. [CrossRef] [PubMed]

14. Cunha, S.C.; Faria, M.A.; Fernandes, J.O. Gas chromatography-mass spectrometry assessment of amines in Port wine and grape juice after fast chloroformate extraction/derivatization. *J. Agric. Food Chem.* **2011**, *59*, 8742–8753. [CrossRef] [PubMed]

15. De Borba, B.M.; Rohrer, J.S. Determination of biogenic amines in alcoholic beverages by ion chromatography with suppressed conductivity detection and integrated pulsed amperometric detection. *J. Chromatogr. A* **2007**, *1155*, 22–30. [CrossRef]

16. Tao, Z.; Sato, M.; Han, Y.; Tan, Z.; Yamaguchi, T.; Nakano, T. A simple and rapid method for histamine analysis in fish and fishery products by TLC determination. *Food Control.* **2011**, *22*, 1154–1157. [CrossRef]

17. Dadáková, E.; Křížek, M.; Pelikánová, T. Determination of biogenic amines in foods using ultra-performance liquid chromatography (UPLC). *Food Chem.* **2009**, *116*, 365–370. [CrossRef]

18. Costa, M.P.; Balthazar, C.F.; Rodrigues, B.L.; Lazaro, C.A.; Silva, A.C.; Cruz, A.G.; Conte Junior, C.A. Determination of biogenic amines by high-performance liquid chromatography (HPLC-DAD) in probiotic cow's and goat's fermented milks and acceptance. *Food Sci. Nutr.* **2015**, *3*, 172–178. [CrossRef]

19. Liu, S.J.; Xu, J.J.; Ma, C.L.; Guo, C.F. A comparative analysis of derivatization strategies for the determination of biogenic amines in sausage and cheese by HPLC. *Food Chem.* **2018**, *266*, 275–283. [CrossRef]

20. Wang, Y.Q.; Ye, D.Q.; Zhu, B.Q.; Wu, G.F.; Duan, C.Q. Rapid HPLC analysis of amino acids and biogenic amines in wines during fermentation and evaluation of matrix effect. *Food Chem.* **2014**, *163*, 6–15. [CrossRef]

21. Mayer, H.K.; Fiechter, G.; Fischer, E. A new ultra-pressure liquid chromatography method for the determination of biogenic amines in cheese. *J. Chromatogr. A* **2010**, *1217*, 3251–3257. [CrossRef]

22. Lázaro, C.A.; Conte-Júnior, C.A.; Cunha, F.L.; Mársico, E.T.; Mano, S.B.; Franco, R.M. Validation of an HPLC Methodology for the Identification and Quantification of Biogenic Amines in Chicken Meat. *Food Anal. Method* **2013**, *6*, 1024–1032. [CrossRef]

23. Jia, S.; Ryu, Y.; Kwon, S.W.; Lee, J. An in situ benzoylation-dispersive liquid-liquid microextraction method based on solidification of floating organic droplets for determination of biogenic amines by liquid chromatography-ultraviolet analysis. *J. Chromatogr. A* **2013**, *1282*, 1–10. [CrossRef] [PubMed]

24. Redruello, B.; Ladero, V.; Cuesta, I.; Alvarez-Buylla, J.R.; Martin, M.C.; Fernandez, M.; Alvarez, M.A. A fast, reliable, ultra high performance liquid chromatography method for the simultaneous determination of amino acids, biogenic amines and ammonium ions in cheese, using diethyl ethoxymethylenemalonate as a derivatising agent. *Food Chem.* **2013**, *139*, 1029–1035. [CrossRef] [PubMed]

25. De Mey, E.; Drabik-Markiewicz, G.; De Maere, H.; Peeters, M.C.; Derdelinckx, G.; Paelinck, H.; Kowalska, T. Dabsyl derivatisation as an alternative for dansylation in the detection of biogenic amines in fermented meat products by reversed phase high performance liquid chromatography. *Food Chem.* **2012**, *130*, 1017–1023. [CrossRef]

26. Kang, L.; You, J.; Sun, Z.; Wang, C.; Ji, Z.; Gao, Y.; Suo, Y.; Li, Y. LC Determination of Trace Biogenic Amines in Foods Samples with Fluorescence Detection and MS Identification. *Chromatographia* **2011**, *73*, 43–50. [CrossRef]

27. Tahmouzi, S.; Khaksar, R.; Ghasemlou, M. Development and validation of an HPLC-FLD method for rapid determination of histamine in skipjack tuna fish (Katsuwonus pelamis). *Food Chem.* **2011**, *126*, 756–761. [CrossRef]

28. Wu, H.; Li, G.; Liu, S.; Ji, Z.; Zhang, Q.; Hu, N.; Suo, Y.; You, J. Simultaneous Determination of Seven Biogenic Amines in Foodstuff Samples Using One-Step Fluorescence Labeling and Dispersive Liquid–Liquid

Microextraction Followed by HPLC-FLD and Method Optimization Using Response Surface Methodology. *Food Anal. Method* **2014**, *8*, 685–695. [CrossRef]

29. Triki, M.; Jiménez-Colmenero, F.; Herrero, A.M.; Ruiz-Capillas, C. Optimisation of a chromatographic procedure for determining biogenic amine concentrations in meat and meat products employing a cation-exchange column with a post-column system. *Food Chem.* **2012**, *130*, 1066–1073. [CrossRef]

30. Kelly, M.T.; Blaise, A.; Larroque, M. Rapid automated high performance liquid chromatography method for simultaneous determination of amino acids and biogenic amines in wine, fruit and honey. *J. Chromatogr. A* **2010**, *1217*, 7385–7392. [CrossRef]

31. Sánchez, J.A.; Ruiz-Capillas, C. Application of the simplex method for optimization of chromatographic analysis of biogenic amines in fish. *Eur. Food Res. Technol.* **2011**, *234*, 285–294. [CrossRef]

32. Zhao, Q.X.; Xu, J.; Xue, C.H.; Sheng, W.J.; Gao, R.C.; Xue, Y.; Li, Z.J. Determination of Biogenic Amines in Squid and White Prawn by High-Performance Liquid Chromatography with Postcolumn Derivatization. *J. Agric. Food Chem.* **2007**, *55*, 3083–3088. [CrossRef] [PubMed]

33. Zotou, A.; Notou, M. Enhancing Fluorescence LC Analysis of Biogenic Amines in Fish Tissues by Precolumn Derivatization with Naphthalene-2,3-dicarboxaldehyde. *Food Anal. Method* **2012**, *6*, 89–99. [CrossRef]

34. Li, G.; Dong, L.; Wang, A.; Wang, W.; Hu, N.; You, J. Simultaneous determination of biogenic amines and estrogens in foodstuff by an improved HPLC method combining with fluorescence labeling. *LWT Food Sci. Technol.* **2014**, *55*, 355–361. [CrossRef]

35. Marks Rupp, H.S.; Anderson, C.R. Determination of putrescine and cadaverine in seafood (finfish and shellfish) by liquid chromatography using pyrene excimer fluorescence. *J. Chromatogr. A* **2005**, *1094*, 60–69. [CrossRef] [PubMed]

36. Lozanov, V.; Petrov, S.; Mitev, V. Simultaneous analysis of amino acid and biogenic polyamines by high-performance liquid chromatography after pre-column derivatization with N-(9-fluorenylmethoxycarbonyloxy)succinimide. *J. Chromatogr. A* **2004**, *1025*, 201–208. [CrossRef]

37. Papageorgiou, M.; Lambropoulou, D.; Morrison, C.; Kłodzińska, E.; Namieśnik, J.; Płotka-Wasylka, J. Literature update of analytical methods for biogenic amines determination in food and beverages. *TrAC Trends Anal. Chem.* **2018**, *98*, 128–142. [CrossRef]

38. Ochi, N. Simultaneous determination of eight underivatized biogenic amines in salted mackerel fillet by ion-pair solid-phase extraction and volatile ion-pair reversed-phase liquid chromatography-tandem mass spectrometry. *J. Chromatogr. A* **2019**, *1601*, 115–120. [CrossRef]

39. Kaufmann, A.; Maden, K. Easy and Fast Method for the Determination of Biogenic Amines in Fish and Fish Products with Liquid Chromatography Coupled to Orbitrap Tandem Mass Spectrometry. *J. AOAC Int.* **2018**, *101*, 336–341. [CrossRef]

40. Zhang, Y.J.; Zhang, Y.; Zhou, Y.; Li, G.H.; Yang, W.Z.; Feng, X.S. A review of pretreatment and analytical methods of biogenic amines in food and biological samples since 2010. *J. Chromatogr. A* **2019**, *1605*, 360361. [CrossRef]

41. Nishikawa, H.; Tabata, T.; Kitani, S. Simple Detection Method of Biogenic Amines in Decomposed Fish by Intramolecular Excimer Fluorescence. *Food Nutr. Sci.* **2012**, *3*, 1020–1026. [CrossRef]

42. Nohta, H.; Satozono, H.; Koiso, K.; Yoshida, H.; Ishida, J.; Yamaguchi, M. Highly Selective Fluorometric Determination of Polyamines Based on Intramolecular Excimer-Forming Derivatization with a Pyrene Labeling Reagent. *Anal. Chem.* **2000**, *72*, 4199–4204. [CrossRef] [PubMed]

43. Yoshitake, T.; Ichinose, F.; Yoshida, H.; Todoroki, K.; Kehr, J.; Inoue, O.; Nohta, H.; Yamaguchi, M. A sensitive and selective determination method of histamine by HPLC with intramolecular excimer-forming derivatization and fluorescence detection. *Biomed. Chromatogr.* **2003**, *17*, 509–516. [CrossRef] [PubMed]

44. Nakano, T.; Todoroki, K.; Ishii, Y.; Miyauchi, C.; Palee, A.; Min, J.Z.; Inoue, K.; Suzuki, K.; Toyo'oka, T. An easy-to-use excimer fluorescence derivatization reagent, 2-chloro-4-methoxy-6-(4-(pyren-4-yl)butoxy)-1,3,5-triazine, for use in the highly sensitive and selective liquid chromatography analysis of histamine in Japanese soy sauces. *Anal. Chim. Acta* **2015**, *880*, 145–151. [CrossRef] [PubMed]

45. DeSilva, K.H.; Vest, F.B.; Kames, H.T. Pyrene Sulphonyl Chloride as a Reagent for Quantitation of Oestrogens in Human Serum Using HPLC with Conventional and Laser-Induced Fluorescence Detection. *Biomed. Chromatogr.* **1996**, *10*, 318–324. [CrossRef]

46. Paiva, B.T.; Tominaga, M.; Paiva, M.C. Ionization of Histamine, N-Acetylhistamine, and Their Iodinated Derivatives. *J. Med. Chem.* **1970**, *13*, 689–692. [CrossRef]
47. PubChem.   Available online: https://pubchem.ncbi.nlm.nih.gov/compound/Cadaverine#section= Dissociation-Constants (accessed on 8 April 2020).
48. PubChem. Available online: https://pubchem.ncbi.nlm.nih.gov/compound/1_4-Diaminobutane#section= pKa (accessed on 8 April 2020).
49. DRUGBANK. Available online: https://www.drugbank.ca/drugs/DB03566 (accessed on 8 April 2020).
50. DRUGBANK. Available online: https://www.drugbank.ca/drugs/DB00127 (accessed on 8 April 2020).
51. Ly, D.; Mayrhofer, S.; Schmidt, J.-M.; Zitz, U.; Domig, K.J. Biogenic Amine Contents and Microbial Characteristics of Cambodian Fermented. *Foods* **2020**, *9*, 198. [CrossRef]

4

Bacterial Production and Control of Biogenic Amines in Asian Fermented Soybean Foods

Jae-Hyung Mah *, Young Kyoung Park, Young Hun Jin, Jun-Hee Lee and Han-Joon Hwang

Department of Food and Biotechnology, Korea University, 2511 Sejong-ro, Sejong 30019, Korea;
eskimo@korea.ac.kr (Y.K.P.); younghoonjin3090@korea.ac.kr (Y.H.J.); bory92@korea.ac.kr (J.-H.L.);
hjhwang@korea.ac.kr (H.-J.H.)
* Correspondence: nextbio@korea.ac.kr

Abstract: Fermented soybean foods possess significant health-promoting effects and are consumed worldwide, especially within Asia, but less attention has been paid to the safety of the foods. Since fermented soybean foods contain abundant amino acids and biogenic amine-producing microorganisms, it is necessary to understand the presence of biogenic amines in the foods. The amounts of biogenic amines in most products have been reported to be within safe levels. Conversely, certain products contain vasoactive biogenic amines greater than toxic levels. Nonetheless, government legislation regulating biogenic amines in fermented soybean foods is not found throughout the world. Therefore, it is necessary to provide strategies to reduce biogenic amine formation in the foods. Alongside numerous existing intervention methods, the use of *Bacillus* starter cultures capable of degrading and/or incapable of producing biogenic amines has been proposed as a guaranteed way to reduce biogenic amines in fermented soybean foods, considering that *Bacillus* species have been known as fermenting microorganisms responsible for biogenic amine formation in the foods. Molecular genetic studies of *Bacillus* genes involved in the formation and degradation of biogenic amines would be helpful in selecting starter cultures. This review summarizes the presence and control strategies of biogenic amines in fermented soybean foods.

Keywords: food safety; biogenic amines; fermented soybean foods; intervention methods; control; starter culture; *Bacillus* spp.

---

1. Introduction

Microbial fermentation is one of the oldest and most practical technologies used in food processing and preservation. However, fermentation of protein-rich raw materials such as fish, meat, and soybean commonly provides abundant precursor amino acids of biogenic amines. Even though most fermented foods have been found to be beneficial to human health, biogenic amines produced through fermentation and/or contamination of protein-rich raw materials by amino acid-decarboxylating microorganisms may cause intoxication symptoms in human unless they are detoxified by human intestinal amine oxidases, viz., detoxification system [1,2]. Thus, the presence of biogenic amines in fermented foods (and non-fermented foods as well) has become one of the most important food safety issues.

According to old documents, the cultivation and use of soybeans, dating back to B.C., were launched in Manchuria on the north side of the Korean Peninsula and have spread to other regions of the world. Hence, a variety of fermented soybean foods have been developed and consumed in north-east Asian countries around the Korean Peninsula, and consequently humans in this region have steadily taken the fermented foods for a long period of time from hundreds to thousands of years, depending on the types of fermented soybean foods consumed [3]. Presently, fermented soybean foods are of public interest and consumed more frequently even in western leading countries because the fermented foods, particularly fermented soybean pastes, not only have been believed by many people, but also have been scientifically proven by researchers to have health-promoting and -protective effects [4]. However, much less attention has been paid to the safety issues of fermented soybean foods [5].

Fermented soybean foods, including various types of fermented soybean pastes and soy sauces, are commonly made from whole soybeans containing abundant amino acids through microbial fermentation. If the fermenting (or sometimes contaminating) microorganisms are significantly capable of decarboxylating amino acids, the resultant fermented soybean foods may contain unignorable amounts of biogenic amines. Indeed, the presence of biogenic amines seems to be quite frequent and inevitable in fermented soybean foods. Therefore, the present review provides information on the presence, bacterial production, and control strategies of biogenic amines in fermented soybean foods, especially focusing on fermented soybean pastes usually considered as heathy foods.

2. A Brief on Biogenic Amines

Biogenic amines are defined as harmful nitrogenous compounds produced mainly by bacterial decarboxylation of amino acids in various foods. The bacterial decarboxylation of amino acids to biogenic amines have been well illustrated in literature and can be found elsewhere [6–8]. Biogenic amines are also endogenous and indispensable components of living cells, and consequently most food materials, including fruit, vegetables, and grains, contain different levels of biogenic amines depending on their variety, maturity and cultivation condition [7]. Usual intake of dietary biogenic amines generally causes no adverse reactions because human intestinal amine oxidases, such as monoamine oxidase (MAO), diamine oxidase (DAO) and polyamine oxidase (PAO), quickly metabolize and detoxify the biogenic amines. If the capacity of amine-metabolizing enzymes is over-saturated and/or the metabolic activity is impaired by specific inhibitors, vasoactive biogenic amines, including histamine, tyramine and β-phenylethylamine, may cause food intoxication and in turn be considered to be toxic substances in humans [1,2]. Furthermore, the toxicity of biogenic amines can be enhanced by putrefactive biogenic amines such as putrescine and cadaverine [9]. The most common symptoms of biogenic amine intoxication in human are nausea, respiratory distress, hot flushes, sweating, heart palpitation, headache, a bright red rash, oral burning, and hypo- or hypertension [10]. Figure 1 schematically illustrates the detoxification and toxicological risks of biogenic amines.

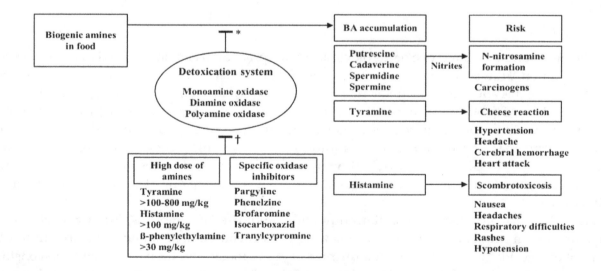

Figure 1. Detoxification and toxicological risks of biogenic amines. *: Metabolic inactivation of biogenic amines through oxidative deamination by oxidases. †: Incapacitation of intestinal detoxication system through saturation by biogenic amines or inhibition by antidepressant medications. BA: Biogenic amines.

The biosynthesis, toxicity and physiological effects have been well reviewed in recent articles [11,12], and will not be summarized here. In addition, it is worth mentioning that in particular, vulnerable people who are immune compromised, such as the elderly, children, and infants, may exhibit intolerance to even low levels of biogenic amines and suffer more severe symptoms [13]. The maximum tolerance levels of vasoactive biogenic amines (mostly histamine in fish and fish products) have established and proposed by government agencies or individual researchers as described below (refer to Table 1), but may need to be further studies and subdivided, considering the vulnerable people.

3. Legal Limits and Toxic Levels of Biogenic Amines in Foods

Early in 1980, the U.S. Food and Drug Administration (FDA) first established regulations for tuna and mahimahi that consider 200 mg histamine/kg as an indication of prior mishandling and 500 mg histamine/kg as an indication of a potential health hazard [14]. Early in 1990, the European Economic Community (EEC) also established regulation for fish species of the *Scombridae* and *Clupeidae* families and fixed a three-class plan for maximum allowable levels of histamine in fresh fish ($n = 9$; $c = 2$; $m = 100$ ppm; $M = 200$ ppm) and enzymatically ripened fish products ($n = 9$; $c = 2$; $m = 200$ ppm; $M = 400$ ppm) where n is the number of units to be analyzed from each lot, m and M are the histamine tolerances, and c is the number of units allowed to contain a histamine level higher than m but lower than M [15]. In 1996, Shalaby [16] suggested the guidelines for histamine content of fish as follows: <50 mg/kg (safe for consumption), 50–200 mg/kg (possibly toxic), 200–1000 mg/kg (probably toxic), >1000 mg/kg (toxic and unsafe for human consumption) based on the review of the regulations and other literature. In the meantime, values of 100–800 mg/kg of tyramine and 30 mg/kg of β-phenylethylamine were reported to be toxic doses in food, respectively, and 100 mg histamine per kg of food and 2 mg histamine per liter of alcoholic beverage were suggested as upper limits for human consumption [6]. The upper limits and toxic doses (stated right above) suggested by Brink et al. [6] have been steadily used by numerous investigators as threshold values to assess human health risks derived from exposure to vasoactive biogenic amines in foods because there have been no other reports describing the guidelines for respective vasoactive biogenic amines in general foods, except for histamine (particularly in fish; not applicable to other foods).

At present, histamine is the only biogenic amine for which the U.S. FDA has set a guidance level, i.e., 50 mg/kg of histamine in the edible portion of fish [17], whereas the European Commission (EC) has established regulatory limits of 100 mg/kg for histamine in fish species and 400 mg/kg for histamine in fish sauce produced by fermentation of fishery products [18]. In the meantime, the European Food Safety Authority [19] reported that although a dose of 50 mg histamine is the no-observed-adverse-effect level (NOAEL), healthy individuals do not experience symptoms unless they ingest a larger amount of histamine than NOAEL. Then, the Food and Agriculture Organization/World Health Organization [20] announced 200 mg histamine/kg as the maximum allowable level for consumption of fish and fish products. According to the Codex standard [21], 200 mg/kg of histamine in fish and fish products and 400 mg/kg of histamine in fish sauce are set as the hygiene and handling indicator levels in the corresponding products, respectively. In addition, the governments of several countries in Asia and Oceania have lately established regulatory limits for histamine in fish and fish products [22–24]. Legal limits and toxic levels set by government agencies or individual researchers for biogenic amines in food products are listed in Table 1. Although several food scientists have referred the suggestion of Brink et al. [6], as described above, there have not been government regulations on maximum allowable levels of biogenic amines, other than histamine, in history. Besides, any government legislation or guidelines on the contents of biogenic amines in fermented soybean foods are not found throughout the world.

Table 1. Legal limits and toxic levels set by agencies for biogenic amines in food products.

Agency	Food	Toxicity Classification	Biogenic Amines (mg/kg) [1]			Governing Entity	Ref.
			PHE	HIS	TYR		
Government	Fish [2] and fish products	Defect action level		50		United States	[17]
		Toxicity level		500			
		Maximum allowable level		200		Australia and New Zealand	[22]
		Maximum allowable level		200		Korea	[23]
		Maximum allowable level		400		China	[24]
	Fish [3] and fish products	Maximum allowable level		200			
International organization	Fresh fish [4]	Defect action level		100		Europe	[15]
		Maximum allowable level		200			
	Enzymaticallyripened fish products [4]	Defect action level		200			
		Maximum allowable level		400			
	Fish [2]	Regulatory limit		100			[18]
	Fish sauce [5]	Regulatory limit		400			
	Fish [2] and fish products	Maximum allowable level		200			[20]
		Decomposition indicator		100			
		Hygiene and handling indicator		200			[21]
	Fish sauce [6]	Hygiene and handling indicator		400			
Independent research	General foods	Toxicity threshold	30	100	100–800		[6]
	Fish [2]	Safe for consumption Possibly toxic Probably toxic Toxic and unsafe for human consumption		<50 50–200 200–1000 >1000			[16]

[1] PHE: β-phenylethylamine, HIS: histamine, TYR: tyramine; [2] *Scombridae, Clupeidae, Engraulidae, Pomatomide, Scombresosidae* and other fish species well known for high histamine content; [3] fish species without high histamine content; [4] *Scombridae* and *Clupeidae* families only; [5] produced by fermentation of fishery products; [6] prepared from fresh fish.

4. Fermented Soybean Foods and Vasoactive Biogenic Amines

Fermented soybean foods have not only been commonly consumed as they are, but have also been frequently used in a variety of processed products, which make them become a necessity in the household in Asian cultures. Moreover, fermented soybean food products have recently gained popularity, crossing from Asian communities to mainstream markets, in many western countries due to the healthy functions of the foods [3,4]. Aside from soy sauces, the most popular fermented soybean foods produced mainly by bacterial fermentation (sometimes with molds) are Natto, Miso (Japanese fermented soybean pastes), Cheonggukjang, Doenjang, Gochujang (Korean fermented soybean pastes), Chunjang, Doubanjiang, Douchi (Chinese fermented soybean pastes) and Tempeh (an Indonesian fermented soybean paste). Some other soybean foods such as Sufu (a Chinese fermented Tofu) and Tauco (an Indonesian fermented yellow soybeans) prepared by mold fermentation are also available in local area (but were excluded from this review due to great differences in the microorganisms involved in fermentation processes as well as little data available in literature). The safety issues of traditional fermented soybean foods have heretofore been overlooked because humans have consistently taken the foods at least for centuries or millennia. However, considering that fermented soybean foods contain not only abundant dietary amino acid precursors of biogenic amines, as mentioned at the beginning of this article, but also significant biogenic amine-producing microorganisms, mainly bacterial species, it is critically important to assess the levels of biogenic amines in the foods.

Based on a critical review of published data (refer to Table 2) [25–37], it seems that the amounts of biogenic amines in most fermented soybean food products are usually within the safe levels for human consumption. It is noteworthy, however, that some specimens of the fermented soybean food products, including both fermented soybean pastes and soy sauces, have been reported to contain vasoactive biogenic amines greater than toxic dose of each amine. For instance, β-phenylethylamine has been detected at concentrations up to 185.6 mg/kg and 239.0 mg/kg in Doubanjiang and

Douchi, respectively [26,36], which are approximately 6-8 times higher than toxic dose of this amine (30 mg/kg) suggested by Brink et al. [6]. In another report, β-phenylethylamine was determined to be 8704.6 mg/kg in a Doenjang sample [34], but which is unreliably larger than those in other articles in which maximum β-phenylethylamine concentrations of 529.2 mg/kg and 544.0 mg/kg have been reported [28,32]; this report was thus excluded from further review. In the meantime, histamine has been detected at concentrations up to 952.0 mg/kg and 808.0 mg/kg in Doenjang and Douchi, respectively [28,36], whereas tyramine has been detected up to 1430.7 mg/kg and 2539.0 mg/kg in Doenjang and Cheonggukjang, respectively [28,33]. The maximum concentrations of histamine and tyramine reported are approximately 8–10 times higher than upper limit of histamine (100 mg/kg) and 14–25 times (on lower toxic dose basis; 2–3 times, upper dose basis) higher than toxic dose of tyramine (100–800 mg/kg), respectively, suggested by Brink et al. [6]. Like the fermented soybean pastes described above, some specimens of soy sauces have been reported to contain high levels of vasoactive biogenic amines, including β-phenylethylamine (up to maximum 121.6 mg/kg), histamine (398.8 mg/kg) and tyramine (794.3 mg/kg), which are much greater than toxic doses of respective amines [28]. As a counter-example, there is a report in which the amounts of respective vasoactive biogenic amines were very low or not detected in all samples (i.e., three batches) of commercial Natto, Miso, Tempeh, and soy sauce products; however, this report seems to insufficiently brief the presence of biogenic amines in the products because samples (batches) of only a single brand for each type of product were available in local stores [4]. The contents of biogenic amines in different types of fermented soybean food products reported in literature have been reviewed once in a book chapter in 2011 [38], and those in the representative fermented soybean food products reviewed herein are compiled in Table 2. After all, it seems likely that there may occasionally be a risk of food poisoning associated with eating fermented soybean pastes, especially when the pastes contain significant amounts of vasoactive biogenic amines, because some types of the pastes, for instance, Natto, Tempeh and sometimes Cheonggukjang, are taken not only as side dishes, but also main dishes. In the case of soy sauces, the risk to consumers may not be so great, considering the small quantity of intake per serve [29].

Table 2. Biogenic amine content in fermented soybean food products.

Fermented Soybean Products	N[1]	Biogenic Amines (mg/kg)[2]								Ref.
		TRP	PHE	PUT	CAD	HIS	TYR	SPD	SPM	
Cheonggukjang	7	6.7–236.4[3]	ND–40.8	4.7–121.3	2.1–20.2	1.3–54.3	0.7–483.1	39.6–59.2	7.1–14.7	[28]
	102	NT[4]	NT	NT	NT	ND[5]–755.40	ND–1913.51	NT	NT	[30]
	13	NT	NT	NT	NT	NT	117.5–2539.0	NT	NT	[33]
Chunjang	4	13.3–19.9	2.2–11.8	9.2–11.7	1.7–6.6	11.6–22.4	29.7–54.6	1.4–12.8	ND–2.9	[28]
	4	19.57–31.35	ND–6.79	3.26–28.59	ND–2.04	1.85–272.55	19.78–131.27	0.24–11.63	ND–1.49	[25]
Doenjang	14	6.1–234.1	ND–529.2	9.9–1453.7	0.3–65.4	1.5–952.0	3.4–1430.7	4.2–23.4	ND–10.2	[28]
	10	ND–449.8	ND–544.0	28.8–1076.6	2.7–144.1	1.4–329.2	12.5–967.6	ND–30.3	ND–9.8	[32]
	23	ND–2808.1	ND–8704.6	ND–4292.3	ND–3235.5	ND–2794.8	ND–6616.1	ND–8804.0	ND–9729.5	[34]
	7	13.5–45.9	3.3–65.0	46.7–168.2	ND–12.9	71.1–382.4	46.4–1190.7	ND–24.7	NT	[29]
Doubanjiang	7	ND–62.43	1.43–185.61	1.15–129.17	ND–0.17	ND	ND–25.75	ND–0.18	ND–1.69	[26]
Douchi	26	ND–440	ND–239	ND–596	ND–191	ND–808	ND–529	ND–719	ND–242	[36]
Gochujang	5	17.9–36.6	0.7–9.1	2.5–3.2	ND–1.1	0.6–1.3	2.1–4.9	1.6–3.4	1.4–1.8	[28]
	7	ND–8.1	1.5–24.8	10.4–36.4	ND–18.1	2.2–59.0	2.9–126.8	ND–14.5	NT	[29]
Miso	5	21.6–23.7	0.7–8.1	16.4–23.2	2.8–3.2	0.8–1.1	2.0–95.3	9.5–21.9	1.3–3.1	[28]
	40	ND–762	ND	ND–12	ND–201	ND–221	ND–49	ND	ND–216	[31]
	22	ND–9.71	2.38–11.76	2.69–14.09	ND–1.31	ND–24.42	ND–66.66	ND–28.31	ND–2.85	[27]
Natto	39	ND–301.0	ND	ND–27.0	ND–42.0	ND–457.0	ND–45.0	ND–124.0	ND–71.0	[35]
	21	ND–45.80	ND–51.50	ND–43.10	ND–36.80	ND–34.40	ND–300.20	246.50–478.10	18.80–80.10	[37]
Soy sauce	11	ND–45.8	1.5–121.6	2.5–1007.5	0.7–32.3	3.9–398.8	26.8–794.3	1.5–53.1	ND–16.1	[28]

[1] Quantity of samples examined; [2] TRP: tryptamine, PHE: β-phenylethylamine, PUT: putrescine, CAD: cadaverine, HIS: histamine, TYR: tyramine, SPD: spermidine, SPM: spermine; [3] the range from minimum to maximum (the same number of digits is used after the decimal point in the values, as was presented in the corresponding references); [4] NT: not tested; [5] ND: not detected.

It is also worth pointing out that some specimens of fermented soybean food products have been found to contain relatively high levels of putrescine and cadaverine (Table 2). The putrefactive biogenic amines have been known to enhance the toxicity of vasoactive biogenic amines in foods [9]. Therefore, comprehensive monitoring and reduction strategies are required to reduce the risk of ingesting putrefactive biogenic amines as well as vasoactive biogenic amines in fermented soybean foods, which may come from the understanding of why there are differences in the amounts and diversity of biogenic amines between the types or batches of the food products. It is probably that the differences may be attributed to (i) the ratio of ingredients used in raw material, (ii) physicochemical and/or microbial contribution, and (iii) conditions and periods of the entire food supply chain [5]. Since fermented soybean foods have their own unique raw materials, physicochemical properties, and production processes, the present review focuses on bacterial contribution to biogenic amine formation conserved across most fermented soybean foods.

5. Bacterial Activity to Produce Biogenic Amines in Fermented Soybean Foods

It has been known that most fermented soybean foods, except for several types of soybean foods prepared by mold fermentation, are mainly fermented (or contaminated) by *Bacillus* species (particularly *B. subtilis*) [5,39,40], which, in turn, leads to biogenic amine formation in the fermented foods, although the abilities of *Bacillus* strains to produce biogenic amines are diverse depending on the types and/or batches of the food products from which the strains are isolated (refer to Table 3) [25,26,31,35–37]. In the studies, the reported ranges (mean ± standard deviation; minimum–maximum) of biogenic amines produced by *Bacillus* spp. in assay media, when cultured for 24 h with proper precursor amino acids, are as follows: histamine 0.22 ± 0.65–29.9 ± 13.4 µg/mL, tyramine 0.3 ± 0.5–30.6 ± 21.7 µg/mL, β-phenylethylamine not detected (ND)—11.2 ± 9.17 µg/mL, tryptamine 0.20 ± 0.45–6.17 ± 3.98 µg/mL, putrescine ND—7.59 ± 3.06 µg/mL, cadaverine ND—1.8 ± 1.1 µg/mL, spermidine 0.40 ± 0.20–9.26 ± 5.73 µg/mL, spermine 1.29 ± 0.86–27.2 ± 12.7 µg/mL. Among the *Bacillus* strains reported, *B. subtilis* strains isolated from Natto exhibited the strongest abilities to produce respective biogenic amines. Table 3 reveals the abilities to produce biogenic amines of different bacterial species isolated from representative types of fermented soybean food products.

Table 3. Production of biogenic amines by bacteria isolated from fermented soybean food products.

Fermented Soybean Products	Isolates	N[1]	Biogenic Amines (µg/mL)[2]								Ref.
			TRP	PHE	PUT	CAD	HIS	TYR	SPD	SPM	
Chunjang	*Bacillus* spp.[3]	89	0.45 ± 0.32[4]	0.85 ± 0.23	0.95 ± 0.55	ND[5]	1.34 ± 1.19	1.41 ± 0.32	9.26 ± 5.73	2.17 ± 1.09	[25]
Doubanjiang	*Bacillus subtilis*	18	0.20 ± 0.45	0.67 ± 1.42	3.45 ± 1.29	1.03 ± 0.46	0.22 ± 0.65	0.59 ± 0.65	0.40 ± 0.20	1.29 ± 0.86	[26]
Douchi	*Bacillus subtilis*	4	NT[6]	2.3 ± 4.5	NT	0.5 ± 0.6	18.7 ± 9.3	0.3 ± 0.5	NT	4.5 ± 5.2	[36]
	Staphylococcus pasteuri	1	NT	ND	NT	1.2	20.0	ND	NT	ND	
	Staphylococcus capitis	3	NT	5.4 ± 9.3	NT	1.1 ± 0.9	375.3 ± 197.0	1.1 ± 1.9	NT	2.7 ± 3.4	
Miso	*Staphylococcus pasteuri*	1	NT	6.4	ND	ND	28.1	NT	ND	NT	[31]
	Bacillus sp.	1	NT	2.7	1.6	2.1	15.3	NT	8.6	NT	
	Bacillus amyloliquefaciens	2	NT	1.6 ± 2.2	ND	1.8 ± 1.1	16.5 ± 8.6	NT	4.2 ± 5.9	NT	
	Bacillus subtilis	2	NT	ND	0.5 ± 0.7	0.6 ± 0.8	29.9 ± 13.4	NT	5.0 ± 7.1	NT	
	Bacillus megaterium	2	NT	7.7 ± 0.4	ND	ND	14.6 ± 2.8	NT	4.7 ± 6.6	NT	
Natto	*Bacillus subtilis*	80	6.17 ± 3.98	11.2 ± 9.17	7.59 ± 3.06	0.94 ± 1.67	9.91 ± 1.61	30.6 ± 21.7	3.34 ± 1.82	27.2 ± 12.7	[37]
	Bacillus subtilis	2	NT	2.4 ± 3.3	NT	1.5 ± 0.1	15.5 ± 2.9	NT	NT	NT	[35]
	Staphylococcus pasteuri	2	NT	ND	NT	1.1 ± 0.1	15.0 ± 1.3	NT	NT	NT	

[1] Quantity of bacterial samples examined; [2] TRP: tryptamine, PHE: β-phenylethylamine, PUT: putrescine, CAD: cadaverine, HIS: histamine, TYR: tyramine, SPD: spermidine, SPM: spermine; [3] *Bacillus* spp. were identified to be *B. subtilis* (91.0%), *B. coagulans* (4.5%), *B. licheniformis* (1.1%) and *B. firmus* (1.1%); [4] mean ± standard deviation (the same number of digits is used after the decimal point in the values, as was presented in the corresponding references); [5] ND: not detected; [6] NT: not tested.

In addition to the aforementioned *Bacillus* spp., *Lactobacillus* sp. and *Enterococcus faecium*, which had been isolated from raw materials of Miso, were proposed to produce histamine and tyramine in Miso, respectively, through a qualitative detection using BCP (Bromo-cresol purple) agar plates and subsequently a quantitative test using liquid media [41,42]. In the quantitative test with incubation for 90 days, the strains of *Lactobacillus* sp. and *E. faecium* produced histamine and tyramine up to approximately 100 μg/mL and 150 μg/mL, respectively. Although *Lactobacillus* species are not commonly involved in the preparation of fermented soybean foods, diverse species of *Lactobacillus* have also been reported to be responsible for the formation of biogenic amines, including histamine, in lactic fermented foods [12]. *E. faecium* and *E. faecalis* have been found to possess *tdc* gene and produce tyramine in fermented foods, including dairy products, fermented sausages, wine and fermented soybean foods [12]. Thus, *E. faecium* strains have been used as target organisms for studies on the reduction of tyramine in fermented soybean foods [33,43], even though *Enterococcus* spp. are present as contaminants at relatively low levels (maximum up to 10^6 CFU/g) in the foods [44–46]. In the meantime, the absence of *hdc* gene encoding histidine decarboxylase was reported in both *E. faecium* and *E. faecalis* in one study [47], while histidine decarboxylase-positive *E. faecium* and *E. faecalis* strains were detected by a PCR (polymerase chain reaction) method in another study [48]. It is interesting to note that the PCR screening method used in the latter study employed the primers developed in the former study, which makes it difficult to conclude whether the species possess *hdc* gene or not.

As shown in Table 4, at present the Gene Bank database of the National Centre for Biotechnology Information (NCBI, National Center for Biotechnology Information, U.S. National Library of Medicine, Bethesda, MD, USA) provides the sequences of *tdc*, *odc*, and *ldc* genes in *E. faecium* and *tdc* and *ldc* genes in *E. faecalis*, while *hdc* gene sequence of both species is not available in the database. In contrast, the sequences of *hdc* gene in *B. licheniformis* and *B. coagulans* (this sequence is completely conserved between the two species) and *ldc* in *B. subtilis* have been deposited in the database, while *tdc* gene sequence is unavailable for the three species of *Bacillus*. Nevertheless, it has lately been suggested that *Bacillus* spp. are as significant as *Enterococcus* spp. for tyramine formation in fermented soybean foods [49]. The deposited genes encoding amino acid decarboxylases of *Bacillus* spp. and *Enterococcus* spp., the most important species related to biogenic amine formation in fermented soybean products, are listed in Table 4 (exceptionally, *odc*-Az encodes an antizyme inhibitor devoid of ornithine decarboxylase activity). All the bacteria and genes mentioned above should be targeted for preventive interventions to reduce biogenic amine formation in fermented soybean foods. Meanwhile, yeasts have been considered to produce only negligible amounts of biogenic amines [50,51]. Fungal distribution to biogenic amine accumulation is remained to be further studied because there appears to be but little literature available dealing with fungal formation of biogenic amines [52].

It is well known that various vasoactive and putrefactive biogenic amines are commonly formed by microbial decarboxylation of amino acids in fermented foods [6,7]. As such, it has been found that soybean fermentation results in an increase in the amount of spermine (and other biogenic amines), but a decrease in that of spermidine [26]. Since spermidine is essential for the growth and development of plants [53,54], this polyamine is abundantly present in soybean and non-fermented soybean foods such as Tofu (a curd product made from soy milk) [26,55,56] and degraded by bacterial enzymes during fermentation [57]. Consequently, fermented soybean foods contain a lower level of spermidine than their raw material, soybean [26,58]. This indicates that development and application of biogenic amine-degrading starter cultures are possible (and necessary) to reduce the contents of biogenic amines in fermented soybean foods. Identifying and understanding the dominant contributors to the formation of biogenic amines may facilitate the development of starter cultures for delaying or avoiding biogenic amine formation in the fermented foods. Taken together, it is clear that distinct and diverse bacterial community and/or capability of producing (and degrading) biogenic amines decisively determine the amounts and diversity of biogenic amines in fermented soybean foods.

Table 4. Genes encoding amino acid decarboxylases in *Bacillus* spp. and *Enterococcus* spp. registered in the NCBI database.

Species	Strain [1]	Source	Gene for Amino Acid Decarboxylase [2]	No. of Amino Acids	Locus Name	Accession (Version)	Size (bp)
B. subtilis	*B. subtilis* subsp. *subtilis* strain 168	Isolated strain	*odc-Az*	331	BACYACA	L77246.1	996
			ldc	490	AF012285	AF012285.1	1473
B. licheniformis / B. coagulans	*B. licheniformis* A5 / *B. coagulans* SL5	Isolated strain	*hdc*	146	AB553282 / AB553281	AB553282.1 / AB553281.1	441
E. faecium	*E. faecium* strain 993	Isolated strains	*odc*	235	PDLZ01000281	PDLZ01000281.1	707
	E. faecium ATCC 700221	ATCC	*ldc*	191	CP014449	CP014449.1	576
	E. faecium ATCC 700221	ATCC	*tdc*	611	CP014449	CP014449.1	1836
E. faecalis	*E. faecalis* ATCC 51299	ATCC	*ldc*	194	JSES01000022	JSES01000022.1	585
	E. faecalis ATCC 19433	Type strain	*tdc*	620	KB944589	KB944589.1	1863

[1] Genes found in a single strain of each *Bacillus* species have been registered, while those of *Enterococcus* spp. found in multiple strains have been separately assigned to different loci, of which a representative locus is presented in the table; [2] *odc-Az*: 37.0% identity over 119 amino acids to the *E. coli* ornithine decarboxylase antizyme, *odc*: gene for ornithine decarboxylase, *ldc*: gene for lysine decarboxylase, *hdc*: gene for histidine decarboxylase, *tdc*: gene for tyrosine decarboxylase; ATCC: the American Type Culture Collection.

6. Control Strategies for Reducing Biogenic Amines in Fermented Soybean Foods

Regarding intervention measures that reduce biogenic amine formation in fermented soybean foods, to date, only a few reports are available in literature as follows: the use of irradiation [59], addition of nicotinic acid as a tyrosine decarboxylase inhibitor [43], and use of *Bacillus* starter cultures [60–63]. However, when extended to other fermented foods, a review of the relevant literature reveals that several types of intervention methods have been developed and used to reduce biogenic amine contents in the foods (mainly fermented sausage and cheese), which involve chemical intervention, such as the use of food additives and natural antimicrobial compounds [43,64–67], physical intervention, such as the use of irradiation [59,68], high hydrostatic pressure [69,70] and modified atmosphere packaging [71,72], and biological intervention, particularly such as the use of starter cultures [60–63,73–76]. The biological intervention methods also involve the control or adjustment of intrinsic and extrinsic factors, such as alterations of temperature, pH, a_w, and Eh, which have been well reviewed in literature [77–79].

Up to this day, thousands of additives have been used to extend shelf life of foods because of their antimicrobials, antioxidants, and antibrowning properties. Natural additives have lately been of great interest in food industry due to consumers' health concerns [80]. Apart from being used as food preservatives, numerous food additives and natural antimicrobial compounds, including glycine [64], nicotinic acid [43], potassium sorbate, sodium benzoate [67], sodium chloride [64,66], clove [65–67], garlic [65], etc., have been found to be effective in suppressing bacterial ability to produce biogenic amines in foods. Among the compounds, nicotinic acid is only one compound proven to practically inhibit the formation of biogenic amines (particularly tyramine) in a fermented soybean food, viz., Cheonggukjang [43]. In the report, the addition of nicotinic acid at concentrations of 0.15% and 0.20% resulted in significant reductions, by approximately 70% and 83%, respectively, compared to the control, of tyramine content in the treated Cheonggukjang samples after 24 h of fermentation. In addition, it is worth noting that even though a successful reduction of biogenic amines in a food product can be achieved by the addition of any of a variety of compounds, some of the additives may cause organoleptic alterations, such as an atypical taste and flavor, in the final food product, especially in the case of fermented soybean foods [5,49]. Therefore, sensory evaluation should be incorporated as an integral part of a program investigating effective inhibitors of biogenic amine formation in fermented soybean foods.

Besides the chemical intervention measures described above, a variety of physical intervention processes have been developed and applied for food preservation, which involve not only well-known classical processes, for instance, heating, refrigeration, and freezing, but also emerging novel processes such as microwave heating, ohmic heating and pulsed electric fields developed during the past 25 to 35 years [81]. Among the physical intervention methods, irradiation, high hydrostatic pressure and modified atmosphere packaging have been relatively recently reported to successfully inhibit biogenic amine formation in fermented foods, which have been achieved mostly by reducing microbial population, for instance, lactic acid bacteria, closely related to the fermentation of foods [59,69–72]. Despite the technological progress that has been made, as for fermented soybean foods, there has been only a single report describing biogenic amine reduction in the food treated by one of the physical intervention processes. In the report, γ-irradiation of raw materials with doses of 5, 10, and 15 kGy significantly reduced the contents of histamine, putrescine, tryptamine and spermidine by approximately 20–50% (but not tyramine, β-phenylethylamine, cadaverine, spermine and agmatine) in the final product of a fermented soybean food, viz., probably Doenjang [59]. However, it needs here to be noted that the irradiation even with the lowest dose resulted in an immediate and significant decrease in the numbers of *Bacillus* spp. and lactic acid bacteria, known as dominant bacteria in the food, by up to about 3 log CFU/g and 2 log CFU/g, respectively. As is well known, many of the physical intervention processes prevent the growth of fermenting microorganisms, as well as of biogenic amine-producing microorganisms, which may in turn not only delay fermentation, but also lead to abnormal fermentation caused by undesirable microorganisms resistant to the treatments [82]. Thus, introducing the processes would be somewhat challenging in the case of fermented soybean foods, considering the presence of fermenting and/or beneficial bioactive microorganisms in the foods.

The use of starter cultures has been suggested to be a successful way to enhance not only the quality and safety, but also the healthy functions, of fermented foods, causing less adverse organoleptic and unhealthy alterations [83–85]. Thus, with that a variety of microorganisms have been compared and screened for the ability to degrade biogenic amines and/or inability to produce biogenic amines in fermented foods, not only at the level of genus, species, or both, but also at the level of individual strain [73–76]. As for the fermentation of soybean, *Bacillus* strains have been steadily proposed as starter cultures to improve the sensory quality, but not the safety of fermented soybean foods [86,87]. On the contrary, less attention has been given to starter cultures for preventing or reducing biogenic amine formation in fermented soybean foods. As mentioned above, *Bacillus* spp. have been known as fermenting (or contaminating) microorganisms responsible for biogenic amine formation in different types of fermented soybean foods. Therefore, it is imperative to screen proper starter cultures (particularly *Bacillus* starter culture) with no or less ability to produce biogenic amines for the production of fermented soybean foods [62]. With respect to this, there have been a few reports in literature in which Doenjang and Cheonggukjang samples prepared with *B. subtilis* and *B. licheniformis* starter cultures, respectively, with low abilities to produce biogenic amines (the data on individual strains were not presented in the reports) contained lower levels of biogenic amines than those of previous studies [61,62]. Alternatively, the use of starter cultures that can degrade biogenic amines may facilitate the reduction of biogenic amines in fermented soybean foods [77]. At present, only two reports of biogenic amine-degrading starter cultures for the production of fermented soybean foods are available in literature as described below. In one study, *B. subtilis* and *B. amyloliquefaciens* strains which had been isolated from traditionally fermented soybean products degraded significant amounts of histamine (up to 71% of its initial concentration by *B. amyloliquefaciens*), tyramine (up to 70% by *B. amyloliquefaciens*), putrescine (up to 92% by *B. subtilis*) and cadaverine (up to 93% by *B. subtilis*) in cooked soybean after 10 days of fermentation [60]. In another study, *B. subtilis* and *B. amyloliquefaciens* strains which had been isolated from commercial fermented soybean products degraded 30–40% of tyramine in a phosphate buffer and probably thereby reduced tyramine content by 40–65% in the final product of Cheonggukjang, as compared to the control [63]. In addition, *B. subtilis* and *B. idriensis* strains isolated from a traditional fermented soybean food have been reported to be not only capable of degrading of, but also incapable of producing histamine and tyramine in vitro (but not applied to practical fermentation of soybean in the study) [88]. Consequently, it is feasible to screen *Bacillus* strains capable of degrading and/or incapable (or less capable) of producing biogenic amines, which would, in turn, make it possible to use them as starter cultures for reducing biogenic amine contents in fermented soybean foods. In the meantime, it is also necessary to fully identify and characterize *Bacillus* genes involved in the formation and degradation of biogenic amines, which would be helpful not only in selecting starter culture candidates but also in providing strategies to efficiently regulate the expression of these genes encoding relevant enzymes. Such molecular genetic studies would be further needed for better understanding of mechanisms by which intervention methods influencing intrinsic and extrinsic factors and/or microbial growth inhibit biogenic amine formation, at the level of gene. It is noteworthy that in addition to the aforementioned *Bacillus* starter cultures, strains of *E. faecium* and *L. plantarum* have also been proposed as starter cultures for fermented soybean foods because of their abilities to produce bacteriocin or to degrade biogenic amines, respectively [46,89]. Considering that the species are present as contaminants at relatively low levels in fermented soybean foods, as described above, further research is required prior to practical application to the fermentation of soybean in food industry.

Aside from the use of starter cultures, the production of biogenic amines has been known to be dependent on intrinsic and extrinsic factors of foods [77–79]. Furthermore, the factors may provide combined effects, especially in connection with technology applied, viz., the chemical and physical intervention measures described above [90]. As of now, however, the alterations of temperature, pH, a_w, and Eh (as another important, but classical, biological intervention strategy), seem to be less preferable for studies on the reduction of biogenic amines in fermented soybean foods than other alternatives, considering that there are no relevant reports available, which might be because of the need to consider strict demands of consumers and governments on unique sensory properties and manufacturing processes of fermented soybean foods. Nonetheless, it is expected that the changes of the intrinsic and extrinsic factors within narrow ranges would be applicable, depending on the types of fermented soybean foods, if organoleptic evaluation is preceded. The intrinsic and extrinsic factors influencing biogenic amine formation in foods have been well reviewed in a recent article [8]. Biogenic amine reduction strategies, including chemical, physical and biological intervention methods, are summarized in Table 5.

Table 5. Biogenic amine reduction strategies for food products.

Parameter Categories		Highly Effective Strategies
Chemical intervention		Nicotinic acid [43], glycine [64], garlic [65], clove [65], clove and sodium chloride [66], clove with potassium sorbate and sodium benzoate [67]
Physical intervention		Irradiation [59,68], high hydrostatic pressure [69,70], modified atmospheric packaging and temperature [71,72]
Biological intervention	Starter cultures	Lactic acid bacteria [84], *Lactobacillus sake* + *Pediococcus pentosaceus* + *Staphylococcus carnosus* + *S. xylosus* [73], *S. carnosus* [74], *S. xylosus* [74–76], *L. plantarum* [89], *Bacillus subtilis* [60,63,88], *B. amyloliquefaciens* [60,63], *B. licheniformis* [62], *B. idriensis* [88], *B. subtilis* + *Aspergillus oryzae* + *Mucor racemosus* [61]
	Intrinsic and extrinsic factors	Temperature, pH, a_w, Eh [77–79]

7. Conclusions

The presence of histamine in fish is of concern in many countries due to its toxic potential and implications. Accordingly, there are specific legislations regarding the histamine content in fish and fish products in US, EU, and other countries. In contrast, the significance of biogenic amines in fermented soybean foods has been overlooked despite the presence not only of abundant precursor amino acids of biogenic amines in soybean, but also of microorganisms capable of producing biogenic amines during the fermentation of soybean. Fortunately, the studies published to date indicate that the amounts of biogenic amines in most fermented soybean food products are within the safe levels for human consumption. However, it should be pointed out that the contents of vasoactive biogenic amines in certain types and/or batches of fermented soybean food products are greater than toxic levels. Nonetheless, lack of both legislation and guidelines on the contents of biogenic amines in fermented soybean food products may lead to serious (or unnecessary) concerns about the safety of the fermented foods. Therefore, it is required to establish guidance levels of biogenic amines in fermented soybean food products based on information about the national daily intake of the fermented foods per person and the amounts of biogenic amines in different types of fermented soybean foods commonly consumed in each country.

Meanwhile, many efforts have been made to reduce biogenic amines in various fermented foods, particularly fermented sausage and cheese, whereas less attention has given to biogenic amines in fermented soybean foods. Consequently, there is at present a little information available regarding intervention methods to reduce biogenic amines in fermented soybean foods. Although empirical data on controlling biogenic amines in fermented soybean foods are not much in literature, several reports have suggested that the use of starter cultures capable of degrading and/or incapable of producing biogenic amines is a preferable way to biocontrol biogenic amines in fermented soybean foods because it probably causes less adverse organoleptic and unhealthy alterations as well as little changes in bacterial communities in the foods. Alterations of intrinsic and extrinsic factors, such as temperature, pH, a_w, and Eh, in fermentation and manufacturing processes are also needed to be taken into consideration when biocontrol strategy is employed. With a successful reduction of biogenic amines in addition to significant health benefits, consumers may place a much higher value on fermented soybean foods.

Author Contributions: Conceptualization, J.-H.M., and H.-J.H.; Literature data collection, J.-H.M., Y.K.P., Y.H.J., and J.-H.L.; Writing-original draft preparation, J.-H.M., Y.K.P.; Writing-review and editing, J.-H.M.; Supervision, J.-H.M.

Acknowledgments: The authors thank Jae Hoan Lee and Junsu Lee of Department of Food and Biotechnology at Korea University for English editing and technical assistance, respectively.

References

1. Joosten, H.M.L.J.; Nuñez, M. Prevention of histamine formation in cheese by bacteriocin-producing lactic acid bacteria. *Appl. Environ. Microbiol.* **1996**, *62*, 1178–1181. [PubMed]
2. Taylor, S.L.; Guthertz, L.S.; Leatherwood, H.; Tillman, F.; Lieber, E.R. Histamine production by foodborne bacterial species. *J. Food Saf.* **1978**, *1*, 173–187. [CrossRef]
3. Shin, D.; Jeong, D. Korean traditional fermented soybean products: *Jang. J. Ethn. Foods* **2015**, *2*, 2–7. [CrossRef]
4. Toro-Funes, N.; Bosch-Fuste, J.; Latorre-Moratalla, M.L.; Veciana-Nogués, M.T.; Vidal-Carou, M.C. Biologically active amines in fermented and non-fermented commercial soybean products from the Spanish market. *Food Chem.* **2015**, *173*, 1119–1124. [CrossRef] [PubMed]
5. Mah, J.-H. Fermented soybean foods: Significance of biogenic amines. *Austin J. Nutr. Food Sci.* **2015**, *3*, 1058.
6. Brink, B.; Damirik, C.; Joosten, H.M.L.J.; Huis In't Veld, J.H.J. Occurrence and formation of biologically active amines in foods. *Int. J. Food Microbiol.* **1990**, *11*, 73–84. [CrossRef]
7. Halász, A.; Baráth, Á.; Simon-Sarkadi, L.; Holzapfel, W. Biogenic amines and their production by microorganisms in food. *Trends Food Sci. Technol.* **1994**, *5*, 42–49. [CrossRef]
8. Gardini, F.; Özogul, Y.; Suzzi, G.; Tabanelli, G.; Özogul, F. Technological factors affecting biogenic amine content in foods: A review. *Front. Microbiol.* **2016**, *7*, 1218. [CrossRef] [PubMed]
9. Stratton, J.E.; Hutkins, R.W.; Taylor, S.L. Biogenic amines in cheese and other fermented foods: A review. *J. Food Prot.* **1991**, *54*, 460–470. [CrossRef]

10. Rice, S.L.; Eitenmiller, R.R.; Koehler, P.E. Biologically active amines in foods: A review. *J. Milk Food Technol.* **1976**, *39*, 353–358. [CrossRef]

11. Fernández-Reina, A.; Urdiales, J.; Sánchez-Jiménez, F. What we know and what we need to know about aromatic and cationic biogenic amines in the gastrointestinal tract. *Foods* **2018**, *7*, 145. [CrossRef] [PubMed]

12. Barbieri, F.; Montanari, C.; Gardini, F.; Tabanelli, G. Biogenic amine production by lactic acid bacteria: A Review. *Foods* **2019**, *8*, 17. [CrossRef] [PubMed]

13. Byun, B.Y.; Bai, X.; Mah, J.-H. Bacterial contribution to histamine and other biogenic amine content in *Juk* (Korean traditional congee) cooked with seafood. *Food Sci. Biotechnol.* **2013**, *22*, 1675–1681. [CrossRef]

14. U.S. Food and Drug Administration. Defect action levels for histamine in tuna: Availability of guide. *Fed. Reg.* **1982**, *47*, 40487–40488.

15. European Economic Commission (EEC). Council Directive 91/493/EEC of 22 July 1991 laying down the health conditions for the production and the placing on the market of fishery products. *Off. J. Eur. Comm.* **1991**, *268*, 15–34.

16. Shalaby, A.R. Significance of biogenic amines to food safety and human health. *Food Res. Int.* **1996**, *29*, 675–690. [CrossRef]

17. U.S. Food and Drug Administration (FDA). *Fish and Fishery Products Hazards and Controls Guidance*, 4th ed.; Center for Food Safety and Applied Nutrition: Rockville, MD, USA, 2011.

18. European Commision (EC). Commission Regulation No. 2073/2005 of 15th November 2005 on microbiological criteria for Foodstuffs. *Off. J. Eur. Union* **2005**, *L338*, 1–25.

19. European Food Safety Authority (EFSA). Scientific opinion on risk based control of biogenic amine formation in fermented foods. *EFSA J.* **2011**, *9*, 2393–2486. [CrossRef]

20. Food and Agriculture Organization of the United Nations/World Health Organization (FAO/WHO). *Joint FAO/WHO Expert Meeting on the Public Health Risk of Histamine and Other Biogenic Amines from Fish and Fishery Products*; FAO/WHO: Rome, Italy, 2013.

21. Codex. Discussion paper histamine. In *Joint FAO/WHO Food Standards Programme, Codex Committee on Fish and Fishery Products, Thirty-Second Session (CX/FFP 12/32/14), Bali, Indonesia, 1–5 October 2012*; FAO/WHO: Rome, Italy, 2012.

22. Food Standards Australia New Zealand (FSANZ). *Imported Food Risk Statement Fish and Fish Products from the Families Specifies and Histamine*; FSANZ: Canberra, Australia, 2016.

23. The Ministry of Food and Drug Safety (MFDS). *Food Code, Notification No. 2017-57*; MFDS: Osong, Korea, 2017.

24. China National Standards. *GB 2733-2015 National Food Safety Standards for Fresh and Frozen Animal Aquatic Products*; Standards Press of China: Beijing, China, 2016.

25. Bai, X.; Byun, B.Y.; Mah, J.-H. Formation and destruction of biogenic amines in *Chunjang* (a black soybean paste) and *Jajang* (a black soybean sauce). *Food Chem.* **2013**, *141*, 1026–1031. [CrossRef] [PubMed]

26. Byun, B.Y.; Bai, X.; Mah, J.-H. Occurrence of biogenic amines in *Doubanjiang* and *Tofu*. *Food Sci. Biotechnol.* **2013**, *22*, 55–62. [CrossRef]

27. Byun, B.Y.; Mah, J.-H. Occurrence of biogenic amines in *Miso*, Japanese traditional fermented soybean paste. *J. Food Sci.* **2012**, *77*, T216–T223. [CrossRef] [PubMed]

28. Cho, T.-Y.; Han, G.-H.; Bahn, K.-N.; Son, Y.-W.; Jang, M.-R.; Lee, C.-H.; Kim, S.-H.; Kim, D.-B.; Kim, S.-B. Evaluation of biogenic amines in Korean commercial fermented foods. *Korean J. Food Sci. Technol.* **2006**, *38*, 730–737.

29. Kim, T.-K.; Lee, J.-I.; Kim, J.-H.; Mah, J.-H.; Hwang, H.-J.; Kim, Y.-W. Comparison of ELISA and HPLC methods for the determination of biogenic amines in commercial *Doenjang* and *Gochujang*. *Food Sci. Biotechnol.* **2011**, *20*, 1747–1750. [CrossRef]

30. Ko, Y.-J.; Son, Y.-H.; Kim, E.-J.; Seol, H.-G.; Lee, G.-R.; Kim, D.-H.; Ryu, C.-H. Quality properties of commercial *Chungkukjang* in Korea. *J. Agric. Life Sci.* **2012**, *46*, 1–11.

31. Kung, H.-F.; Tsai, Y.-H.; Wei, C.-I. Histamine and other biogenic amines and histamine-forming bacteria in miso products. *Food Chem.* **2007**, *101*, 351–356. [CrossRef]

32. Lee, H.T.; Kim, J.H.; Lee, S.S. Analysis of microbiological contamination and biogenic amines content in traditional and commercial Doenjang. *J. Food Hyg. Saf.* **2009**, *24*, 102–109.

33. Oh, S.-J.; Mah, J.-H.; Kim, J.-H.; Kim, Y.-W.; Hwang, H.-J. Reduction of tyramine by addition of *Schizandra chinensis* Baillon in Cheonggukjang. *J. Med. Food* **2012**, *15*, 1109–1115. [CrossRef] [PubMed]

34. Shukla, S.; Park, H.-K.; Kim, J.-K.; Kim, M. Determination of biogenic amines in Korean traditional fermented soybean paste (Doenjang). *Food Chem. Toxicol.* **2010**, *48*, 1191–1195. [CrossRef] [PubMed]

35. Tsai, Y.-H.; Chang, S.-C.; Kung, H.-F. Histamine contents and histamine-forming bacteria in natto products in Taiwan. *Food Control* **2007**, *28*, 1026–1030. [CrossRef]

36. Tsai, Y.-H.; Kung, H.-F.; Chang, S.-C.; Lee, T.-M.; Wei, C.-I. Histamine formation by histamine-forming bacteria in douchi, a Chinese traditional fermented soybean product. *Food Chem.* **2007**, *103*, 1305–1311. [CrossRef]

37. Kim, B.; Byun, B.Y.; Mah, J.-H. Biogenic amine formation and bacterial contribution in *Natto* products. *Food Chem.* **2012**, *135*, 2005–2011. [CrossRef] [PubMed]

38. Shukla, S.; Kim, J.-K.; Kim, M. Occurrence of biogenic amines in soybean food products. In *Soybean and Health*; El-Shemy, H., Ed.; InTech: Rijeka, Croatia, 2011; pp. 182–184.

39. Onda, T.; Yanagida, F.; Shinohara, T.; Yokotsuka, K. Time series analysis of aerobic bacterial flora during Miso fermentation. *Lett. Appl. Microbiol.* **2003**, *37*, 162–168. [CrossRef] [PubMed]

40. Tamang, J.P. Diversity of fermented foods. In *Fermented Foods and Beverages of the World*; Tamang, J.P., Kailasapathy, K., Eds.; CRC Press: Boca Raton, FL, USA, 2010; pp. 41–84.

41. Ibe, A.; Nishima, T.; Kasai, N. Formation of tyramine and histamine during soybean paste (Miso) fermentation. *Jpn. J. Toxicol. Environ. Health* **1992**, *38*, 181–187. [CrossRef]

42. Ibe, A.; Nishima, T.; Kasai, N. Bacteriological properties of and amine-production conditions for tyramine-and histamine-producing bacterial strains isolated from soybean paste (Miso) starting materials. *Jpn. J. Toxicol. Environ. Health* **1992**, *38*, 403–409. [CrossRef]

43. Kang, H.-R.; Kim, H.-S.; Mah, J.-H.; Kim, Y.-W.; Hwang, H.-J. Tyramine reduction by tyrosine decarboxylase inhibitor in *Enterococcus faecium* for tyramine controlled *cheonggukjang*. *Food Sci. Biotechnol.* **2018**, *27*, 87–93. [CrossRef] [PubMed]

44. Oh, E.J.; Oh, M.-H.; Lee, J.M.; Cho, M.S.; Oh, S.S. Characterization of microorganisms in *Eoyukjang*. *Korean J. Food Sci. Technol.* **2008**, *40*, 656–660.

45. Sarkar, P.; Tamang, J.P.; Cook, P.E.; Owens, J. Kinema-a traditional soybean fermented food: Proximate composition and microflora. *Food Microbiol.* **1994**, *11*, 47–55. [CrossRef]

46. Yoon, M.Y.; Kim, Y.J.; Hwang, H.-J. Properties and safety aspects of *Enterococcus faecium* strains isolated from *Chungkukjang*, a fermented soy product. *LWT-Food Sci. Technol.* **2008**, *41*, 925–933. [CrossRef]

47. Coton, E.; Coton, M. Multiplex PCR for colony direct detection of Gram-positive histamine-and tyramine-producing bacteria. *J. Microbiol. Methods* **2005**, *63*, 296–304. [CrossRef] [PubMed]

48. Komprda, T.; Sládková, P.; Petirová, E.; Dohnal, V.; Burdychová, R. Tyrosine-and histidine-decarboxylase positive lactic acid bacteria and enterococci in dry fermented sausages. *Meat sci.* **2010**, *86*, 870–877. [CrossRef] [PubMed]

49. Jeon, A.R.; Lee, J.H.; Mah, J.-H. Biogenic amine formation and bacterial contribution in *Cheonggukjang*, a Korean traditional fermented soybean food. *LWT-Food Sci. Technol.* **2018**, *92*, 282–289. [CrossRef]

50. Caruso, M.; Fiore, C.; Contursi, M.; Salzano, G.; Paparella, A.; Romano, P. Formation of biogenic amines as criteria for the selection of wine yeasts. *World J. Microbiol. Biotechnol.* **2002**, *18*, 159–163. [CrossRef]

51. Izquierdo-Pulido, M.; Font-Fábregas, J.; Vidal-Carou, C. Influence of *Saccharomyces cerevisiae* var. *uvarum* on histamine and tyramine formation during beer fermentation. *Food Chem.* **1995**, *54*, 51–54.

52. Nout, M.K.R.; Ruiker, M.M.W.; Bouwmeester, H.M.; Beljaars, P.R. Effect of processing conditions on the formation of biogenic amines and ethyl carbamate in soybean tempe. *J. Food Saf.* **1993**, *33*, 293–303. [CrossRef]

53. Fuell, C.; Elliott, K.A.; Hanfrey, C.C.; Franceschetti, M.; Michael, A.J. Polyamine biosynthetic diversity in plants and algae. *Plant Physiol. Biochem.* **2010**, *48*, 513–520. [CrossRef] [PubMed]

54. Kumar, A.; Taylor, M.A.; Arif, S.A.M.; Davies, H.V. Potato plants expressing antisense and sense S-adenosylmethionine decarboxylase (SAMDC) transgenes show altered levels of polyamines and ethylene: Antisense plants display abnormal phenotypes. *Plant J.* **1996**, *9*, 147–158. [CrossRef]

55. Kalaè, P.; Krausová, P. A review of dietary polyamines: Formation, implications for growth, and health and occurrence in foods. *Food Chem.* **2005**, *90*, 219–230.

56. Liu, Z.-F.; Wei, Y.-X.; Zhang, J.-J.; Liu, D.-H.; Hu, Y.-Q.; Ye, X.-Q. Changes in biogenic amines during the conventional production of stinky tofu. *Int. J. Food Sci. Technol.* **2011**, *46*, 687–694. [CrossRef]

57. Woolridge, D.P.; Vazquez-Laslop, N.; Markham, P.N.; Chevalier, M.S.; Gerner, E.W.; Neyfakh, A.A. Efflux of the natural polyamine spermidine facilitated by the *Bacillus subtilis* multidrug transporter Blt. *J. Biol. Chem.* **1997**, *272*, 8864–8866. [CrossRef] [PubMed]

58. Righetti, L.; Tassoni, A.; Bagni, N. Polyamines content in plant derived food: A comparison between soybean and Jerusalem artichoke. *Food Chem.* **2008**, *111*, 852–856. [CrossRef]

59. Kim, J.-H.; Ahn, H.-J.; Kim, D.-H.; Jo, C.; Yook, H.-S.; Park, H.-J.; Byun, M.-W. Irradiation effects on biogenic amines in Korean fermented soybean paste during fermentation. *J. Food Sci.* **2003**, *68*, 80–84. [CrossRef]

60. Kim, Y.S.; Cho, S.H.; Jeong, D.Y.; Uhm, T.-B. Isolation of biogenic amines-degrading strains of *Bacillus subtilis* and *Bacillus amyloliquefaciens* from traditionally fermented soybean products. *Korean J. Microbiol.* **2012**, *48*, 220–224. [CrossRef]

61. Shukla, S.; Lee, J.S.; Park, H.-K.; Yoo, J.-A.; Hong, S.-Y.; Kim, J.-K.; Kim, M. Effect of novel starter culture on reduction of biogenic amines, quality improvement, and sensory properties of *Doenjang*, a traditional Korean soybean fermented sauce variety. *J. Food Sci.* **2015**, *80*, M1794–M1803. [CrossRef] [PubMed]

62. Kim, S.-Y.; Kim, H.-E.; Kim, Y.-S. The potentials of *Bacillus licheniformis* strains for inhibition of *B. cereus* growth and reduction of biogenic amines in *cheonggukjang* (Korean fermented unsalted soybean paste). *Food Control* **2017**, *79*, 87–93.

63. Kang, H.-R.; Lee, Y.-L.; Hwang, H.-J. Potential for application as a starter culture of tyramine-reducing strain. *J. Korean Soc. Food Sci. Nutr.* **2017**, *46*, 1561–1567. [CrossRef]

64. Mah, J.-H.; Hwang, H.-J. Effects of food additives on biogenic amine formation in *Myeolchi-jeot*, a salted and fermented anchovy (*Engraulis japonicus*). *Food Chem.* **2009**, *114*, 168–173. [CrossRef]

65. Mah, J.-H.; Kim, Y.J.; Hwang, H.-J. Inhibitory effects of garlic and other spices on biogenic amine production in *Myeolchi-jeot*, Korean salted and fermented anchovy product. *Food Control* **2009**, *20*, 449–454. [CrossRef]

66. Wendakoon, C.N.; Sakaguchi, M. Combined effect of sodium chloride and clove on growth and biogenic amine formation of *Enterobacter aerogenes* in mackerel muscle extract. *J. Food Prot.* **1993**, *56*, 410–413. [CrossRef]

67. Wendakoon, C.N.; Sakaguchi, M. Combined effects of cloves with potassium sorbate and sodium benzoate on the growth and biogenic amine production of *Enterobacter aerogenes*. *Biosci. Biotechnol. Biochem.* **1993**, *57*, 678–679. [CrossRef]

68. Rabie, M.A.; Siliha, H.; el-Saidy, S.; el-Badawy, A.A.; Malcata, F.X. Effects of γ-irradiation upon biogenic amine formation in Egyptian ripened sausages during storage. *Innov. Food Sci. Emerg. Technol.* **2010**, *11*, 661–665. [CrossRef]

69. Novella-Rodriguez, S.; Veciana-Nogués, M.T.; Saldo, J.; Vidal-Carou, M.C. Effects of high hydrostatic pressure treatments on biogenic amine contents in goat cheeses during ripening. *J. Agric. Food Chem.* **2002**, *50*, 7288–7292. [CrossRef] [PubMed]

70. Ruiz-Capillas, C.; Colmenero, F.J.; Carrascosa, A.V.; Muñoz, R. Biogenic amine production in Spanish dry-cured "chorizo" sausage treated with high-pressure and kept in chilled storage. *Meat Sci.* **2007**, *77*, 365–371. [CrossRef] [PubMed]

71. Patsias, A.; Chouliara, I.; Paleologos, E.K.; Savvaidis, I.; Kontominas, M.G. Relation of biogenic amines to microbial and sensory changes of precooked chicken meat stored aerobically and under modified atmosphere packaging at 4 °C. *Eur. Food Res. Technol.* **2006**, *223*, 683–689. [CrossRef]

72. Ruiz-Capillas, C.; Pintado, T.; Jiménez-Colmenero, F. Biogenic amine formation in refrigerated fresh sausage "chorizo" keeps in modified atmosphere. *J. Food Biochem.* **2012**, *36*, 449–457. [CrossRef]

73. Ayhan, K.; Kolsarici, N.; Özkan, G.A. The effects of a starter culture on the formation of biogenic amines in Turkish soudjoucks. *Meat Sci.* **1999**, *53*, 183–188. [CrossRef]

74. Bover-Cid, S.; Izquierdo-Pulido, M.; Vidal-Carou, M.C. Effect of proteolytic starter cultures of *Staphylococcus* spp. on biogenic amine formation during the ripening of dry fermented sausages. *Int. J. Food Microbiol.* **1999**, *46*, 95–104. [CrossRef]

75. Gardini, F.; Martuscelli, M.; Crudele, M.A.; Pararella, A.; Suzzi, G. Use of *Staphylococcus xylosus* as a starter culture in dried sausages: Effect on the biogenic amine content. *Meat Sci.* **2002**, *61*, 275–283. [CrossRef]

76. Mah, J.-H.; Hwang, H.-J. Inhibition of biogenic amine formation in a salted and fermented anchovy by *Staphylococcus xylosus* as a protective culture. *Food Control* **2009**, *20*, 796–801. [CrossRef]

77. Alvarez, M.A.; Moreno-Arribas, M.V. The problem of biogenic amines in fermented foods and the use of potential biogenic amine-degrading microorganisms as a solution. *Trends Food Sci. Technol.* **2014**, *39*, 146–155. [CrossRef]

78. Naila, A.; Flint, S.; Fletcher, G.; Bremer, P.; Meerdink, G. Control of biogenic amines in food-existing and emerging approaches. *J. Food Sci.* **2010**, *75*, R139–R150. [CrossRef] [PubMed]

79. Stadnik, J. Significance of biogenic amines in fermented foods and methods of their control. In *Beneficial Microbes in Fermented and Functional Foods*; Rai, V.R., Bai, J.A., Eds.; CRC Press: Boca Raton, FL, USA, 2014; pp. 149–163.

80. Carocho, M.; Barreiro, M.F.; Morales, P.; Ferreira, I.C. Adding molecules to food, pros and cons: A review on synthetic and natural food additives. *Compr. Rev. Food Sci. Food Saf.* **2014**, *13*, 377–399. [CrossRef]

81. Floros, J.D.; Newsome, R.; Fisher, W.; Barbosa-Cánovas, G.V.; Chen, H.; Dunne, C.P.; German, J.B.; Hall, R.L.; Heldman, D.R.; Karwe, M.V.; et al. Feeding the world today and tomorrow: The importance of food science and technology. *Compr. Rev. Food Sci. Food Saf.* **2010**, *9*, 572–599. [CrossRef]

82. Fang, S.-H.; Lai, Y.-J.; Chou, C.-C. The susceptibility of *Streptococcus thermophilus* 14085 to organic acid, simulated gastric juice, bile salt and disinfectant as influenced by cold shock treatment. *Food Microbiol.* **2013**, *33*, 55–60. [CrossRef] [PubMed]

83. Hansen, E.B. Commercial bacterial starter cultures for fermented foods of the future. *Int. J. Food Microbiol.* **2002**, *78*, 119–131. [CrossRef]

84. Leroy, F.; De Vuyst, L. Lactic acid bacteria as functional starter cultures for the food fermentation industry. *Trends Food Sci. Technol.* **2004**, *15*, 67–78. [CrossRef]

85. Leroy, F.; Verluyten, J.; De Vuyst, L. Functional meat starter cultures for improved sausage fermentation. *Int. J. Food Microbiol.* **2006**, *106*, 270–285. [CrossRef] [PubMed]

86. Omafuvbe, B.O.; Abiose, S.H.; Shonukan, O.O. Fermentation of soybean (*Glycine max*) for soy-*daddawa* production by starter cultures of *Bacillus*. *Food Microbiol.* **2002**, *19*, 561–566. [CrossRef]

87. Tamang, J.P.; Nikkuni, S. Selection of starter cultures for the production of kinema, a fermented soybean food of the Himalaya. *World J. Microbiol. Biotechnol.* **1996**, *12*, 629–635. [CrossRef] [PubMed]

88. Eom, J.S.; Seo, B.Y.; Choi, H.S. Biogenic amine degradation by *Bacillus* species isolated from traditional fermented soybean food and detection of decarboxylase-related genes. *J. Microbiol. Biotechnol.* **2015**, *25*, 1519–1527. [CrossRef] [PubMed]

89. Yi-Chen, L.; Hsien-Feng, K.; Ya-Ling, H.; Chien-Hui, W.; Yu-Ru, H.; Yung-Hsiang, T. Reduction of biogenic amines during miso fermentation by *Lactobacillus plantarum* as a starter culture. *J. Food Prot.* **2016**, *79*, 1556–1561.

90. Ruiz-Capillas, C.; Herrero, A.M. Impact of biogenic amines on food quality and safety. *Foods* **2019**, *8*, 62. [CrossRef] [PubMed]

5

Effect of Fermentation, Drying and Roasting on Biogenic Amines and Other Biocompounds in Colombian Criollo Cocoa Beans and Shells

Johannes Delgado-Ospina [1,2,*], **Carla Daniela Di Mattia** [1], **Antonello Paparella** [1], **Dino Mastrocola** [1], **Maria Martuscelli** [1,*] **and Clemencia Chaves-Lopez** [1]

[1]   Faculty of Bioscience and Technology for Food, Agriculture and Environment, University of Teramo, Via R. Balzarini 1, 64100 Teramo, Italy; cdimattia@unite.it (C.D.D.M.); apaparella@unite.it (A.P.); dmastrocola@unite.it (D.M.); cchaveslopez@unite.it (C.C.-L.)

[2]   Grupo de Investigación Biotecnología, Facultad de Ingeniería, Universidad de San Buenaventura Cali, Carrera 122 # 6-65, Cali 76001, Colombia

*   Correspondence: jdelgado1@usbcali.edu.co (J.D.-O.); mmartuscelli@unite.it (M.M.)

Abstract: The composition of microbiota and the content and pattern of bioactive compounds (biogenic amines, polyphenols, anthocyanins and flavanols), as well as pH, color, antioxidant and reducing properties were investigated in fermented Criollo cocoa beans and shells. The analyses were conducted after fermentation and drying (T1) and after two thermal roasting processes (T2, 120 °C for 22 min; T3, 135 °C for 15 min). The fermentation and drying practices affected the microbiota of beans and shells, explaining the great variability of biogenic amines (BAs) content. Enterobacteriaceae were counted in a few samples with average values of 10^3 colony forming units per gram (CFU g^{-1}), mainly in the shell, while *Lactobacillus* spp. was observed in almost all the samples, with the highest count in the shell with average values of 10^4 CFU g^{-1}. After T1, the total BAs content was found to be in a range of 4.9÷127.1 mg kg^{-1}DFW; what was remarkable was the presence of cadaverine and histamine, which have not been reported previously in fermented cocoa beans. The total BAs content increased 60% after thermal treatment *T2*, and of 21% after processing at *T3*, with a strong correlation ($p < 0.05$) for histamine (ß = 0.75) and weakly correlated for spermidine (ß = 0.58), spermine (ß = 0.50), cadaverine (ß = 0.47) and serotonine (ß = 0.40). The roasting treatment of *T3* caused serotonin degradation (average decrease of 93%) with respect to unroasted samples. However, BAs were detected in a non-alarming concentration (e.g., histamine: n.d ÷ 59.8 mg kg^{-1}DFW; tyramine: n.d. ÷ 26.5 mg kg^{-1}DFW). Change in BAs level was evaluated by principal component analysis. PC1 and PC2 explained 84.9% and 4.5% of data variance, respectively. Antioxidant and reducing properties, polyphenol content and BAs negatively influenced PC1 with both polyphenols and BA increasing during roasting, whereas PC1 was positively influenced by anthocyanins, catechin and epicatechin.

Keywords: biogenic amines; polyphenols; histamine; microbiota; roasting

---

1. Introduction

In the last years, an increase in global cocoa production has been observed with a market demand of high-quality cocoa products [1]. Colombian cocoa was declared by the International Cocoa Organization as "fine" and "flavour" due to the agro-ecological characteristics of the areas in which it is cultivated and the adequate fermentation and drying processes that are carried out. In particular, Criollo is a variety of Colombia and other Latin American countries, known for its high quality.

Cocoa is produced from cocoa beans that undergo several processes such as fermentation, drying, roasting, dutching, conching, and tempering. In the first stages, cocoa pods (fruits) are picked from the trees (*Theobroma cacao*), collected in piles and immediately opened or left to stand for a few days (pod storage) to obtain positive effects on the quality of the final products. After harvesting, beans together with mucilage are removed from the pod, fermented, dried, and roasted [2].

Fermentation is essential for the degradation of mucilage thanks to the production of ethanol, which kills cocoa bean cotyledons, and to the production of different organic acids and important volatile compounds that diffuse into the interior of the beans and react with substances responsible for the flavour of final products during the subsequent roasting process. In addition, fermentation influences some functional properties such as antiradical activity and reduces the power of cocoa beans [3,4]. However, biogenic amines (BAs) can be formed during this step, with detrimental effects on cocoa quality and human health. The occurrence of BAs in food originates from decarboxylation of free amino acids, amination and transamination of ketones and aldehydes or during thermal processes. In fermented products, the concentration of BAs is the result of a balance between formation and degradation reactions in which several microorganisms are involved. In fact, cocoa microbiota may present strains with decarboxylase activity [5–8] and amino-oxidase activity [9].

A decrease of bioactive compounds, such as BAs and polyphenolic compounds, occurs in different steps of the cocoa beans processing, affecting their final content and functional properties in cocoa derivatives [10–12]. During roasting, physical and chemical changes occur in the beans, such as differences in colour, removal of undesirable volatile compounds, formation of desirable aroma and flavour, reduction of water content (up to 2%), and formation of a brittle structure, as well as changes in flavanols, proanthocyanidins and antioxidant activity [13,14]. In addition, peculiar cocoa volatile compounds are generated by Maillard reactions and their release is favoured by modifications of the matrix structure [15]. In spite of this, during roasting critical changes may also take place such as the formation of water-insoluble melanoidins, the degradation of catechin-containing compounds [16], the reduction of polyphenol content and antioxidant activity [17], and an increase of the biogenic amines content [12]. If some Maillard Reaction products, such as melanoidins, are required for the development of the peculiar cocoa sensory characteristics and brown colour, some furanic compounds are supposed to have negative effects on human health, as they can show cytotoxicity at high concentration and are "possibly carcinogenic to humans" [18]. Furthermore, since the presence of cocoa shell in cocoa beans derivatives adversely affects the final product quality [19], beans should be peeled before or after roasting [20,21].

The present research was aimed to study the effect of fermentation, drying and roasting on the microbiological, physical and chemical characteristics of Colombian Criollo cocoa (bean and shell), with a particular focus on the content of bioactive compounds such as BAs and polyphenols.

2. Materials and Methods

2.1. Origin of the Samples

Criollo cocoa bean samples were collected in spring 2018 directly from 18 farms (identified with a numerical code) located in three Departments of Colombia, with different environmental conditions and different fermentation and drying systems (Table 1). Thirteen samples were from Valle del Cauca, located in the western part of the country, between 3°05′ and 5°01′ N latitude, 75°42′ and 77°33′ W longitude; four samples from Cauca, located in the southwest of the country on the Andean and Pacific regions, between 0°58′54″ N and 3°19′04″ N latitude, 75°47′36″ W and 77°57′05″ W longitude; one sample from Nariño, located in the west of the country (1°16′0.01″ N latitude and 77°22′0.12″ W longitude) and, despite low altitude, affected by cold winds from the south of the continent.

Table 1. Geographical characteristics of the area of origin and fermentation and drying condition of the cocoa samples.

	Farm Location	Altitude (masl *)	Fermentation				Drying			
			T_{day} (°C)	T_{night} (°C)	Time (d)	Box	T_{day} (°C)	T_{night} (°C)	Time (d)	Drying Surface
1	Valle del Cauca	1000	27–31	18–20	4	plastic	28–30	17–18	4	wooden trays
2	Valle del Cauca	1000	27–31	18–20	3	plastic	28–30	17–18	5	wooden trays
3	Valle del Cauca	1000	25–27	17–18	6	wooden	25–27	18–19	3	wooden trays
4	Valle del Cauca	1000	27–31	18–20	6	plastic	28–30	17–18	3	wooden trays
5	Cauca	990	29–30	19–20	4	wooden	29–30	18–19	5	wooden trays
6	Valle del Cauca	1000	27–31	18–20	6	wooden	28–30	17–18	6	floors
7	Valle del Cauca	1000	25–27	17–18	6	plastic	25–27	18–19	4	wooden trays
8	Valle del Cauca	1000	25–27	17–18	6	plastic	25–27	18–19	4	metal trays
9	Cauca	990	29–30	19–20	4	wooden	29–30	18–19	6	floors
10	Cauca	990	28–33	19–20	4	wooden	28–33	17–20	7	floors
11	Valle del Cauca	1000	27–31	18–20	6	wooden	28–30	17–18	4	floors
12	Valle del Cauca	1000	27–31	18–20	6	plastic	28–30	17–18	3	floors
13	Valle del Cauca	1000	25–27	17–18	6	plastic	25–27	18–19	3	wooden trays
14	Cauca	990	29–30	19–20	4	wooden	29–30	18–19	5	wooden trays
15	Valle del Cauca	1000	27–31	18–20	6	plastic	28–30	17–18	5	wooden trays
16	Nariño	30	21–25	12–22	4	wooden	20–25	11–15	4	wooden trays
17	Valle del Cauca	1000	27–31	18–20	6	plastic	28–30	17–18	4	floors
18	Valle del Cauca	1000	25–27	17–18	6	plastic	25–27	18–19	3	wooden trays

* meters above sea level.

2.2. Samples Preparation and Defatting

The samples were collected at the final stage of fermentation and drying (step T1). Moreover, dried cocoa beans were divided into two batches and treated in a convection oven (Memmert UN110, Büchenbach, Germany) in two different commercial roasting conditions: T2, 120 °C for 22 min; T3, 135 °C for 15 min.

After cooling, the shell was removed manually from the cocoa beans. After the removal of the external skin and grinding (IKA M20, Staufen, Germany), cocoa samples were defatted by three cycles of hexane washing (8 g of cocoa sample in 50 mL of hexane), following the method described by Di Mattia et al. [4]. Four grams of sample were weighed and 25 mL of hexane added, then the mixture was vortexed for 1 min and centrifuged (2325× g for 10 min), each time discharging the supernatant. To completely remove the hexane from the sample, the lipid-free solids were air-dried at room temperature. The fat-free samples were then used for the extraction of the phenolic fraction and other chemical determinations.

2.3. Moisture and pH Determination

The pH of defatted cocoa nibs was measured by diluting in distilled water (1:1) by using an electrode probe connected to a pHmeter (FE20, Mettler Toledo, Columbus, OH, USA).

Moisture content was determined according to the official procedure adopted by the Association of Official Analytical Chemists (AOAC) [22]. In particular, 1 g of sample was dried in a forced-air drying oven at 105 °C up to a constant weight.

2.4. Microbiological Analyses

Microbiological analyses were performed according to Chaves et al. [23]. From samples of dried cocoa beans, the beans (from here they are beans without shell) and the shells were obtained by manual separation. Twenty grams of cocoa nibs and separate shells were homogenized in a Stomacher Lab-blender (Thomas Scientific, Swedesboro, NJ, USA) in 90 mL phosphate buffer solution (PBS, Biolife, Milan, Italy) sterile solution, pH 7.4. Decimal dilutions of the suspension were prepared in PBS, plated and incubated as follows: Enterobacteriaceae were counted and isolated in Violet Red Bile Glucose Agar (Oxoid, Basingstoke, UK) at 37 °C in anaerobiosis for 48 h; mesophilic aerobic bacteria in Plate Count Agar (PCA) at 30 °C for 48 h; thermophilic aerobic bacteria in PCA and incubated at 45 °C for 48 h; lactobacilli in De Man Rogose and Sharp (MRS) Broth (Oxoid, Basingstoke, UK) at 37 °C in anaerobiosis for 72 h; lactic streptococci in M17 agar (Oxoid, Basingstoke, UK) at 37 °C in anaerobiosis for 72 h; yeasts in Yeast Extract-Peptone-Dextrose (YPD) agar medium and Walerstein Laboratory (WL) medium agar (Biolife, Milan, Italy) at 25 °C for 48 h; moulds in DG18 Agar (Oxoid, Basingstoke, UK) and Czapec-Agar (Biolife, Milan, Italy) added with 150 ppm chloramphenicol (Sigma-Aldrich Italy, Milan, IT) for 5 days.

2.5. Biogenic Amines Determination

Defatted samples were subjected to BAs extraction, detection, identification and quantification by high-performance liquid chromatography (HPLC) using an Agilent 1200 Series (Agilent Technologies, Milano, Italy), optimizing the method described by Chaves-Lopez et al. [23]. Shortly after, 1.0 g of sample was added of 5.0 mL of 0.1 N HCl and stirred in vortex (1 min) and ultrasound (20 min). It was centrifuged (Hettich Zentrifugen, Tuttlingen, Germany) at relative centrifugal force of 2325× g for 10 min and the supernatant recovered. Then, 150 μL of saturated $NaHCO_3$ was added to 0.5 mL of the supernatant, adjusting the pH to 11.5 with 0.1 N NaOH. For derivatization, 2.0 mL of dansyl chloride/acetone (10 mg mL^{-1}) was added and incubated at 40 °C for 1 h under agitation (195 stokes) (Dubnoff Bath-BSD/D, International PBI, Milano, Italy). To remove excess of dansyl chloride, 200 μL of 30% ammonia was added, allowed to stand for 30 min at room temperature, and diluted with 1950 μL of acetonitrile.

In a Spherisorb S30ODS Waters C18-2 column (3 μm, 150 mm × 4.6 mm ID), 10 μl of sample were injected with gradient elution, acetonitrile (solvent A) and water (solvent B) as follows: 0–1 min 35% B isocratic; 1–5 min, 35%–20% B linear; 5–6 min, 20%–10% linear B; 6–15 min, 10% B isocratic; 15–18 min, 35% linear B; 18–20 min, 35% B isocratic. Identification and quantification of cadaverine (CAD), dopamine (DOP), ethylamine (ETH), histamine (HIS), 2-phenylethylamine (PHE), putrescine (PUT), serotonin (SER), spermidine (SPD), spermine (SPM), and tyramine (TYR) was performed by comparing retention times and calibration curves of pure standards. The results were reported as mg of BA kg^{-1} of defatted dry weight (of DFW).

2.6. Colour Analysis

Colour analysis of the cocoa samples was carried out by a Minolta Bench-top Colorimeter CR-5 (Konica Minolta, Tokyo, Japan) CM-500 spectrophotometer. Before analysis, two calibrations were carried out, one with black standard and the other one with white standard. For each measurement, a single layer of grounded cocoa beans was spread on a Petri dish. The analysis was repeated three times on each sample.

The instrument gave the results in terms of CIE L* a* b* parameters (CIELAB is a colour space specified by the International Commission on Illumination, French Commission Internationale de l'éclairage, CIE), where L* indicates the lightness within the range from 0 (black) to 100 (white); a* ranges from −60 (green) to +60 (red); b* ranges from −60 (blue) to +60 (yellow). a* and b* indicate colour direction and from these values we obtained the Hue angle (h°), calculated as h° = arctan(b*/a*)

2.7. Anthocyanin Determination

The total anthocyanins content was determined by the method described by do Carmo Brito et al. [11]. In brief, 1.0 mL of 95% ethanol and 1.5 N hydrochloric acid solution (85:15 v/v) were added to 0.1 g of defatted cocoa sample, then stirred in vortex for two min and allowed to stand overnight. The sample was centrifuged at 10,000× g, and the supernatant was suitably diluted to measure its absorbance at 535 nm by a spectrometer (Eppendorf Biospectrometer kinetic, Hamburg, Germany. The results were reported as mg of anthocyanins g^{-1} of sample.

2.8. Extraction of the Phenolic Fraction

The defatted samples were further ground with mortar and pestle to reduce the powder size and to allow better contact of the extracting solvent with the sample. The sample extraction was carried out according to Di Mattia et al. [4] with some modifications. One gram of defatted sample was added to 5 mL of 70:29.5:0.5 acetone/water/acetic acid; the mixture was vortexed for 1 min, then sonicated in an ultrasonic bath (Labsonic LBS 1, Falc, Treviglio, Bergamo, Italy) at 20 °C for 10 min and finally centrifuged (2325× g for 10 min). The surnatant liquid was recovered and filtrated through cellulose filters. The extracted polyphenols were then stored in the freezer at −32 °C until analyses. This extract was used for the evaluation of total polyphenol content, radical scavenging activity and ferric reducing properties. For flavanols analysis, samples were extracted, kept at −80 °C and analysed on the same day or at the latest a few days after extraction.

2.8.1. Total Polyphenols Content (TPC)

The total polyphenol content (TPC) was determined according to a procedure modified from Di Mattia et al. [10]. To a volume of 0.1 mL of diluted defatted sample extract, water was added up to a volume of 5 mL, and 500 μL of Folin–Ciocalteu reagent was added. After 3 min, 1.5 mL of a 25% (w/v) Na_2CO_3 solution was added and then deionized water up to 10 mL of the final volume. Solutions were maintained at room temperature under dark conditions for 60 min and the total polyphenols content was determined at 765 nm using a spectrophotometer (Lambda Bio 20 Perkin Elmer, Waltham, MA, USA) Gallic acid standard (Fluka, Buchs, Switzerland) solutions were used for calibration. Results were expressed as milligrams of gallic acid equivalents (GAE) per gram of defatted and dry weight.

2.8.2. Flavanols Identification and Quantification

HPLC (high-performance liquid chromatography) was used for separation, quantitative determination and identification of flavonoids. The chromatographic analyses were performed on a 1200 Agilent Series HPLC (Agilent Technologies, Milano, Italy) equipped with a quaternary pump, a degasser, a column thermostat, an autosampler injection system and a diode array detector (DAD). The system was controlled by Agilent ChemStation for Windows (Agilent Technologies). Flavanols determination was carried out according to Ioannone et al. [13]. The sample (20 μL) was injected into a C18 reversed-phase column. Separation of phenolic compounds was carried out at a flow rate of 1 mL/min with a non-linear gradient from A (1% acetic acid solution) to B (ACN). Gradient elution was as follows: from 6% to 18% B from 0 to 40 min, from 18% to 100% B from 40 to 45 min, from 100% to 6% B from 45 to 50 min, isocratic from 50 to 53 min. The DAD acquisition range was set from 200 to 400 nm. The calibration curves were made with epicatechin and catechin, and the results were expressed as mg per gram of defatted and dry weight.

2.8.3. ABTS (2,2'-azino-bis(3-ethylbenzthiazoline-6-sulfuric acid)) Assay

The radical scavenging activity was measured by ABTS (2,2'-azino-bis (3-ethylbenzthiazoline-6-sulfuric acid)) radical cation decoloration assay, as described by Re et al. [24]. The ABTS radical stock solution was prepared by dissolving ABTS in water to a 7 mM concentration and by making this solution react with 2.45 mM of potassium persulfate. The mixture was then left in the dark at room temperature for 12–16 h before use. The ABTS+• stock solution was diluted in water to an Abs of 0.70 ± 0.02 for the analysis, and the reaction was started by the addition of 30 μL of cocoa extract to 2.97 mL of ABTS+• radical solution. The bleaching rate of ABTS+• in the presence of the sample was monitored at 25 °C at 734 nm using a spectrophotometer (Lambda Bio 20, Perkin Elmer, Boston, MA, USA) and decoloration after 5 min was used as the measure of antioxidant activity. Radical scavenging activity was expressed as Trolox Equivalents Antioxidant Capacity (TEAC-μmol of Trolox equivalents per g of defatted and dry weight), calculated by the ratio between the correlation coefficient of the dose–response curve of the sample and the correlation coefficient of the dose–response curve of Trolox, the standard compound.

2.8.4. Ferric Reducing Antioxidant Power (FRAP)

The reducing activity of the samples was determined according to the method described by Benzie and Strain [25] with some modifications. One hundred microliters of suitably diluted sample extract were added to 2900 μL of the FRAP reagent obtained by mixing acetate buffer (300 mM, pH 3.6), TPTZ (2,4,6-tripyridyl-s-triazine), 10 mM solubilized in HCl 40 mM and $FeCl_3$ 20 mM in the ratio 10:1:1. The absorbance change was followed at 593 nm for 6 min. A calibration plot based on $FeSO_4 \cdot 7H_2O$ was used, and results were expressed as mmols of Fe^{2+} per gram of defatted and dry weight.

2.9. Statistical Analyses

All determinations were done in triplicate, except where differently indicated. Means and relative standard deviations were calculated. Analysis of variance (ANOVA) was performed to test the significance of the effects of the factor variables (processing steps); differences among means were separated by the least significant differences (LSD) test. Statistical analysis of data was performed using XLSTAT software version 2019.1 for Microsoft Excel (Addinsoft, New York, NY, USA). All results were considered statistically significant at $p < 0.05$.

The multivariate descriptive analysis was used to understand the presence of the main descriptors related to the BAs content of cocoa. The principal components analysis (PCA) started with the analysis of a matrix (18 × 55) that consisted of 18 samples of Criollo cocoa. The analyses were performed in triplicate. The 55 conformations of the values of the evaluated variables were gathered by the following tests: roasting temperature (raw, 120 °C for 22 min and 135 °C for 15 min), pH, content of total polyphenols (TPC), anthocyanins, antioxidant activity (FRAP and TEAC), flavonols (catechin and epicatechin), levels of the main microorganisms groups in cocoa beans and shell, ethylamine (ETH), dopamine (DOP), 2-phenylethylamine (PHE), putrescine (PUT), cadaverine (CAD), serotonin (SER), histamine (HIS), tyramine (TYR), spermidine (SPD), and spermine (SPM).

3. Results

3.1. Characterization of Fermented and Dried Cocoa Beans

Several indicators are used to measure the quality of cocoa beans. These include, in addition to microbiota, composition, colour and acidity of the beans [26].

3.1.1. Microbiota

Some researchers found that variations in the content of BAs in cocoa are mainly affected by fermentation, which is directly correlated to the type and quantity of microbial populations [11]. Figure 1a,b shows the distribution of microorganism groups enumerated both in cocoa beans and cocoa shells of investigated samples. A large variability was observed, confirming that postharvest practices carried out in the different Colombian farms affected microbiota, which in turn can explain the great diversity of decarboxylation products, such as BAs content at the T1 step. The microbial load of the shell was determined because many of the metabolites produced in the shell during fermentation and drying can migrate to the beans, causing a pH decrease due to the accumulation of organic acids. The microbiological analyses showed the presence of enterobacteria, total aerobic mesophiles, total aerobic thermophiles, acetic bacteria, spore-forming bacteria, lactobacilli, lactococci, fungi, and yeasts that are mainly involved in fermentation and drying. Variations in microbial counts and species were observed in the different samples, likely due to different fermentation and drying practices (pod ripeness, postharvest pod storage, variations in pulp/bean ratio, fermentation method, batch size, frequency of bean mixing or turning, and fermentation time), as well as due to some characteristics of the environment where the cultivation takes place (farm, weather conditions, pod diseases) [27,28].

A great difference was also observed between the microbial community found in the shell and inside the beans. As expected, the shell contained a greater number of microorganisms because sugars and other rapidly degradable nutrients are concentrated here, while a smaller number of microbial populations could adapt to the conditions of the beans. According to Lima et al. [29], average levels of total aerobic microorganisms and aerobic total spores are reduced in the beans, while Enterobacteriaceae and fungi were not detected.

Generally, the production of BAs is attributed to certain species of Enterobacteriaceae, mainly *Clostridium* spp., *Lactobacillus* spp., *Streptococcus* spp., *Micrococcus* spp., and *Pseudomonas* spp. [30]. Two of these groups were found in cocoa samples; Enterobacteriaceae were counted in a few samples with average values of 10^3 CFU g^{-1}, mainly in the shell and probably due to contamination during outdoor drying, while *Lactobacillus* spp. was observed in almost all the samples, with the highest count in the shell with average values of 10^4 CFU g^{-1}.

3.1.2. pH, Moisture and Colour

The characteristics of Colombian Criollo cocoa samples at the end of fermentation and drying (T1) are shown in Table 2. The great variability found in the samples depends on several factors, namely fermentation and drying, as well as some intrinsic characteristics of the farming system. The organic acids produced by lactic and acetic bacteria during fermentation diffuse within the beans and cause a pH decrease; low pH values are considered an index of appropriate fermentation while pH values above 5.5 may indicate an inadequate or incomplete fermentation [31]. The pH of the samples ranged between 4.43 and 6.17 (C.V. 10%). With the exception of sample 18, it can be stated that the samples coming from Valle de Cauca were generally characterized by lower pH values compared to Cauca samples.

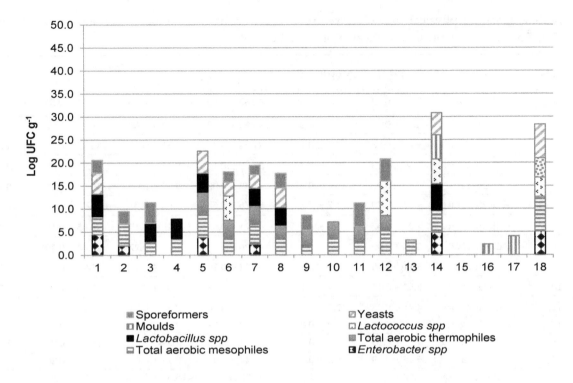

(a)

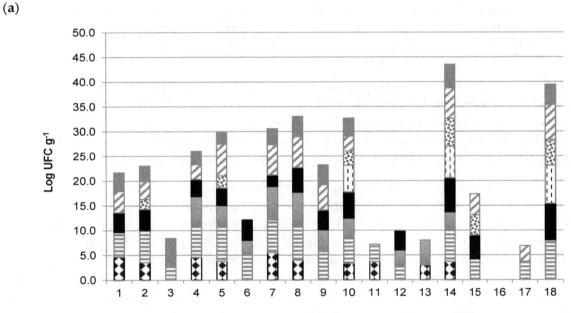

(b)

Figure 1. Levels of the main microorganisms groups in beans (**a**) and shells (**b**) of Colombian Criollo cocoa samples (after fermentation and drying, step T1).

Table 2. Chemico-physical and colour parameters (L*, a*, b* and h°) for fermented and dried samples (T1). The data are expressed as mean ± standard deviation.

Samples	Origin	pH	Moisture (%)	Colour			
				L*	a*	b*	h°
1	Valle de Cauca	5.01 ± 0.02	3.4 ± 0.3	40.83 ± 0.80	7.06 ± 0.08	11.17 ± 0.35	57.67± 0.01
2	Valle de Cauca	4.79 ± 0.11	3.9 ± 0.2	48.92 ± 0.27	3.64 ± 0.08	9.38 ± 0.10	68.77 ± 0.02
3	Valle de Cauca	4.43 ± 0.08	4.2 ± 0.4	41.78 ± 0.43	9.87 ± 0.12	12.35 ± 0.22	51.35 ± 0.06
4	Valle de Cauca	4.49 ± 0.03	3.8 ± 0.3	38.40 ± 0.16	9.49 ± 0.12	9.57 ± 0.13	45.24 ± 0.02
5	Cauca	5.85 ± 0.06	5.0 ± 0.2	41.42 ± 0.30	9.21 ± 0.05	10.25 ± 0.16	48.07 ± 0.01
6	Valle de Cauca	4.54 ± 0.10	1.2 ± 0.1	38.76 ± 0.67	9.05 ± 0.32	6.70 ± 0.54	36.49 ± 0.02
7	Valle de Cauca	4.99 ± 0.07	3.5 ± 0.2	41.71 ± 0.65	8.29 ± 0.15	10.85 ± 0.16	52.62 ± 0.01
8	Valle de Cauca	5.05 ± 0.05	4.8 ± 0.4	43.26 ± 0.37	8.25 ± 0.27	13.43 ± 0.20	58.45 ± 0.01
9	Cauca	5.11 ± 0.18	2.5 ± 0.1	36.84 ± 0.30	7.30 ± 0.16	11.79 ± 0.41	58.22 ± 0.01
10	Cauca	5.52 ± 0.15	1.7 ± 0.1	38.89 ± 0.68	4.37 ± 0.02	8.30 ± 0.09	62.22 ± 0.01
11	Valle de Cauca	4.41 ± 0.08	2.5 ± 0.2	39.85 ± 0.34	9.60 ± 0.24	10.27 ± 0.38	46.90 ± 0.01
12	Valle de Cauca	4.62 ± 0.03	2.2 ± 0.0	44.50 ± 0.38	7.31 ± 0.16	11.34 ± 0.18	57.19 ± 0.00
13	Valle de Cauca	4.67 ± 0.33	4.9 ± 0.4	37.54 ± 0.18	10.04 ± 0.05	15.77 ± 0.03	57.53 ± 0.00
14	Cauca	5.45 ± 0.09	6.2 ± 0.3	38.44 ± 1.02	8.18 ± 0.09	12.07 ± 0.26	55.87 ± 0.01
15	Valle de Cauca	5.07 ± 0.12	2.5 ± 0.2	38.23 ± 0.91	8.33 ± 0.14	9.30 ± 0.08	48.15 ± 0.00
16	Nariño	4.68 ± 0.13	3.9 ± 0.1	43.78 ± 0.07	5.34 ± 0.09	12.25 ± 0.25	66.43 ± 0.01
17	Valle de Cauca	4.44 ± 0.10	2.8 ± 0.2	48.89 ± 0.73	5.34 ± 0.09	12.25 ± 0.25	66.43 ± 0.01
18	Valle de Cauca	6.17 ± 0.27	4.7 ± 0.7	32.96 ± 0.89	9.39 ± 0.32	12.93 ± 0.46	54.00 ± 0.00

The moisture in bean samples ranged between 1.2% (sample 6) and 6.2% (sample 14), with differences depending on process conditions (solar or artificial dryers) and processing time. However, for all the samples, moisture content was below 12% which is considered the threshold value for optimal beans storage, corresponding to inhibition of both enzymatic reactions and fungal growth that can produce undesired metabolites during storage, such as mycotoxins.

The lightness (L*) of the 18 samples had a mean value of 40.8 (±4.03), ranging from 48.92, observed in sample 2, to 32.96 in sample 18. For redness values (a*), a mean of 7.78 (±1.94) was detected with 10.04 as the maximum value (in sample 13) and 3.64 as the minimum (in sample 2). For yellowness (b*), we observed the highest value in sample 13 (11.11 ± 2.07) and the lowest value in sample 6 (6.70 ± 0.54). Finally, for hue angle (h°), a mean value of 55.12 ± 8.26 was determined with a range from 68.77 (sample 2) to 36.49 (sample 6). Other authors reported L* values quite different from those obtained in the present study, while results for parameters of a* and b* were similar [32]. The values obtained for h° were similar to those reported by Sacchetti et al. [14].

3.2. Biogenic Amines Profile

The BAs profile of the Criollo cocoa beans under investigation is described in Table 3. In unroasted samples (T1), the total BAs amount was found to be 57.5 (±37.5) mg kg^{-1}DFW, with a minimum value of 4.9 mg kg^{-1}DFW (in sample 16, from Nariño region) and a maximum of 127.1 mg kg^{-1}DFW (in sample 4, from Valle de Cauca). As far as the BAs pattern is concerned, the most represented BAs in unroasted beans (T1) were CAD, SER, HIS, SPD, and SPM (Table 3); DOP was also detected in unroasted sample 15 (from Cauca).

The Pearson correlation coefficient between total BAs content and each single BA was calculated. A strong correlation was only found with HIS (ß = 0.75); tot BAs correlated weakly with SPD (ß = 0.58), SPM (ß = 0.50), CAD (ß = 0.47), and SER (ß = 0.40), while no significant correlation was found with other amines.

Table 3. Biogenic amines (mg kg$^{-1}_{DFW}$) in Criollo cocoa bean samples after fermentation and drying (T1) and after roasting (T2, 120 °C for 22 min; T3, 135 °C for 15 min).

Sample		ETH *	DOP	PHE	PUT	CAD	SER	HIS	TYR	SPD	SPM
						Biogenic Amines Content (mg kg$^{-1}_{DFW}$)					
1	T1	nd	nd	nd	nd	49.75 ± 1.4	3.06 ± 0.0	5.96 ± 1.2	nd	30.85 ± 3.6	nd
	T2	14.72 ± 2.3	nd	12.2 ± 0.1	1.32 ± 0.2	nd	17.18 ± 4.2	0.21 ± 3.1	17.02 ± 0.2	nd	14.30 ± 2.2
	T3	15.56 ± 5.2	nd	26.23 ± 0.3	59.02 ± 3.5	5.38 ± 0.6	nd	13.25 ± 1.1	11.11 ± 0.0	nd	36.84 ± 5.2
2	T1	nd	nd	nd	nd	48.62 ± 0.2	0.60 ± 0.0	nd	nd	4.67 ± 0.3	nd
	T2	24.77 ± 0.7	33.27 ± 0.2	11.18 ± 0.1	nd	nd	17.70 ± 0.0	nd	26.51 ± 4.2	nd	46.42 ± 8.4
	T3	19.22 ± 0.4	1.99 ± 0.0	25.98 ± 1.2	62.58 ± 6.2	nd	2.95 ± 0.2	17.13 ± 0.9	11.27 ± 0.8	2.50 ± 0.0	52.82 ± 3.9
3	T1	nd	nd	nd	nd	nd	nd	3.56 ± 1.1	nd	13.91 ± 0.2	0.37 ± 0.0
	T2	0.89 ± 0.3	nd	12.55 ± 0.5	nd	nd	9.11 ± 0.6	0.44 ± 0.5	25.22 ± 1.7	nd	nd
	T3	1.08 ± 0.4	nd	9.69 ± 136	nd	nd	nd	0.80 ± 1.2	8.29 ± 0.3	nd	nd
4	T1	nd	nd	nd	nd	22.60 ± 0.6	11.65 ± 1.1	41.9 ± 2.2	nd	42.06 ± 2.8	8.86 ± 3.2
	T2	2.74 ± 0.1	nd	11.77 ± 0.1	nd	2.22 ± 0.1	3.07 ± 0.0	0.82 ± 0.5	14.96 ± 1.1	nd	2.03 ± 1.8
	T3	1.68 ± 0.2	nd	17.22 ± 0.0	nd	nd	1.01 ± 0.0	1.01 ± 1.2	15.98 ± 0.4	nd	nd
5	T1	nd	nd	nd	nd	66.57 ± 0.7	0.36 ± 0.0	39.8 ± 7.1	nd	5.99 ± 0.0	nd
	T2	29.17 ± 0.3	65.12 ± 0.1	10.36 ± 0.5	6.46 ± 0.3	3.28 ± 0.0	25.63 ± 2.3	0.46 ± 0.7	20.80 ± 0.5	nd	60.04 ± 8.4
	T3	17.06 ± 1.2	95.03 ± 0.1	10.67 ± 0.1	2.51 ± 0.1	3.87 ± 0.0	nd	0.38 ± 0.2	9.87 ± 0.0	1.15 ± 0.0	nd
6	T1	nd	nd	nd	nd	1.50 ± 0.0	6.98 ± 0.1	38.90 ± 5.3	nd	48.66 ± 2.1	13.61 ± 2.2
	T2	2.95 ± 1.2	nd	11.36 ± 0.1	nd	3.14 ± 0.0	nd	17.13 ± 2.2	17.03 ± 0.7	nd	5.38 ± 0.4
	T3	2.17 ± 0.7	nd	14.75 ± 0.1	nd	nd	nd	nd	13.50 ± 0.4	nd	nd
7	T1	nd	nd	nd	nd	1.80 ± 0.0	0.11 ± 0.0	5.96 ± 1.2	nd	18.53 ± 6.2	1.12 ± 0.0
	T2	0.98 ± 0.1	nd	13.85 ± 0.0	nd	nd	nd	13.55 ± 3.0	20.51 ± 0.2	nd	nd
	T3	1.24 ± 0.1	nd	15.72 ± 0.2	nd	nd	nd	nd	13.12 ± 0.8	nd	nd
8	T1	nd	nd	nd	nd	15.86 ± 0.1	nd	nd	nd	12.81 ± 2.5	0.12 ± 0.1
	T2	3.61 ± 0.4	nd	12.70 ± 0.1	nd	nd	nd	48.18 ± 1.9	14.88 ± 0.1	nd	1.05 ±
	T3	3.36 ± 0.3	nd	12.81 ± 0.1	nd	nd	nd	nd	10.82 ± 0.0	nd	nd
9	T1	nd	nd	nd	nd	51.77 ± 0.5	0.85 ± 0.1	nd	nd	11.93 ± 1.1	0.37 ± 0.3
	T2	13.67 ± 0.5	nd	14.14 ± 0.1	nd	nd	nd	37.73 ± 2.8	18.16 ± 7.7	0.51 ± 0.0	nd
	T3	11.3 ± 0.1	nd	14.98 ± 0.1	1.50 ±	2.31 ± 0.0	nd	nd	15.85 ± 1.6	0.03 ± 0.0	nd

Table 3. *Cont.*

Sample		ETH*	DOP	PHE	PUT	CAD	SER	HIS	TYR	SPD	SPM
10	T1	nd	nd	nd	nd	nd	nd	7.76 ± 4.1	nd	12.37 ± 0.0	2.37 ± 0.2
	T2	21.65 ± 0.1	76.92 ± 0.1	12.04 ± 0.1	3.99 ± 0.4	2.22 ± 0.0	3.71 ± 0.1	0.27 ± 0.7	14.97 ± 1.1	1.87 ± 0.0	nd
	T3	12.55 ± 0.2	56.68 ± 0.1	8.42 ± 0.5	nd	nd	nd	1.25 ± 0.2	11.5 ± 2.7	nd	nd
11	T1	nd	nd	nd	nd	5.54 ± 0.1	0.24 ± 0.0	5.04 ± 0.1	nd	23.15 ± 1.2	4.37 ± 0.6
	T2	2.63 ± 0.1	nd	13.35 ± 0.0	nd	nd	nd	21.53 ± 3.1	14.20 ± 0.7	nd	nd
	T3	2.89 ± 0.1	nd	14.83 ± 0.6	nd	nd	nd	0.27 ± 0.5	14.49 ± 5.7	nd	0.37 ± 0.2
12	T1	nd	nd	nd	nd	7.56 ± 0.1	4.28 ± 0.1	16.74 ± 3.1	nd	44.48 ± 4.2	9.61 ± 1.1
	T2	4.55 ± 0.4	nd	13.94 ± 3.2	nd	nd	nd	59.78 ± 6.4	19.68 ± 0.3	nd	nd
	T3	3.19 ± 0.2	nd	16.16 ± 4.1	nd	nd	nd	1.24 ± 0.2	8.54 ± 0.6	nd	nd
13	T1	nd	nd	nd	nd	0.61 ± 0.0	4.41 ± 1.1	4.76 ± 2.6	nd	40.75 ± 2.7	4.86 ± 0.2
	T2	8.38 ± 0.1	nd	11.81 ± 2.9	nd	nd	nd	0.60 ± 0.3	16.24 ± 0.5	nd	nd
	T3	7.46 ± 0.1	nd	13.43 ± 1.6	nd	nd	nd	0.17 ± 0.4	14.42 ± 0.8	nd	nd
14	T1	nd	nd	nd	nd	nd	15.33 ± 1.2	2.96 ± 0.3	nd	18.31 ± 3.1	0.62 ± 0.1
	T2	6.47 ± 0.2	nd	14.65 ± 2.8	nd	nd	5.62 ± 0.1	0.79 ± 0.7	18.51 ± 1.0	nd	nd
	T3	6.33 ± 0.3	nd	12.50 ± 3.1	nd	2.87 ± 0.0	nd	2.70 ± 0.5	12.41 ± 3.4	nd	nd
15	T1	nd	57.35 ± 01	nd	nd	0.83 ± 0.0	3.55 ± 0.0	nd	0.19 ± 0.0	24.47 ± 2.2	8.86 ± 0.8
	T2	11.92 ± 0.0	nd	16.27 ± 2.1	nd	nd	4.90 ± 0.0		13.65 ± 0.3	nd	nd
	T3	6.102 ± 0.2	nd	10.81 ± 1.0	nd	nd	nd		12.14 ± 0.6	nd	nd
16	T1	nd	nd	nd	nd	2.85 ± 0.0	1.34 ± 0.0	nd	nd	0.28 ± 0.0	0.42 ± 0.0
	T2	27.00 ± 0.2	92.17 ± 0.1	18.78 ± 1.2	nd	nd	nd	29.87 ± 5.6	9.64 ± 0.0	2.62 ± 0.2	3.35 ± 0.7
	T3	9.41 ± 0.8	55.76 ± 0.1	6.02 ± 0.0	nd	nd	nd	1.18 ± 9.4	0.79 ± 0.0	nd	nd
17	T1	nd	nd	nd	nd	5.77 ± 0.3	9.68 ± 0.4	5.36 ± 0.7	nd	14.79 ± 2.1	1.19 ± 0.5
	T2	23.52 ± 0.4	77.52 ± 0.1	11.81 ± 0.8	nd	5.45 ± 0.2	nd	27.27 ± 4.9	11.37 ± 0.4	2.30 ± 0.0	nd
	T3	22.27 ± 0.0	16.27 ± 0.1	11.73 ± 0.3	4.44 ± 1.1	5.45 ± 0.7	nd	8.12 ± 0.9	12.51 ± 0.0	2.60 ± 0.0	2.85 ± 0.0
18	T1	nd	nd	nd	nd	nd	nd	nd	nd	15.45 ± 0.0	0.87 ± 0.0
	T2	25.66 ± 0.1	nd	17.24 ± 2.1	nd	nd	nd	nd	15.14 ± 3.2	nd	nd
	T3	11.77 ± 0.1	nd	20.60 ± 1.6	nd	nd	nd	nd	16.94 ± 1.9	nd	nd

Biogenic Amines Content (mg kg^{-1}$_{DFW}$)

* Legend: ethylamine (ETH), dopamine (DOP), 2-phenylethylamine (PHE), putrescine (PUT), cadaverine (CAD), histamine (HIS), serotonin (SER), tyramine (TYR), spermidine (SPD), and spermine (SPM); nd, not detectable.

To the best of our knowledge, there are no studies reporting the occurrence of CAD and HIS in raw cocoa beans, although there are few studies where BAs are identified in cocoa. Some authors [12] found tyramine, 2-phenylethylamine, tryptamine, serotonin, and dopamine in different varieties of raw cocoa beans; other authors [11] also found spermidine and spermine in Brazilian samples during fermentation.

Most of the analysed samples presented similar profiles of BAs that might be explained by the fact that they belong to the same variety. However, variations in their concentration were found and can be explained by the difference between cultivars, different growth, fermentation and drying conditions, as well as the microbiota of beans and shell (see Figure 1a,b).

Polyamines can also occur naturally due to the large proliferation of cells that occur in the early stages of growth or germination caused by physiological changes in tissues [33,34]. In fact, being that the cocoa bean is a seed and germination does not start if the optimal fermentation conditions are not present, a consequence of this could be that secondary metabolites such as the aliphatic amines (PUT, CAD, SPM and SPD) could be accumulated in cells. To our best knowledge, very few studies have been published on the relation between the physiological conditions and the BAs content in cocoa seeds, thus these aspects should be thoroughly investigated.

Although the development of microorganisms with amino acid decarboxylases activity occurs in environments with optimal pH between 4.0 and 5.5, no correlation was observed between BAs content and low pH.

No direct relationship was found between the content of polyphenols and the content of BAs in cocoa beans (see below). However, it is possible to hypothesize that the presence of metabolites as polyamines may influence antioxidant activity in cocoa samples or exhibit pro-oxidant properties [35].

3.3. Effect of Roasting on the BAs Content

A significant effect of both the roasting processes on total BAs content was found in all the samples; in particular, the beans treated at T2 (120 °C for 22 min) showed an increase of 60% with respect to the raw beans samples (T1), whilst the roasting process T3 (135 °C, 15 min) caused an increase of 21% compared to T1 samples.

In our experiments, we observed a large variability in the behaviour of each BA in the samples (Table 3); after the high temperature treatment, we determined the presence of TYR, 2-PHE, ETH, and PUT that were not detected in unroasted beans (T1). On the other hand, the roasting treatment increased the concentration of DOP and SPM with the increase of the temperature, while CAD and SPD levels decreased dramatically.

Several factors could affect the final accumulation of BAs. In particular, some authors have reported that Strecker degradation is responsible for the formation of BAs during the thermal decarboxylation of amino acids in the presence of α-dicarbonyl compounds formed during the Maillard reaction [12,36] or lipid peroxidation [37].

After treatment at 120 °C (T2), total BAs concentration correlated significantly with SPM ($\beta = 0.77$), SPD ($\beta = 0.67$) and PUT ($\beta = 0.60$), while at 135 °C (T3) there was a strong correlation between tot BAS and SPD ($\beta = 0.85$), HIS ($\beta = 0.81$), and PUT, CAD and SPM ($\beta > 0.70$).

Some authors suggested that serotonin could be formed as a result of the transformation of its precursors (tryptophan and 5-hydroxytryptophan) at very high temperatures [38]. In this study, we detected an increase in the concentration of serotonin only in three samples after T2 treatment, while this monoamine neurotransmitter in most of the samples decreased considerably after roasting at 135 °C (T3) with respect to unroasted samples (T1); a similar behaviour was observed for histamine in 50% of investigated samples. These results are in contrast with other authors who demonstrated the histamine thermostability during cooking processes [39,40].

It was also observed that the histamine level increased in foods after frying and grilling [41]. However, other authors elucidated the mechanism by which certain cooking ingredients and common organic acids destroy histamine [42], so it could be very interesting to deepen this aspect by considering the occurrence of bio-compounds that develop following the roasting process of cocoa and their possible role in the control of the BAs levels in food.

3.4. Anthocyanins, Total Polyphenols and Flavanols Content

The results on the content of anthocyanins, total polyphenols and flavanols of the eighteen cocoa samples at different process steps (T1, T2 and T3) are reported in Table 4.

After fermentation and drying treatment (T1), the anthocyanin concentration was between 0.17 and 3.36 mg g^{-1}_{DFW}, with an average value of 1.02 mg g^{-1}_{DFW}; these pigments disappeared during fermentation [11], reaching low values on the sixth day of fermentation, and they are a good parameter to determine the progress or status of the fermentation. The contents found are similar to other cocoa varieties from Colombia [43], but inferior to those found in other studies conducted on Ghana cocoa varieties [44]. In unroasted samples (T1), the average content of total polyphenols was 45.50 mg GAE g^{-1}_{DFW}, values that are similar to other cocoa varieties planted in Colombia [43,45], with the only exception being sample 4 which presented higher contents (over 80 mg GAE g^{-1}_{DFW}). These are more similar to the values found in other studies carried out on varieties planted in Ghana, as well as in other varieties planted in Colombia [44,46]. It is important to point out that these results may have been affected by the fact that each single phenol shows a different response to the Folin-Ciocalteau reagent [14].

According to Carrillo et al. [45], the cocoa-producing region can have a significant effect on the total polyphenol content, as a proportional relationship was found between polyphenols content and altitude of plant crops. Their results suggest that plants grown at lower altitude accumulate more polyphenols compared to plants grown at higher altitude. In the present study, the TPC determined for sample 16 (from geographical area at 30 m.a.s.l) was lower than other cocoa samples so it seems that the theory proposed by Carillo et al. is not confirmed by our data, although this aspect would be worth investigating with a large number of samples.

Roasting did not cause a statistically significant decrease in anthocyanin content in samples from all the three regions, with the following exception: a decrease of 50%–60% was observed in roasted cocoa beans in the sample of the Narino region (sample 16) due to its highest values at the end of fermentation. The decrease in anthocyanin content is in accordance with data observed by other authors [12,43] for different roasting temperatures. Regarding the TPC, a not statistically significant increase was found from 45.50 mg GAE g^{-1}_{DFW} for T1 to 55.26 mg GAE g^{-1}_{DFW} (+21%) for T2 and 62.01 mg GAE g^{-1}_{DFW} (+14%) for T3. However, an increase in TPC values after the roasting process is consistent with the data reported by Ioannone et al. [13]; these authors suggested that an increase in TPC is dependent on temperature and exposure time, as a series of condensation and polymerization reactions occur with the formation of complex molecules such as pro-anthocyanidins from lower molecular weight compounds such as phenols and anthocyanins. Additionally, through Maillard reactions, melanoidins can be formed from reducing sugars and free amino acids; as a consequence, melanoidins can have reducing properties that affect the response to the Folin-Ciocalteu reagent, thus causing an overestimation of the TPC values [14].

The occurrence of flavanols before and after roasting was also investigated in Criollo cocoa samples and the results are reported in Table 3. Moreover, Table S1 shows the epicatechin to catechin ratio (epi/cat) for both unroasted and roasted samples. Catechin was found in all unroasted samples (ranging from n.d. to 4.43 ± 0.13) with the exception of the samples 1, 5 and 9. Epicatechin was detected in all the samples with a maximum value of 5.7 ± 0.17 mg g^{-1} (in sample 12) and a minimum of 0.45 ± 0.01 mg g^{-1} (in sample 11). Similar catechin contents were found by Loureiro et al. in dried cocoa beans from Latin America [47].

Table 4. Anthocyanins, total polyphenols (TPC) and flavanols content of the Criollo cocoa samples after fermentation and drying (T1) and after roasting (T2, 120 °C for 22 min; T3, 135 °C for 15 min). The data are expressed as mean of triplicate analysis.

Sample	Anthocyanins (mg g^{-1} DFW)				TPC (mg GAE g^{-1} DFW)				Catechin (mg g^{-1} DFW)				Epicatechin (mg g^{-1} DFW)			
	T1	T2	T3	Sign.	T1	T2	T3	Sign.	T1	T2	T3	Sign.	T1	T2	T3	Sign.
1	2.28 [a]	1.36 [c]	2.08 [b]	*	43.00 [c]	68.79 [b]	83.85 [a]	*	nd	8.41 [b]	11.49 [a]	**	0.5 [b]	0.87 [a]	0.51 [b]	*
2	3.36 [a]	1.77 [b]	2.00 [b]	***	22.97 [b]	100.73 [a]	110.17 [a]	***	0.18 [b]	4.36 [a]	nd	***	0.76	0.9	0.54	**
3	0.24 [c]	0.31 [b]	0.37 [a]	*	47.38 [b]	33.54 [c]	45.56 [b]	*	1.47 [b]	1.64 [b]	2.17 [a]	*	1.72 [a]	nd	0.7 [b]	**
4	0.76 [a]	0.59 [b]	0.51 [b]	*	85.75 [a]	48.56 [c]	56.42 [b]	*	1.82 [b]	2.23 [b]	3.72 [a]	*	1.41	nd	nd	*
5	2.33 [a]	1.79 [b]	1.80 [b]	*	40.66 [b]	79.91 [a]	83.76 [a]	*	n.d	3.36 [b]	4.66 [a]	*	1.16 [a]	nd	1.1 [a]	*
6	0.36 [b]	0.50 [a]	0.56 [c]	*	65.12 [a]	68.36 [a]	60.39 [b]	*	2.16 [b]	5.93 [a]	6.12 [a]	**	1.07 [a]	0.47 [b]	nd	*
7	0.35 [b]	0.33 [b]	0.45 [a]	*	70.61 [a]	33.29 [b]	44.41 [b]	*	0.16 [b]	1.69 [a]	0.16 [b]	**	1.29	nd	nd	*
8	0.29 [c]	0.33 [b]	0.45 [a]	*	34.04 [b]	29.33 [b]	43.24 [a]	*	0.49 [a]	0.17 [b]	0.18 [b]	*	1.22	nd	nd	*
9	1.42 [a]	0.97 [c]	0.86 [b]	*	47.34 [c]	83.17 [a]	57.77 [b]	*	nd	0.65	nd	*	1.26	nd	nd	*
10	1.94 [c]	1.86 [b]	2.29 [a]	*	54.73 [c]	75.15 [b]	89.98 [a]	***	0.03 [b]	13.02 [a]	0.19 [b]	***	0.45 [b]	nd	0.83 [a]	***
11	0.57 [a]	0.49 [b]	0.52 [ab]	*	53.02 [a]	40.38 [c]	46.23 [b]	*	4.35 [a]	3.06 [b]	0.13 [c]	*	5.7 [a]	nd	3.58 [b]	**
12	0.46	0.48	0.44	n.s.	49.59 [b]	55.46 [a]	58.36 [a]	**	3.42 [b]	4.39 [a]	0.42 [c]	*	1.61 [b]	nd	5.26 [a]	***
13	0.17 [c]	0.23 [b]	0.31 [a]	*	22.62 [b]	18.99 [b]	38.49 [a]	*	4.43 [a]	0.74 [b]	0.05 [c]	*	0.59 [b]	nd	2.07 [a]	**
14	0.92 [b]	0.48 [c]	1.15 [a]	*	42.02 [b]	28.78 [c]	48.02 [a]	*	0.66	0.13	0.33	n.s.	0.79 [b]	nd	2.62 [a]	**
15	0.52 [b]	0.64 [a]	0.66 [a]	**	32.90 [c]	46.56 [b]	57.72 [a]	**	4.43 [b]	6.61 [a]	nd	**	2.34 [a]	1.62 [b]	nd	**
16	1.39 [a]	0.63 [c]	0.84 [b]	**	23.89 [b]	60.64 [a]	62.91 [a]	**	4.00 [b]	0.50 [b]	nd	**	0.97 [b]	5.87 [a]	0.78 [b]	***
17	0.72 [a]	0.71 [a]	0.35 [b]	***	33.83 [b]	87.43 [a]	88.57 [a]	***	0.65 [b]	0.27 [b]	nd	*	0.7 [c]	2.27 [a]	1.14 [b]	*
18	0.35 [a]	0.52 [a]	0.60 [a]	*	49.53 [a]	35.56 [c]	40.32 [b]	*	0.09 [b]	0.11 [b]	1.65 [a]	*	1.58	0.95 [b]	0.53 [c]	*

Legend: nd, not detectable; data followed by different superscript letters, in the same line, are significantly different (LSD test, $p < 0.05$); asterisks indicate significance at * $p < 0.05$; ** $p < 0.01$; *** $p < 0.001$; n.s. not significant.

The epi/cat ratio is a widely used index as it may be associated with the degree of cocoa processing [48,49] (Table S1). Generally, with the increase of temperature the epi/cat ratio tends to decrease due to isomerization reactions and the faster degradation of epicatechin with respect to catechin [50]. The major flavanol present in unroasted samples was (−)-epicatechin. According to Hurst et al. [51], the high temperatures may induce the epimerization of this flavanol to (−)-catechin, and (+)-catechin to (+)-epicatechin. This behaviour was noticed in many samples, even though in other cases the opposite was observed. Moreover, in many cases the ratio could not be calculated since either catechin or epicatechin was not detected. Finally, it can be said that both roasting conditions had a similar effect on flavanols.

3.5. Trolox Equivalent Antioxidant Capacity (TEAC) and the Ferric Reducing Antioxidant Power (FRAP) Assays

Table 5 shows the effect of the different roasting treatments on the radical scavenging activity (TEAC) and the reducing activity (FRAP) assays with respect to unroasted cocoa bean samples. The Pearson correlation coefficient was calculated: a strong correlation was found between TPC content and TEAC (ß = 0.88, $p < 0.05$) and between TPC content and FRAP (ß = 0.92, $p < 0.05$).

Table 5. Results of Trolox Equivalent Antioxidant Capacity (TEAC) and the Ferric Reducing Antioxidant Power (FRAP) assays on the Criollo cocoa samples after fermentation and drying (T1) and after roasting (T2, 120 °C for 22 min; T3, 135 °C for 15 min). The data are expressed as mean of triplicate analysis.

Sample	TEAC (μmol TE g^{-1})			Sign.	FRAP (μmol Fe^{2+} g^{-1})			Sign.
	T1	T2	T3		T1	T2	T3	
1	293.6 [b]	270.6 [c]	374.0 [a]	*	374.9 [c]	594.7 [a]	439.4 [b]	*
2	125.0 [c]	529.8 [b]	578.6 [a]	***	144.6 [c]	714.8 [b]	790.6 [a]	**
3	380.1 [a]	100.9 [c]	193.5 [b]	*	382.4 [a]	217.5 [c]	261.9 [b]	*
4	304.7 [a]	181.7 [b]	158.2 [c]	*	486.6 [a]	301.5 [c]	405.6 [b]	*
5	200.4 [c]	313.5 [b]	384.6 [a]	*	249.2 [b]	521.5 [a]	488.0 [a]	*
6	268.4 [a]	220.7 [b]	260.9 [a]	*	510.1	481.2	488.3	n.s.
7	411.0 [a]	94.5 [c]	170.6 [b]	***	562.3 [a]	242.3 [b]	243.0 [b]	*
8	101.5 [b]	91.4 [b]	187.9 [a]	*	259.2 [a]	187.0 [b]	269.9 [a]	*
9	161.4 [c]	396.9 [a]	188.9 [b]	*	282.6 [c]	321.0 [b]	431.2 [a]	*
10	217.1 [c]	389.0 [b]	459.3 [a]	*	432.2 [c]	562.7 [b]	703.5 [a]	**
11	238.4 [a]	135.2 [c]	168.3 [b]	*	401.9 [a]	282.0 [b]	279.5 [b]	*
12	246.3 [b]	284.8 [a]	279.2 [a]	*	313.8 [c]	352.0 [b]	440.1 [a]	*
13	98.2 [b]	69.6 [c]	141.3 [a]	**	174.5 [b]	135.7 [c]	222.1 [a]	***
14	195.5 [a]	136.1 [c]	161.1 [b]	*	264.3 [b]	252.8 [b]	334.3 [a]	*
15	161.3 [b]	202.5 [a]	111.3 [c]	*	205.6 [c]	349.9 [b]	466.3 [a]	**
16	184.1 [c]	224.8 [b]	351.0 [a]	*	151.1 [c]	388.4 [b]	498.8 [a]	***
17	191.0 [c]	391.9 [b]	441.5 [a]	***	214.6 [c]	636.9 [b]	789.1 [a]	***
18	231.0 [a]	167.2 [b]	122.3 [c]	*	275.3	229.1	273.8	n.s.

Legend: data followed by different superscript letters, in the same line, are significantly different (LSD test, $p < 0.05$); asterisks indicate significance at * $p < 0.05$; ** $p < 0.01$; *** $p < 0.001$; n.s. not significant.

Regarding the reducing capacity as evaluated by FRAP assay (Table 5), a mean value of 395 μmol Fe^{2+}/g was determined in unroasted samples, which is in agreement with other authors [4,52].

The trend of values of FRAP was similar to those obtained in the ABTS assay. Moreover, in roasted samples results of TEAC and FRAP were comparable with values found by Ioannone et al. [13]. Generally, the antioxidant activity was higher in roasted samples compared with unroasted ones, with the exception of some cases (samples 3, 4, 7, 11, 14 and 18).

The improvement in the antioxidant and reducing properties after the roasting process may be related to the formation of reducing molecules, not quantified in the present work, as well as to the occurrence of condensation reactions among polyphenols, as evidenced by the results reported in Table 4.

3.6. Principal Component Analysis

A principal component analysis (PCA) was performed to highlight how factors or variables can influence the changes in the BA level in raw cocoa beans after roasting. Figure 2 shows the distribution of the variables analysed in the two first principal components which represent 90.1% of data variance. Usually, the two first principal components are sufficient to explain the maximum variation in all data [31]. PC1 and PC2 explained 84.9% and 5.2% of date variance related to the BA content of cacao. In order to better describe the data set, the following results and information were included: microbial counts, polyphenols (TPC), anthocyanins, antioxidant activity (FRAP and TEAC), flavonols (catechin and epicatechin), BA content at different processing conditions (T1, T2, and T3) and origin of the samples.

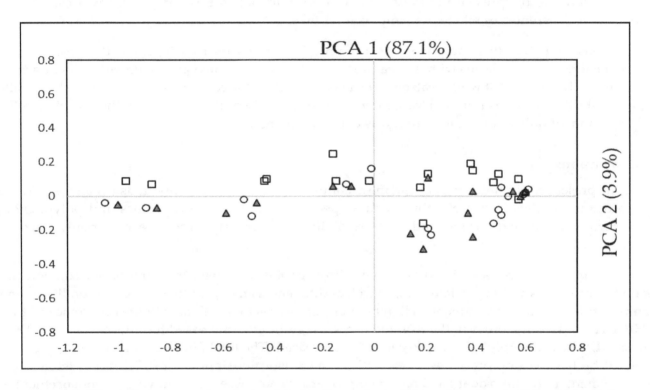

Figure 2. Principal component analysis related to the content of biogenic amines in Colombian criollo cocoa samples. Legend: empty square, T1 (raw cocoa beans); gray triangle, T2 (120 °C for 22 min); empty circle, T3 (135 °C for 15 min).

Concerning PC1, antioxidant activity (FRAP and TEAC) (−1.23 to −1.01), polyphenol content (TPC), and BA (−0.72 to −0.56) showed a negative influence on this component, while FRAP, TEAC, and TPC showed a significant increase in concentration in the same way as BA during roasting conditions. On the other side, anthocyanins, catechin, and epicatechin (0.43 to 0.55) showed a positive influence on this component. The anthocyanin content is a good parameter to determine if fermentation is carried out properly since they decrease as the fermentation progresses; therefore, the correlation found between a high anthocyanin content in raw cocoa (T1) and a high BA content may be related to a non-ideal fermentation process in which, for different reasons, the enzymatic activity of the grains remained active, generating metabolic intermediates such as BAs.

As for the individual BAs, a positive influence was found for DOP (0.51) under initial conditions (T1) after treatment at 120 °C (T2) for PUT, CAD, and SPD (0.49 to 0.54), and at 135 °C (T3) for CAD, SPD, SER, HIS, and SPM (0.44 to 0.56), while the other BAs showed no influence on this component.

The variables pH (0.1), region (0.41), shell microbiota (−0.26), and bean microbiota (−0.12) showed a weak correlation with each other. The pH is important to select the type of microorganisms that can grow and therefore quantity, and on the type of BAs they can generate [53]; in the present study, the values found for pH were not low enough to inhibit Enterobacteriaceae, which is one of the main groups that can produce BAs [31]. Moreover, pH values were in the optimal range that can favour BAs accumulation. The synthesis of polyamines, such as spermine and spermidine, occurs in response to high pH environments; these BAs act as inhibitors of carbonic anhydrase enzymes that catalyse the interconversion of carbon dioxide and water into bicarbonate and protons and vice versa [54].

According to Lima et al. [29], average levels of microorganisms are lower in the beans compared to those found in the shell due to the lower availability of nutrients, which can cause the activation of metabolic pathways in some groups of microorganisms that can lead to the accumulation of decarboxylation products such as BAs; however, no influence was observed in this component. Regarding the origin, the difference between cultivars, different growth, and postharvest conditions may be related to the presence of these BAs, but no influence in this component was established between the different sites where the samples were taken.

Concerning PC2, this component was mainly influenced by roasting. On the positive axis, the characteristics of the beans without heat treatment (T1) were located predominantly, differing from the samples T2 and T3 that were located on the negative axis of the component. Although a statistically significant difference was found in the content of BAs at T1, T2 and T3 in most of the samples, in the PC2 component only a small correlation was evident among them.

4. Conclusions

The present study aimed to evaluate the accumulation of bioactive compounds in eighteen Criollo cocoa beans samples from Colombia, with a special focus on biogenic amines and polyphenols, after fermentation and drying and after two different roasting processes commonly used in cocoa factories.

The samples showed a similar BAs profile, with a variability in their concentration as a consequence of both cocoa beans and shell microbiota, as well as differences among cultivars, growth conditions and fermentation and drying treatments. High temperature seems to correlate with the occurrence of TYR, PHE, ETH and PUT; moreover, the roasting process significantly increased the concentration of DOP and SPM, whilst CAD and SPD levels generally decreased. The total phenolic content was positively affected by the roasting processes; even without a statistically significant difference a remarkable improvement in the antioxidant and reducing properties were observed, showing an enhancement of their functionality.

No direct relationship was found between the content of polyphenols and the content of BAs in cocoa beans, even if it can be speculated that polyamines could have a role by influencing the antioxidant activity or exhibiting pro-oxidant properties in cocoa beans. Therefore, the correlation found between a high anthocyanin content and a high BAs content in unroasted cocoa samples (T1) could be attributable to a non-ideal fermentation process. One important result that it is worth pointing out is that the quantities of BAs found in the unroasted cocoa beans were not alarming, especially with regard to HIS and TYR, the amines of toxicological interest.

Moreover, low BAs amounts were also found in roasted samples, which is of crucial importance considering that such values that were calculated for defatted samples will be further processed and used as ingredients in complex formulations.

Author Contributions: Conceptualization, C.C.-L., M.M., C.D.D.M. and J.D.O.; methodology, M.M., C.D.D.M. and C.C.-L.; software, M.M. and J.D.-O.; formal analysis, J.D.-O., C.D.D.M., M.M.; investigation, J.D.-O., C.D.D.M., and M.M.; resources, J.D.-O., C.D.D.M. and M.M.; data curation, M.M.; writing—original draft preparation, J.D.-O., M.M.; writing—review and editing, M.M. and C.D.D.M.; visualization, M.M. and C.C.-L.; supervision, D.M., A.P.; project administration, C.C.-L.; funding acquisition, J.D.-O. All authors have read and agreed to the published version of the manuscript.

References

1.   International Cocoa Organization. *Quarterly Bulletin of Cocoa Statistics*; International Cocoa Organization: London, UK, 2019; Volume 45, p. 101. Available online: http:www.icco.org (accessed on 12 December 2019).

2.   Beg, M.S.; Ahmad, S.; Jan, K.; Bashir, K. Status, supply chain and processing of cocoa—A review. *Trends Food Sci. Technol.* **2017**, *66*, 108–116. [CrossRef]

3.   Afoakwa, E.O. The chemistry of flavour development during Cocoa processing and chocolate manufacture. In *Chocolate Science and Technology*, 1st ed.; Afoakwa, E.O., Ed.; Wiley online library: Hoboken, NJ, USA, 2016; p. 296.

4.   Di Mattia, C.; Martuscelli, M.; Sacchetti, G.; Scheirlinck, I.; Beheydt, B.; Mastrocola, D.; Pittia, P. Effect of fermentation and drying on procyanidins, antiradical activity and reducing properties of cocoa beans. *Food Bioprocess. Technol.* **2013**, *6*, 3420. [CrossRef]

5.   Coton, E.; Coton, M. Evidence of horizontal transfer as origin of strain to strain variation of the tyramine production trait in *Lactobacillus brevis*. *Food Microbiol.* **2009**, *26*, 52–57. [CrossRef] [PubMed]

6.   Lucas, P.M.; Blancato, V.S.; Claisse, O.; Magni, C.; Lolkema, J.S.; Lonvaud-Funel, A. Agmatine deiminase pathway genes in *Lactobacillus brevis* are linked to the tyrosine decarboxylation operon in a putative acid resistance locus. *Microbiology* **2007**, *153*, 2221–2230. [CrossRef] [PubMed]

7.   Lucas, P.M.; Wolken, W.A.M.; Claisse, O.; Lolkema, J.S.; Lonvaud-Funel, A. Histamine-producing pathway encoded on an unstable plasmid in *Lactobacillus hilgardii* 0006. *Appl. Environ. Microbiol.* **2005**, *71*, 1417–1424. [CrossRef] [PubMed]

8.   Marcobal, Á.; De Las Rivas, B.; Moreno-Arribas, M.V.; Muñoz, R. Evidence for horizontal gene transfer as origin of putrescine production in *Oenococcus oeni* RM83. *Appl. Environ. Microbiol.* **2006**, *72*, 7954–7958. [CrossRef] [PubMed]

9.   Martuscelli, M.; Crudele, M.A.; Gardini, F.; Suzzi, G. Biogenic amine formation and oxidation by *Staphylococcus xylosus* strains from artisanal fermented sausages. *Lett. Appl. Microbiol.* **2000**, *31*, 228–232. [CrossRef] [PubMed]

10. Di Mattia, C.; Martuscelli, M.; Sacchetti, G.; Beheydt, B.; Mastrocola, D.; Pittia, P. Effect of different conching processes on procyanidin content and antioxidant properties of chocolate. *Food Res. Int.* **2014**, *63*, 367–372. [CrossRef]

11. Do Carmo Brito, B.D.N.; Campos Chisté, R.; da Silva Pena, R.; Abreu Gloria, M.B.; Santos Lopes, A. Bioactive amines and phenolic compounds in cocoa beans are affected by fermentation. *Food Chem.* **2017**, *228*, 484–490. [CrossRef]

12. Oracz, J.; Nebesny, E. Influence of roasting conditions on the biogenic amine content in cocoa beans of different *Theobroma cacao* cultivars. *Food Res. Int.* **2014**, *55*, 1–10. [CrossRef]

13. Ioannone, F.; Di Mattia, C.D.; De Gregorio, M.; Sergi, M.; Serafini, M.; Sacchetti, G. Flavanols, proanthocyanidins and antioxidant activity changes during cocoa (*Theobroma cacao* L.) roasting as affected by temperature and time of processing. *Food Chem.* **2015**, *174*, 263–269. [CrossRef] [PubMed]

14. Sacchetti, G.; Ioannone, F.; De Gregorio, M.; Di Mattia, C.; Serafini, M.; Mastrocola, D. Non enzymatic browning during cocoa roasting as affected by processing time and temperature. *J. Food Eng.* **2016**, *169*, 44–52. [CrossRef]

15. Hinneh, M.; Abotsi, E.E.; Van De Walle, D.; Tzompa-Sosa, D.A.; De Winne, A.; Simonis, J.; Messens, K.; Van Durme, J.; Afoakwa, E.O.; De Cooman, L.; et al. Pod storage with roasting: A tool to diversifying the flavor profiles of dark chocolates produced from 'bulk' cocoa beans? (part I: Aroma profiling of chocolates). *Food Res. Int.* **2019**, *119*, 84–98. [CrossRef] [PubMed]

16. Oliviero, T.; Capuano, E.; Cämmerer, B.; Fogliano, V. Influence of roasting on the antioxidant activity and HMF formation of a cocoa bean model systems. *J. Agric. Food Chem.* **2009**, *57*, 147–152. [CrossRef]

17. Djikeng, F.T.; Teyomnou, W.T.; Tenyang, N.; Tiencheu, B.; Morfor, A.T.; Touko, B.A.H.; Houketchang, S.N.; Boungo, G.T.; Karuna, M.S.L.; Ngoufack, F.Z.; et al. Effect of traditional and oven roasting on the physicochemical properties of fermented cocoa beans. *Heliyon* **2018**, *4*, e00533. [CrossRef]

18. International agency for research on cancer (IARC). Monographs on the evaluation of carcinogenic risks to humans. In *Dry Cleaning, Some Chlorinated Solvents and Other Industrial Chemicals*; IARC Publications: Lyon, France, 1995; Volume 63, pp. 3194–3407.

19. Codex Alimentarius Commission. Standard for Cocoa (Cacao) Mass (Cocoa/Chocolate Liquor) and Cocoa Cake Codex Stan 141-1983; 2014. Available online: http://www.fao.org/input/download/standards/69/CXS_141e.pdf (accessed on 1 March 2020).

20. Okiyama, D.C.G.; Navarro, S.L.B.; Rodrigues, C.E.C. Cocoa shell and its compounds: Applications in the food industry. *Trends Food Sci. Technol.* **2017**, *63*, 103–112. [CrossRef]

21. Quelal-Vásconez, M.A.; Lerma-García, M.J.; Pérez-Esteve, É.; Arnau-Bonachera, A.; Barat, J.M.; Talens, P. Fast detection of cocoa shell in cocoa powders by near infrared spectroscopy and multivariate analysis. *Food Control* **2019**, *99*, 68–72. [CrossRef]

22. Association of Official Analytical Chemists. *Official Methods of Analysis*, 17th ed.; AOAC: Gaithersburg, MD, USA, 2002.

23. Chaves-López, C.; Serio, A.; Montalvo, C.; Ramirez, C.; Peréz-Álvarez, J.A.; Paparella, A.; Mastrocola, D.; Martuscelli, M. Effect of nisin on biogenic amines and shelf life of vacuum packaged rainbow trout (*Oncorhynchus mykiss*) fillets. *J. Food Sci. Technol.* **2017**, *54*, 3268–3277. [CrossRef] [PubMed]

24. Re, R.; Pellegrini, N.; Proteggente, A.; Pannala, A.; Yang, M.; Rice-Evans, C. Antioxidant activity applying an improved ABTS radical cation decolorization assay. *Free Radic. Biol. Med.* **1999**, *26*, 1231–1237. [CrossRef]

25. Benzie, I.F.F.; Strain, J.J. Ferric reducing/antioxidant power assay: Direct measure of total antioxidant activity of biological fluids and modified version for simultaneous measurement of total antioxidant power and ascorbic acid concentration. *Methods Enzymol.* **1999**, *299*, 15–27.

26. Kongor, J.E.; Hinneh, M.; Van de Walle, D.; Afoakwa, E.O.; Boeckx, P.; Dewettinck, K. Factors influencing quality variation in cocoa *Theobroma cacao* bean flavour profile: A review. *Food Res. Int.* **2016**, *82*, 44–52. [CrossRef]

27. Schwan, R.F.; Pereira, G.D.M.; Fleet, G.H. Microbial activities during cocoa fermentation. In *Cocoa and Coffee Fermentations*, 1st ed.; Shwan, R.F., Fleet, G.H., Eds.; Taylor & Francis: Boca Raton, FL, USA, 2014; pp. 129–192.

28. Schwan, R.F.; Wheals, A.E. The microbiology of cocoa fermentation and its role in chocolate quality. *Crit. Rev. Food Sci. Nutr.* **2004**, *44*, 205–221. [CrossRef] [PubMed]

29. Lima, L.J.R.; van der Velpen, V.; Wolkers-Rooijackers, J.; Kamphuis, H.J.; Zwietering, M.H.; Rob Nout, M.J. Microbiota dynamics and diversity at different stages of industrial processing of cocoa beans into cocoa powder. *Appl. Environ. Microbiol.* **2012**, *78*, 2904–2913. [CrossRef] [PubMed]

30. Baraggio, N.G.; Velázquez, N.S.; Simonetta, A.C. Aminas biógenas generadas por cepas bacterianas provenientes de alimentos lácteos y cárnicos. *Rev. Cienc. Tecnol.* **2010**, *13*. Available online: http://www.scielo.org.ar/scielo.php?script=sci_arttext&pid=S1851-75872010000100012 (accessed on 20 March 2020).

31. Rodriguez-Campos, J.; Escalona-Buendía, H.B.; Orozco-Avila, I.; Lugo-Cervantes, E.; Jaramillo-Flores, M.E. Dynamics of volatile and non-volatile compounds in cocoa (*Theobroma cacao* L.) during fermentation and drying processes using principal components analysis. *Food Res. Int.* **2011**, *44*, 250–258. [CrossRef]

32. García-Alamilla, P.; Lagunes-Gálvez, L.M.; Barajas-Fernández, J.; García-Alamilla, R. Physicochemical changes of cocoa beans during roasting process. *J. Food Qual.* **2017**, *12*, 1–11. [CrossRef]

33. Gloria, M.B.A.; Tavares-Neto, J.; Labanca, R.A.; Carvalho, M.S. Influence of cultivar and germination on bioactive amines in soybeans (*Glycine max* L. Merril). *J. Agric. Food Chem.* **2005**, *53*, 7480–7485. [CrossRef]

34. Bandeira, C.M.; Evangelista, W.P.; Gloria, M.B.A. Bioactive amines in fresh, canned and dried sweet corn, embryo and endosperm and germinated corn. *Food Chem.* **2012**, *131*, 1355–1359. [CrossRef]

35. Albertini, B.; Schoubben, A.; Guarnaccia, D.; Pinelli, F.; Della Vecchia, M.; Ricci, M.; Di Renzo, G.C.; Blasi, P. Effect of fermentation and drying on cocoa polyphenols. *J. Agric. Food Chem.* **2015**, *63*, 9948–9953. [CrossRef]

36. Granvogl, M.; Bugan, S.; Schieberle, P. Formation of amines and aldehydes from parent amino acids during thermal processing of cocoa and model systems: New insights into pathways of the Strecker reaction. *J. Agric. Food Chem.* **2006**, *54*, 1730–1739. [CrossRef]

37. Zamora, R.; Delgado, R.M.; Hidalgo, F.J. Formation of β-phenylethylamine as a consequence of lipid oxidation. *Food Res. Int.* **2012**, *46*, 321–325. [CrossRef]

38. Martins, A.C.C.L.; Gloria, M.B.A. Changes on the levels of serotonin precursors - tryptophan and 5-hydroxytryptophan-during roasting of Arabica and Robusta coffee. *Food Chem.* **2010**, *118*, 529–533. [CrossRef]

39. Luten, J.B.; Bouquet, W.; Seuren, L.A.J.; Burggraaf, M.M.; Riekwel-Booy, G.; Durand, P.; Etienne, M.; Gouyou, J.P.; Landrein, A.; Ritchie, A.; et al. Biogenic amines in fishery products: Standardization methods

within E.C. In *Quality Assurance in the Fish Industry*; Elsevier Science Publishers, B.V.: Amsterdam, The Netherlands, 1992; pp. 427–439.

40. Wendakoon, C.N.; Sakaguchi, M. Combined effect of sodium chloride and clove on growth and biogenic amine formation of *Enterobacter aerogenes* in Mackerel muscle extract. *J. Food Prot.* **1993**, *56*, 410–413. [CrossRef] [PubMed]

41. Chung, B.Y.; Park, S.Y.; Byun, Y.S.; Son, J.H.; Choi, Y.W.; Cho, H.S.; Kim, H.O.; Park, C.W. Effect of different cooking methods on Histamine levels in selected foods. *Ann. Dermatol.* **2017**, *296*, 706–714. [CrossRef]

42. Thadhani, V.M.; Jansz, E.R.; Peiris, H. Destruction of histamine by cooking ingredients-an artifact of analysis. *J. Natl. Sci. Found. Sri Lanka* **2001**, *29*, 129–135. [CrossRef]

43. Zapata-Bustamante, S.; Tamayo-Tenorio, A.; Rojano, B.A. Effect of roasting on the secondary metabolites and antioxidant activity of Colombian cocoa clones. *Rev. Fac. Nac. Agron. Medellín* **2015**, *68*, 7497–7507. [CrossRef]

44. Afoakwa, E.O.; Ofosu-Ansah, E.; Budu, A.S.; Mensah-Brown, H.; Takrama, J.F. Roasting effects on phenolic content and free-radical scavenging activities of pulp preconditioned and fermented cocoa (*Theobroma cacao*) beans. *Afr. J. Food Agric. Nutr. Dev.* **2015**, *15*, 9635–9650.

45. Carrillo, L.C.; Londoño-Londoño, J.; Gil, A. Comparison of polyphenol, methylxanthines and antioxidant activity in *Theobroma cacao* beans from different cocoa-growing areas in Colombia. *Food Res. Int.* **2014**, *60*, 273–280. [CrossRef]

46. Porras Barrientos, L.D.; Torres Oquendo, J.D.; Gil Garzón, M.A.; Martínez Álvarez, O.L. Effect of the solar drying process on the sensory and chemical quality of cocoa (*Theobroma cacao* L.) cultivated in Antioquia, Colombia. *Food Res. Int.* **2019**, *115*, 259–267. [CrossRef]

47. Loureiro, G.A.H.A. Qualidade de solo e Qualidade de Cacau. Master's Thesis, Universidade Estadual de Santa Cruz, Ilhéus, Bahia, Brazil, 2014.

48. Janszky, I.; Mukamai, K.J.; Ljung, R.; Ahnve, S.; Ahlbom, A.; Hallqvist, J. Chocolate consumption and mortality following a first acute myocardial infarction: The Stockholm Heart Epidemiology Study. *J. Intern. Med.* **2009**, *266*, 248–257. [CrossRef]

49. Payne, M.J.; Hurst, W.J.; Miller, K.B.; Rank, C.; Stuart, D.A. Impact of fermentation, drying, roasting, and dutch processing on epicatechin and catechin content of cacao beans and cocoa ingredients. *J. Agric. Food Chem.* **2010**, *58*, 10518–10527. [CrossRef] [PubMed]

50. Caligiani, A.; Cirlini, M.; Palla, G.; Ravaglia, R.; Arlorio, M. GC-MS detection of chiral markers of cocoa beans of different quality and geographic origin. *Chirality* **2007**, *19*, 329–334. [CrossRef]

51. Hurst, W.J.; Krake, S.H.; Bergmeier, S.C.; Payne, M.J.; Miller, K.B.; Stuart, D.A. Impact of fermentation, drying, roasting and Dutch processing on flavan-3-ol stereochemistry in cacao beans and cocoa ingredients. *Chem. Cent. J.* **2011**, *5*, 53. [CrossRef] [PubMed]

52. Bauer, D.; de Abreu, J.P.; Oliveira, H.S.S.; Goes-Neto, A.; Koblitz, M.G.B.; Teodoro, A.J. Antioxidant activity and cytotoxicity effect of cocoa beans subjected to different processing conditions in human lung carcinoma cells. *Oxid. Med. Cell. Longev.* **2016**, *2016*, 7428515. [CrossRef] [PubMed]

53. Tabanelli, G.; Montanari, C.; Gardini, F. Biogenic amines in food: A review of factors affecting their formation. In *Encyclopedia of Food Chemistry*; Melton, L., Shahidi, F., Varelis, P., Eds.; Academic Press: New York, NY, USA, 2019; pp. 337–343.

54. Carta, F.; Temperini, C.; Innocenti, A.; Scozzafava, A.; Kaila, K.; Supuran, C.T. Polyamines inhibit carbonic anhydrases by anchoring to the zinc-coordinated water molecule. *J. Med. Chem.* **2010**, *53*, 511–522. [CrossRef] [PubMed]

6

Biogenic Amines in Plant-Origin Foods: Are they Frequently Underestimated in Low-Histamine Diets?

Sònia Sánchez-Pérez [1,2,3], Oriol Comas-Basté [1,2,3], Judit Rabell-González [1,2,3],
M. Teresa Veciana-Nogués [1,2,3], M. Luz Latorre-Moratalla [1,2,3] and M. Carmen Vidal-Carou [1,2,3,*]

[1] Departament de Nutrició, Ciències de l'Alimentació i Gastronomia, Facultat de Farmàcia i Ciències de l'Alimentació, Universitat de Barcelona (UB), Av. Prat de la Riba 171, 08921 Santa Coloma de Gramenet, Spain; soniasanchezperez@ub.edu (S.S.-P.); oriolcomas@ub.edu (O.C.-B.); juditrabellgonzalez@hotmail.com (J.R.-G.); veciana@ub.edu (M.T.V.-N.); mariluzlatorre@ub.edu (M.L.L.-M.)
[2] Institut de Recerca en Nutrició i Seguretat Alimentària (INSA·UB), Universitat de Barcelona (UB), Av. Prat de la Riba 171, 08921 Santa Coloma de Gramenet, Spain
[3] Xarxa de Referència en Tecnologia dels Aliments de la Generalitat de Catalunya (XaRTA), C/ Baldiri Reixac 4, 08028 Barcelona, Spain
* Correspondence: mcvidal@ub.edu

Abstract: Low-histamine diets are currently used to reduce symptoms of histamine intolerance, a disorder in histamine homeostasis that increases plasma levels, mainly due to reduced diamine-oxidase (DAO) activity. These diets exclude foods, many of them of plant origin, which patients associate with the onset of the symptomatology. This study aimed to review the existing data on histamine and other biogenic amine contents in nonfermented plant-origin foods, as well as on their origin and evolution during the storage or culinary process. The only plant-origin products with significant levels of histamine were eggplant, spinach, tomato, and avocado, each showing a great variability in content. Putrescine has been found in practically all plant-origin foods, probably due to its physiological origin. The high contents of putrescine in certain products could also be related to the triggering of the symptomatology by enzymatic competition with histamine. Additionally, high spermidine contents found in some foods should also be taken into account in these diets, because it can also be metabolized by DAO, albeit with a lower affinity. It is recommended to consume plant-origin foods that are boiled or are of maximum freshness to reduce biogenic amine intake.

Keywords: histamine; putrescine; tyramine; cadaverine; biogenic amines; histamine intolerance; low-histamine diet; plant-origin foods; culinary process; storage conditions

1. Introduction

In recent years, various diets have been proposed for the treatment of histamine intolerance [1–8]. These diets, known as low- or free-histamine diets, usually exclude foods that patients associate with the onset of intolerance symptoms. Such foods tend to be rich in histamine, but some, surprisingly, are not usually regarded as sources of this amine.

As described in the literature and scientific reports issued by the European Food Safety Authority (EFSA) and a joint Food and Agriculture Organization of the United Nations (FAO)/World Health Organization (WHO) committee, histamine intolerance (also called food histaminosis or food histamine sensitivity) is a disorder associated with increased plasma histamine levels and is recognized as clinically different from the more established histamine intoxication [9,10]. Although in both cases, histamine is the causative agent, the etiology of the disorders differs. Intoxication appears after the consumption of foods with unusually high histamine concentrations, while intolerance is due to a deficiency in histamine metabolism, so that symptoms may be triggered even by the intake of low amounts [1,9–11].

Diamine oxidase (DAO) is the main enzyme responsible for the metabolism of histamine and other amines at the intestinal level, and impaired DAO activity is one of the main causes of histamine intolerance [1,12,13]. This enzymatic deficit may have its origins in genetic mutations. Different polymorphisms of a single nucleotide in the gene that encodes this enzyme (*AOC1* on chromosome 7) have been associated with lower DAO activity [14–16]. The deficit may also be due to acquired causes such as inflammatory bowel diseases that block the secretion of DAO [1,3,12], or to the inhibitory action of drugs, some of them with a very widespread use (e.g., acetylcysteine, clavulanic acid, metoclopramide, verapamil) [1,17]. Another enzyme involved in histamine metabolization is monoamine oxidase (MAO) [13]. Therefore, MAO inhibitor drugs, such as selegiline or rasagiline, could also favor the plasmatic accumulation of histamine and the onset of symptoms of histamine intolerance. In addition, the presence of other biogenic amines, mainly putrescine and cadaverine, may compromise the intestinal degradation of histamine by enzymatic competition with DAO [9].

The symptoms of histamine intolerance are numerous and highly variable, due to the effects and functions of histamine in multiple organs and systems of the body. They include gastrointestinal (abdominal pain, diarrhea, vomiting), dermatological (urticaria, dermatitis, or pruritus), respiratory (rhinitis, nasal congestion, and asthma), cardiovascular (hypotonia and arrhythmias), and neurological (headaches) symptoms, and it is common for more than one disorder to occur simultaneously [1,11,12]. Several clinical studies have shown that patients with a potential diagnosis of histamine intolerance or with a diagnosis of migraine, intestinal, or dermatological diseases (atopic dermatitis, eczema, or chronic urticaria) have a higher prevalence of DAO deficits compared to the control population [3,6,18–28].

In order to carry out a correct dietary treatment of histamine intolerance, it is necessary to know what foods may contain this amine and what factors influence its accumulation. Likewise, it is also important to consider the occurrence of other amines that are also metabolized by the DAO enzyme. In contrast to plant-origin foods, there is more available information on the contents of histamine and other amines in fish and fish derivatives and all types of fermented products (cheeses, sausages, sauerkraut, wines, beer), in which their presence is attributed to the aminogenic activity of spoilage microorganisms and also to fermentative microorganisms [9,10,29]. Therefore, the freshness of the food and the hygienic conditions of the raw materials and manufacturing processes, as well as the adequate selection of starter cultures without decarboxylase activity, are of vital importance to avoid or reduce the formation of these compounds [9,29–31].

Due to the information available on the contents of biogenic amines in nonfermented plant-origin foods being scarce, the aim of this study was to review the existing data on the contents of histamine and other biogenic amines in these types of products, as well as their origin and evolution during storage or cooking.

2. Methods

A selective search of scientific literature dealing with biogenic amine contents in nonfermented plant-origin foods, including vegetables, fruits, and cereals, was performed. The bibliographic search was carried out in the PubMed and Web of Science databases using the following keywords: "histamine", "biogenic amines", "tyramine", "putrescine", "cadaverine", "plant-origin food", "food samples", "storage", "cooking", "fruit", "vegetable", "legume", "cereal", "spinach", "eggplant", "tomato", "citrus", "modified atmosphere packaging", and "microbial decarboxylase activity". Original analytical studies, reviews, and table compilations of content in food were included. Articles published before 1990 were excluded from this review.

Apart from data obtained from the literature, data on the biogenic amine content of plant-origin foods from our own database of Spanish market products were also used. Specifically, histamine, tyramine, putrescine, and cadaverine contents of 25 types of vegetables, 19 fruits, and 8 cereals were included.

3. Content of Biogenic Amines in Plant-Origin Foods

In this section, the contents of biogenic amines (histamine, tyramine, putrescine, and cadaverine) in different plant-origin foods are reviewed, using our own database and data from studies published by other authors. A total of 20 studies reporting data on biogenic amine contents in such foods were found. Most provided data on putrescine contents (normally together with the polyamines spermine and spermidine, not dealt with in this section), and only a few included other amines, such as histamine, tyramine, and cadaverine.

3.1. Vegetables and Legumes

Table 1 shows the contents of biogenic amines in different types of vegetables and legumes (nonfermented).

The only products found to contain significant levels of histamine were eggplant, spinach, and tomato, each showing a great variability in content, both in samples from the same study and among different studies. Histamine values ranged from 4.2 to 100.6 mg/kg in eggplant, from 9.5 to 69.7 mg/kg in spinach, and from not detected to 17.1 mg/kg in tomato. In the case of asparagus, pumpkin, and chard, histamine was found in only a few samples and at very low levels (<2 mg/kg).

Histamine occurs naturally in certain foods [29,32], which explains why it was recorded in practically all samples of spinach, eggplant, and tomato. The variability observed may have been due to botanical variety, as reported by Kumar et al. [33] for eggplant. However, as occurs in foods of animal origin, the presence of high contents of histamine and other amines in plant-origin products could also be associated with microbial activity [29,32,34]. Lavizzari et al. [32] attributed the high contents of histamine in spinach to the activity of contaminating bacteria during storage, belonging mainly to the groups Enterobacteriaceae and Pseudomonadaceae. There is currently a need for more research to understand in more detail the origin of histamine in plant foods such as spinach, eggplant, and tomatoes.

Tyramine has been found in more foods than histamine, although in lower concentrations, in no case exceeding 10 mg/kg. It should be noted that histamine-containing foods also contained tyramine (eggplants, tomatoes, spinach, chard, and asparagus). Although there is very little information about the origin of tyramine in nonfermented vegetables, its presence seems to be associated with microbial aminogenic activity. The ability to form tyramine has been reported for bacteria of the genus *Enterococcus* isolated from plants and fruits, mainly *E. faecium*, *E. mundtii*, and *E. casseliflavus* [35].

Putrescine has been detected in all the studied vegetables and legumes, although its content varied greatly among foods and sometimes also within the same product. In most vegetables and legumes, the average values ranged from 1 to 25 mg/kg. However, some samples of green pepper, eggplant, sweet corn, green and purple beans, spinach, tomato ketchup, soybeans, and peas had strikingly high putrescine contents, in some cases exceeding 200 mg/kg (Table 1). The putrescine found in food can have a dual origin. In plant-origin foods, low contents of this amine generally have a physiological source, as it performs different functions in plants, as do the polyamines spermidine and spermine, ranging from the activation of organogenesis to protection against stress [34,36,37]. On the other hand, the presence of putrescine is also associated with the decarboxylase activity of different groups of spoilage bacteria, mainly Enterobacteriaceae and *Clostridium* spp. [36]. According to Kalač et al. [38], the high amounts of putrescine found in frozen peas are due to bacterial activity in the period between harvesting and freezing or during thawing. However, high putrescine contents cannot always be attributed to bacterial decarboxylase activity. Toro-Funes et al. [39] have suggested that the considerable levels of putrescine found in soybean sprouts arise from the germination process, as this amine is a plant growth factor. In general, based on the available information, and due to the great variability in the reported contents, it is difficult to establish to what degree the presence of putrescine in plant-origin products can be considered physiological or the result of bacterial activity.

Cadaverine, like tyramine, has been described in few vegetables and legumes and in relatively low concentrations, with average values that in no case exceeded 8 mg/kg. The values reported by Nishimura et al. [40] in onion (29 mg/kg) and tofu (18 mg/kg) were an exception.

3.2. Fruits and Nuts

Table 2 shows the content of biogenic amines in different types of fresh fruits, fruit juices, and nuts. There were fewer publications reporting amine data for this type of food than for vegetables and legumes. In general, the contents were low, putrescine being in many cases the only amine found (in addition to the polyamines spermidine and spermine).

Avocado and kiwi, and grapefruit, orange, and pineapple juices, are the only products in this category for which the presence of histamine has been reported, but not in all studies. The 23 mg/kg of histamine in avocado reported by Jarisch et al. [12] stands out, although no relevant information about its possible origin was provided. A study conducted by Preti et al. [41] concluded that the presence of histamine in grapefruit, orange, and pineapple juices is due to a lack of hygienic quality during processing or storage, since this amine is not found in the original fresh fruit.

Similarly, very few fruits contained tyramine, and levels have always been low (Table 2). Avocado and plum stand out for their content of this amine, although in no case has it exceeded 7 mg/kg.

Putrescine has been found in practically all the fruits and nuts, with the highest levels in orange, orange juice, mandarin, grapefruit, grapefruit juice, banana, passion fruit, and pistachio. The range of contents of this amine in citrus fruits and their juices has been very broad, varying from not detected to as high as 200 mg/kg. Suggested explanations for this variability have included different origins, cultivation, and transport and storage conditions [41–44]. As reported by Gonzalez-Aguilar et al. [45], the contents of putrescine in mandarin (flavedo) can be increased by a drop in temperature before harvesting and by damage of mechanical origin. Its presence in most of the samples, unaccompanied by high levels of other amines (related to bacterial activity), seemed to indicate that, with some exceptions, putrescine in fruits has a physiological origin. To confirm this, it would be necessary to carry out more studies analyzing the fruit at the moment of collection. The only fruits reported as having no putrescine were avocado and plum, although interestingly, these did contain histamine and tyramine.

The only fruits with a notable content of cadaverine were bananas and sunflower seeds, for which Nishimura et al. [40] reported average levels of 11 and 22 mg/kg, respectively, although these data were from the analysis of only two samples.

3.3. Cereals and Derivatives

Table 3 shows the contents of biogenic amines in cereals and some derivatives such as breakfast cereals, pasta, and bread. The quantitative information available on amines in cereals is very limited. In principle, these foods do not contain amines other than putrescine, which has a physiological origin [36]. The only standout source of putrescine is wheat germ, which, like soya bean sprouts, has a high rate of cell division, in which putrescine and polyamines play a significant role [36].

Table 1. Biogenic amine contents (mg/kg fresh weight) found in vegetables and legumes. Data are presented as average (standard deviation) and range (minimum–maximum).

Food Categories	n	Occurrence of Biogenic Amines (mg/kg)								Reference
		Histamine		Tyramine		Putrescine		Cadaverine		
		Mean (SD)	Range	Mean (SD)	Range	Mean (SD)	Range	Mean (SD)	Range	
Vegetables and Vegetable Products										
Asparagus (wheat)	5	0.34 (0.62)	nd–1.42	0.69 (0.88)	nd–2.1	13.08 (2.81)	8.58–16.27	0.12 (0.19)	nd–0.43	#
Beans (green)	12	nd	-	2.46 (3.24)	nd–9.86	10.30 (8.61)	2.98–28.81	nd	-	#
	-	nd	-	-	-	34.9 (6.2)	-	-	-	[46]
Beans (purple)	-	nd	-	-	-	77.8 (7.7)	-	nd	-	[46]
Beans (yellow)	-	nd	-	-	-	14.8 (1.4)	-	nd	-	[46]
Broccoli	4	-	-	-	-	9	7–10.5	-	-	[47]
	5	-	-	-	-	6.4 (2.9)	3.4–10.8	-	-	[42]
	10	-	-	-	-	5.7	-	-	-	[48]
	5	-	-	-	-	1.9	1.3–2.6	-	-	[49]
	2	-	-	-	-	-	8.37–20.53	0.1	-	[40]
Cabbage	3	-	-	<0.5	-	-	-	-	-	[47]
	2	-	-	-	-	16	-	-	-	[34]
Cabbage (white)	10	-	-	-	-	14.18	-	6.03	-	[40]
	8	-	-	-	-	6.6	0.7–6.2	-	-	[48]
	3	-	-	-	-	2.7	-	<0.5	-	[49]
Cauliflower	3	nd	-	1.05 (0.91)	nd–1.71	3 (1.49)	1.61–4.58	0.16 (0.27)	nd–0.48	#
	7	-	-	-	-	4.9	3.1–4.5	-	-	[50]
	5	-	-	-	-	5.3 (2.1)	2.2–7.6	-	-	[47]
	2	-	-	-	-	3.7	3.3–8.9	-	-	[42]
	-	-	-	<0.5	-	9	-	-	-	[40]
Carrot	13	nd	-	nd	-	2.27 (2.20)	0.35–8.92	nd	-	[34]
	3	-	-	1	-	2.8	1.2–1.8	-	-	#
	4	-	-	-	-	3.5	2–3.9	-	-	[50]
	2	-	-	-	-	1.5 (0.7)	-	-	-	[47]
	6	-	-	-	-	0.7	0.7–2.7	-	-	[51]
	10	-	-	-	-	12.1 (3.9)	-	-	-	[42]
	4	-	-	-	-	14.8	7–18	-	-	[48]
	5	-	-	-	-	5	8–24.7	-	-	[52]
	-	-	-	-	-	-	-	-	-	[49]
Celeriac	2	-	-	-	-	6.1	5.73–14.10	2.55	-	[34]
	3	-	-	-	-	-	3.7–7.7	-	-	[40]
Chard	8	0.79 (0.41)	nd–1.33	1.90 (0.98)	0.74–3.48	6.38 (3.22)	2.4–11.94	0.13 (0.11)	nd–0.24	[47]
Courgette	41	nd	-	nd	-	7.94 (4.12)	2.74–24.81	0.32 (0.7)	nd–2.07	#
	-	-	-	2	-	4	-	-	-	#
Cucumber	10	nd	-	0.61 (0.89)	nd–2.33	5.42 (3.13)	1.32–10.62	nd	-	[34]
	3	-	-	-	-	3.2	-	-	-	[50]
	5	-	-	-	-	6.9 (1.4)	5.5–8.7	-	-	[42]
	10	-	-	-	-	8.7	-	-	-	[48]

Table 1. *Cont.*

Food Categories	n	Occurrence of Biogenic Amines (mg/kg)								Reference
		Histamine Mean (SD)	Range	Tyramine Mean (SD)	Range	Putrescine Mean (SD)	Range	Cadaverine Mean (SD)	Range	
Vegetables and Vegetable Products										
	5	-	-	-	-	13.1	2.1–28.8	-	-	[49]
	-	-	-	1	-	25	-	-	-	[34]
	2	-	-	-	-	9.87	-	2.96	-	[40]
Eggplant	23	39.42 (30.66)	4.17–100.6	0.60 (0.90)	nd–2.27	34.30 (6.98)	24.10–48.63	nd	-	#
	-	26	-	-	-	18.35	-	-	-	[12]
	2	-	-	-	-	18.35	-	5	-	[40]
Lettuce	4	nd	-	nd	-	2.85 (0.75)	2.20–3.90	nd	-	#
	3	-	-	-	-	-	3.3–4.8	-	-	[50]
	3	-	-	-	-	5.6 (1.3)	4.5–7.3	-	-	[42]
	10	-	-	-	-	7.9	-	-	-	[48]
	7	-	-	-	-	20.7	10.2–42.3	-	-	[49]
	2	-	-	-	-	6.87	-	3.27	-	[40]
Mushroom	11	nd	-	nd	-	1.29 (1.20)	0.02–3.65	0.10 (0.4)	nd–1.59	#
	10	-	-	-	-	11.7	-	-	-	[48]
	213	-	-	-	-	-	nd–156	-	-	[53]
Onion	4	nd	-	nd	-	nd	-	nd	-	#
	3	-	-	-	-	0.5	5.5–7.2	-	-	[50]
	10	-	-	-	-	0.6	-	-	-	[48]
	6	-	-	-	-	2	0.2–1	-	-	[49]
	-	-	-	3	-	-	-	-	-	[34]
	2	-	-	-	-	3.96	-	-	-	[40]
Pepper (green)	9	nd	-	nd	-	90.04 (41.65)	11.7–148.9	29.32	-	#
	2	-	-	-	-	-	104–237	0.05 (0.14)	nd–0.41	[40]
	5	-	-	-	-	70 (31)	13.2–96.9	5.62	-	[52]
Pepper (red)	8	nd	-	nd	-	2.42 (2.21)	0.59–5.35	nd	-	#
Potato	10	nd	-	0.58 (0.64)	nd–2.2	4.14 (3.06)	1.05–11.68	0.22 (0.54)	nd–1.75	#
	3	-	-	-	-	9.7	-	-	-	[50]
	3	-	-	-	-	17.6	-	-	-	[51]
	6	-	-	-	-	9.7 (2.1)	5.8–12.8	-	-	[42]
	2	-	-	-	-	-	0.1–22.4	-	-	[40]
	10	-	-	-	-	2.8	-	-	-	[48]
	6	-	-	-	-	7.2	1.1–10.5	-	-	[49]
	-	-	-	5	-	8	-	<0.5	-	[34]
Pumpkin	12	0.28 (0.54)	nd–1.90	nd	-	9.87 (6.19)	2.95–24.23	0.54 (0.76)	nd–2.15	#
Spinach	18	31.77 (17.02)	9.46–69.71	2.05 (0.83)	0.785–4.28	4.48 (2.46)	0.14–9.19	nd	-	#
	5	16	-	-	-	4.8	1.8–13.5	-	-	[49]
	-	-	-	6	-	6.0	-	1	-	[34]
	-	-	30–60	-	-	-	-	-	-	[12]
	2	37.5	-	-	-	-	-	-	-	[5]
	-	-	-	-	-	4.41	-	8.48	-	[40]
	-	61 (1.5)	-	nd	-	7.8 (0.1)	-	nd	-	[54]

Table 1. *Cont.*

Occurrence of Biogenic Amines (mg/kg)

Food Categories	n	Histamine Mean (SD)	Histamine Range	Tyramine Mean (SD)	Tyramine Range	Putrescine Mean (SD)	Putrescine Range	Cadaverine Mean (SD)	Cadaverine Range	Reference
Vegetables and Vegetable Products										
Sweet corn	32	-	-	-	-	12.9	nd–119	-	-	[52]
Tomato	5	nd	-	nd	-	38.44 (9.50)	30.5–119.2	0.18 (0.25)	nd–0.46	#
	53	2.51 (4.08)	nd–17.07	0.49 (0.92)	nd–6.38	16.48 (6.93)	6.29–35.55	0.50 (0.48)	nd–2.33	#
	3	-	-	-	-	10.6	9.3–122	-	-	[50]
	2	-	-	-	-	-	-	-	-	[51]
	5	6	-	<0.5	-	10	5.3–20.7	<0.5	-	[49]
	-	22	-	-	-	-	-	-	-	[34]
	2	-	-	-	-	23	-	-	-	[5]
		-	-	-	-	23.96	-	1.63	-	[40]
Tomato (concentrated)	19	-	-	-	-	25.9 (8.2)	7.9–41.1	-	-	[38]
Tomato (crushed)	3	1.22 (1.69)	0.24–3.17	0.14 (0.02)	0.12–0.16	9.66 (8.78)	4.5–19.80	0.18 (0.06)	0.12–0.23	#
	-	2	-	2	-	20	-	-	-	[34]
Tomato (ketchup)	3	0.37 (0.64)	nd–1.11	nd	-	1.07 (0.08)	1–1.15	nd	-	#
	24	22	-	-	-	52.5 (54.1)	nd–165	-	-	[38]
	-	-	-	-	-	-	-	-	-	[12]
Legumes and Derivatives										
Beans (white)	6	nd	-	nd	-	0.66 (0.64)	0.35–1.96	nd	-	#
Beans (red kidney)	3	-	-	2	-	3	-	-	-	[34]
	5	-	-	-	-	-	0.3–0.4	-	-	[50]
		-	-	-	-	-	nd–4	-	-	[52]
Chickpeas	4	-	-	3	-	1	-	-	-	[34]
		nd	-	nd	-	3.63 (2.49)	0.90–6.39	nd	-	#
Lentils	7	nd	-	<0.50	-	2	-	<0.50	-	[34]
	5	-	-	-	-	8.19 (8.36)	1.96–21.81	-	-	[52]
		-	-	-	-	3	nd–20.2	-	-	[34]
Peanuts	7	nd	-	nd	-	0.87 (1.01)	nd–2.56	nd	-	#
Peas	9	nd	-	nd	-	34.28 (13.50)	8.74–54.44	nd	-	#
	10	-	-	-	-	17.3	-	-	-	[48]
	6	-	-	-	-	32.3	5.5–51.1	-	-	[49]
	14	-	-	-	-	46.3 (27)	11.7–107	-	-	[38]
Peas (frozen)	3	-	-	-	-	17	1.6–6.5	-	-	[50]
Soybean, dried	1	-	-	-	-	41	-	-	-	[47]
	2	-	-	-	-	-	-	-	-	[51]
	13	-	-	-	-	30.9 (15.5)	3.7–16.8	-	-	[55]
	4	-	-	-	-	-	16.3–57	-	-	[52]
	2	-	-	-	-	-	35.2–57.2	-	-	[40]
	5	-	-	-	-	17.1	6.4–24.2	-	-	[49]
Soybean milk	3	nd	-	nd	-	1.02 (0.73)	0.39–1.81	0.28 (0.24)	nd–0.42	[39]
	2	-	-	-	-	2.11	-	13.9	-	[40]

Table 1. *Cont.*

Food Categories	n	Histamine		Tyramine		Putrescine		Cadaverine		Reference
		Mean (SD)	Range	Mean (SD)	Range	Mean (SD)	Range	Mean (SD)	Range	
Vegetables and Vegetable Products										
Soybean sprouts	3	-	-	-	-	44.71 (3.21)	41.13–47.43	0.21 (0.18)	nd–0.33	[39]
Tofu	6	nd	-	nd	-	0.76 (0.55)	nd–1.49	0.67 (0.49)	nd–1.42	#
	4	-	-	-	-	nd	-	-	-	[49]
	19	-	-	-	-	2.6 (1.4)	nd–5	-	-	[56]
	2	-	-	-	-	1.76	-	18.4	-	[40]

Here, *n*: number of samples; SD: standard deviation; nd: not detected; -: values not reported by the study; #: data on the biogenic amine content from our own database of Spanish market products.

Table 2. Biogenic amine contents (mg/kg fresh weight) found in fruits and nuts. Data are presented as average (standard deviation) and range (minimum–maximum).

Food Categories	n	Histamine		Tyramine		Putrescine		Cadaverine		Reference
		Mean (SD)	Range	Mean (SD)	Range	Mean (SD)	Range	Mean (SD)	Range	
Fruit and Fruit Products										
Apple	3	-	-	-	-	-	0.4–1.7	-	-	[50]
	2	-	-	-	-	nd	-	-	-	[51]
	2	-	-	-	-	1.5	-	nd	-	[40]
Apple juice	10	nd	-	0.67 (0.50)	nd–1.6	1.02 (0.35)	0.59–1.68	2.30 (1.53)	0.55–4.27	[41]
Avocado	5	nd	-	1.81 (2.06)	0.58–5.44	nd	-	nd	-	#
	-	23	-	-	-	-	-	-	-	[12]
	2	-	-	-	-	nd	-	nd	-	[40]
Banana	8	nd	-	0.53 (0.79)	nd–1.85	37.94 (8.32)	25.50–49.49	nd	-	#
	2	-	-	-	-	-	15.86–41.05	10.83	-	[40]
Cherry	5	nd	-	nd	-	3.08 (0.51)	2.35–3.46	nd	-	#
	2	-	-	-	-	4.67	-	nd	-	[40]
Grape	2	-	-	-	-	9.34	-	5.93	-	[40]
Grapefruit	2	nd	-	nd	-	55.55 (12.8)	46.52–64.57	nd	-	#
	2	-	-	-	-	51.1	-	nd	-	[40]
	3	-	-	-	-	98.6	-	-	-	[50]
Grapefruit Juice	10	0.31 (0.58)	nd–1.74	nd	-	10.08 (4.11)	7.17–20.8	1 (0.64)	0.38–2.28	[41]
Guava	21	-	-	-	-	1	0.4–1.8	-	-	[57]
Kiwi	13	nd	-	nd	-	2.49 (3.96)	0.5–15.57	nd	-	#
	2	-	-	-	-	1.06	-	nd	-	[40]
Lemon	-	1.9 (0.1)	-	-	-	3.1 (0.1)	-	nd	-	[54]
	3	nd	-	nd	-	2.33 (2.02)	nd–3.67	nd	-	#
Mandarin	21	nd	-	0.94 (1.31)	nd–5.76	90.16 (36.6)	12.29–173.8	nd	-	#
	10	-	-	-	-	122 (44.2)	67.3–200	-	-	[42]

Table 2. *Cont.*

Occurrence of Biogenic Amines (mg/kg)

Food Categories	n	Histamine Mean (SD)	Histamine Range	Tyramine Mean (SD)	Tyramine Range	Putrescine Mean (SD)	Putrescine Range	Cadaverine Mean (SD)	Cadaverine Range	Reference
Fruit and Fruit Products										
Mango	21	-	-	-	-	0.9	nd-2.7	-	-	[57]
Orange	12	nd	-	nd	-	91.24 (41.7)	11.34-151.1	nd	-	#
	3	-	-	-	-	117	95.1-140	-	-	[50]
	2	-	-	-	-	-	-	-	-	[51]
	5	-	-	-	-	137 (11.3)	119-153	-	-	[42]
	2	-	-	-	-	-	54.62-119.82	2.04	-	[40]
Orange juice	3	nd	-	nd	-	45.51 (10.5)	37.35-57.3	nd	-	#
	3	-	-	-	-	85 (11.4)	76.6-100	-	-	[42]
Papaya	11	0.46 (0.41)	nd-1.32	nd	-	45.51 (8.35)	34.70-60.97	nd	-	[41]
	21	-	-	-	-	4.67	5.3-19.3	-	-	[57]
Passion fruit	21	nd	-	nd	-	17.9	6.5-40.5	nd	-	[40]
Pear	3	-	-	-	-	-	23.6-24.2	-	-	[57]
	2	-	-	-	-	1.5	-	0.41	-	[50]
Peach	2	nd	-	nd	-	1.92 (0.14)	1.82-2.02	nd	-	[40]
	2	-	-	-	-	0.35	-	<0.10	-	#
Pineapple	6	nd	-	nd	-	4.20 (2.17)	1.39-7.96	nd	-	[40]
	2	-	-	-	-	4.05	-	3.07	-	#
	21	-	-	-	-	1.1	nd-2.5	-	-	[57]
Pineapple juice	12	2.44 (1.59)	nd-4.61	0.87 (0.86)	nd-1.93	1.79 (0.16)	1.53-1.98	1.21 (1.22)	nd-3.14	[41]
Plum	2	nd	-	4.02 (4.32)	0.96-7.07	nd	-	nd	-	#
Strawberry	9	nd	-	nd	-	3.77 (1.52)	2.04-6.42	nd	-	#
	2	-	-	-	-	0.97	-	4.29	-	[40]
Nuts										
Almonds	7	nd	-	nd	-	2.47 (1.24)	nd-4.36	nd	-	#
	2	-	-	-	-	4.32	-	5.57	-	[40]
Chestnuts	2	nd	-	nd	-	4.53 (3.40)	2.12-6.93	nd	-	#
	2	-	-	-	-	5.2	-	1.33	-	[40]
Hazelnuts	9	nd	-	0.49 (0.85)	nd-2.63	1.18 (1.09)	nd-3.19	nd	-	#
Nuts	6	nd	-	nd	-	5.64 (4.17)	2.82-13.79	nd	-	#
Pistachios	7	nd	-	nd	-	14.84 (14.0)	4.31-39.51	1.65 (4.37)	nd-11.58	#
	2	-	-	-	-	43	-	3.27	-	[40]
Sunflower seeds	2	nd	-	nd	-	0.50 (0.19)	0.36-0.63	nd	-	#
	2	-	-	-	-	3	-	22.58	-	[40]

Here, *n*: number of samples; SD: standard deviation; nd: not detected; -: values not reported by the study; #: data on the biogenic amine content from our own database of Spanish market products.

Table 3. Biogenic amine contents (mg/kg fresh weight) found in cereals and cereal-based products. Data are presented as average (standard deviation) and range (minimum–maximum).

Food Categories	n	Occurrence of Biogenic Amines (mg/kg)								Reference
		Histamine		Tyramine		Putrescine		Cadaverine		
		Mean (SD)	Range	Mean (SD)	Range	Mean (SD)	Range	Mean (SD)	Range	
Barley	2	nd	-	nd	-	2.19 (1.55)	1.09–3.28	nd	-	#
Bread, white	2	nd	-	nd	-	nd	-	nd	-	#
	3	-	-	-	-	-	1.5–1.8	-	-	[50]
	10	-	-	-	-	1.1	-	-	-	[48]
	2	-	-	-	-	1.32	-	2.35	-	[40]
Bread, wholemeal	6	nd	-	nd	-	1.96 (1.45)	nd–4.32	nd	-	#
Cereal (corn, chocolate)	8	nd	-	nd	-	0.32 (0.44)	nd–0.93	nd	-	#
	10	-	-	-	-	-	2–2.2	-	-	[50]
Oats	2	nd	-	nd	-	0.67 (0.32)	0.44–0.89	nd	-	#
Pasta (wheat)	7	nd	-	nd	-	1.56 (1.65)	0.81–4.52	nd	-	#
Rice	2	nd	-	nd	-	2.4 (0.03)	2.38–2.42	nd	-	#
	2	-	-	-	-	<0.9	-	-	-	[51]
	2	-	-	-	-	1.2	-	-	-	[40]
	6	-	-	-	-	0.2	0.2–0.3	-	-	[49]
	10	-	-	-	-	0.2	-	-	-	[48]
Wheat germ	2	nd	-	nd	-	31.64 (0.35)	31.39–31.9	0.63 (0.08)	0.57–0.69	#
	2	-	-	-	-	62.1	-	-	-	[40]

Here, n: number of samples; SD: standard deviation; nd: not detected; -: values not reported by the study; #: data on the biogenic amine content from our own database of Spanish market products.

Putrescine contents in wholemeal bread were slightly higher than in bread made with refined flour. In white bread, low contents of cadaverine have also been reported, although only in one study, and from the analysis of two samples.

4. Evolution of Amine Contents during Storage and Cooking

The variability of amine contents observed among samples of the same product can be attributed mainly to conditions of production, transport, and storage [42].

The storage temperature is one of the most important factors in the formation of biogenic amines [11,29]. Refrigeration delays or reduces the aminogenic potential of microorganisms, although the formation of amines at refrigeration temperatures (4–10 °C) has been reported. The influence of the conservation temperature has been widely studied in foods such as meats, fish, and fermented products [29,30,58], but scarcely in plant-origin foods.

A study conducted by Simon-Sarkadi et al. [59] showed a clear increase in putrescine in different types of green leafy vegetables (lettuce, endives, Chinese cabbage, and radicchio) during six days of storage at 5 °C. The authors concluded that there was a positive correlation between putrescine contents and the hygienic state of these foods (total microorganism counts). Tyramine contents also showed a tendency to increase slightly. Histamine was present only in Chinese cabbage and in very low concentrations, remaining stable throughout the study period. In contrast, when Moret et al. [34] studied the effect of storage temperature on the amine content in various vegetables (parsley, zucchini, broccoli, and cucumber), no significant changes in histamine, tyramine, putrescine, and cadaverine were observed after three weeks of refrigeration.

Lavizzari et al. [32] also reported an increase in histamine in different spinach samples over 12–15 days of storage at 6 °C, noting that the relatively high pH of this vegetable favored the growth of Gram-negative bacteria, which could have been responsible for the formation of this amine during storage. The contents of tyramine and putrescine did not undergo significant changes under these storage conditions. It should be noted that in two of the five trials carried out in this study, histamine levels decreased in the last days of storage. The authors suggested that this histamine degradation could have been due to the action of bacteria with DAO activity, as well as the effect of the pH, which reached values above 8 [32]. Another study also recently reported the complete degradation of histamine in a spinach sample (61 mg/kg) after three weeks of storage at 4 °C [54].

Modified atmosphere packaging, together with low storage temperatures, is commonly used to extend the life of fresh vegetables and fruits. This type of packaging can influence the capacity of microorganisms to form amines [30,58,60]. Esti et al. [43] monitored the contents of amines during the ripening of cherries and apricots packaged in modified atmospheres and stored at 0 °C, and found that after 20 days of storage the contents of amines (mainly putrescine) had decreased by 20% compared to the initial value. Although the authors did not provide an explanation for this reduction, it could have been due to putrescine serving as a substrate for polyamine formation [36].

Another factor that can affect the content of biogenic amines in foods of plant origin, especially vegetables, is the culinary process. Again, the results reported in the literature were variable, depending on the type of cooking and the amine in question.

Latorre-Moratalla et al. [61] evaluated the effect of cooking spinach in water, with or without salt. The cooking process reduced the histamine content in all the samples by an average of 83% with respect to the raw product (after a correction for the dilution effect of the cooking). Analysis confirmed a transfer of histamine to the cooking water, which was not enhanced by the addition of salt. Likewise, Kumar et al. [33] observed the loss of 11–14% histamine in eggplants boiled at 100 °C for 10 min. Veciana et al. [62] also concluded that the putrescine content in certain vegetables (spinach, cauliflower, Swiss chard, potato, and green beans) is reduced by transfer to the cooking water. However, this heat treatment had no effect on the putrescine content in other vegetables such as pepper, pea, and asparagus. Eliassen et al. [42] also found no significant differences in putrescine levels among different types of raw and boiled vegetables (carrot, broccoli, cauliflower, and potato), although they acknowledged that the low number of samples analyzed (two per food) was a limitation when trying to reach a conclusion.

Conversely, three recent studies have shown an increase in amine levels after a cooking process. According to Lo Scalzo et al. [63], boiling and grilling enhanced the putrescine content in a specific variety of eggplant by 55% and 32%, respectively. In the other two varieties of eggplant tested, the cooking had no effect. Similarly, Preti et al. [46] reported a significant increase in putrescine in green beans after boiling, whereas steaming did not modify the contents. According to the work performed by Chung et al. [64], frying brought about a 2.5- and 4-fold increase in histamine in carrots and seaweed, respectively. The authors attributed this increase to the loss of water caused by the high heat treatment. The same process had no effect on spinach and onions. However, it should be noted that in this study, the contents of histamine in all foods were well below 1 mg/kg, both before and after frying.

Amines are thermostable compounds, so in principle changes in contents can only be due to their transfer to the cooking water or by dilution or concentration effects of the culinary process, in which the food gains or loses water.

5. Plant-Origin Foods in Low-Histamine Diets

At present, the main strategy to prevent the onset of histamine intolerance symptoms is to follow a low-histamine diet. Its efficacy has been demonstrated in different clinical studies, which have always described an improvement or remission of gastrointestinal, dermatological, and neurological symptoms [3,6,18–20,22,24,27,65–67] if the diet was followed.

Current low-histamine diets exclude foods that patients associate with the onset of symptoms [1–8], such as blue fish and their preserves, and all kinds of fermented products (cheeses, sausages, wine, beer, sauerkraut, and fermented soy derivatives), all of which are susceptible to having high contents of histamine and other amines. A high number of nonfermented plant-origin foods are also excluded: The average contents of biogenic amines and polyamines in these foods are shown in Table 4. As can be seen, with the exception of spinach, eggplant, tomatoes, and avocado, for which high amounts of histamine have been described, the rest contained very little or no histamine, so a priori should not be responsible for triggering symptoms. However, some of them had relatively high contents of other biogenic amines and polyamines.

Table 4. Content of histamine and other biogenic amines (mg/kg fresh weight) in plant-origin foods excluded from different low-histamine diets [1–8]. Data obtained from own database and from different scientific studies [5,12,34,38–42,47–55,57].

Food Items	Histamine	Putrescine	Cadaverine	Tyramine	Spermidine	Spermine
Spinach [a]	9–70	nd–119	nd–9	1–10	14–53	nd–9
Eggplant [a]	4–101	24–49	nd–5	nd–2	2–12	nd–6
Tomato [a]	nd–17	5–122	nd–2	nd–6	2–16	nd–2
Ketchup [a]	nd–22	nd–165	nd	nd	nd–33	nd–12
Avocado [a]	nd–23	nd	nd	0.5–5	nd–7	2–8
Citrus (fresh and juices) [b]	nd–2	7–200	nd–2	nd–5	nd–12	nd–5
Mushroom [b]	nd	nd–156	nd	nd	9–155	nd–13
Banana [b]	nd	15–50	nd–10	nd–2	8–16	nd–3
Soybean or soybean sprouts [b]	nd	2–57	nd–0.3	nd	33–389	7–114
Nuts [b]	nd	nd–40	nd–23	nd–3	6–40	2–33
Pears [b]	-	2–25	nd–0.4	-	30–76	8–49
Lentils [b]	nd	nd–21	nd	nd	15–107	5–18
Chickpeas [b]	nd	1–6	nd–0.5	nd–0.5	15–85	4–32
Peanuts [c]	nd	nd–3	nd	nd	23–48	5–13
Kiwi [c]	nd–2	nd–15	nd	nd	3–6	nd–2
Papaya [c]	-	5–20	nd	-	4–8	nd–2
Strawberry [c]	nd	2–6	nd–4	nd	5–10	nd–2
Pineapple [c]	nd	nd–8	nd–3	nd	nd–3	nd–1
Plum [c]	nd	nd	nd	1–7	2–3	nd–4

Here, nd: not detected; -: values not reported by the studies; [a] plant-origin foods with histamine; [b] plant-origin foods without histamine but with high contents of other amines; [c] plant-origin foods with low levels of all amines.

Putrescine, cadaverine, and tyramine are all substrates of the DAO enzyme, so if present in high amounts they may increase the adverse effects of histamine by competing as rival substrates or for binding sites in the intestinal mucosa [1,9,68,69]. The high putrescine contents found in citrus fruits, mushrooms, soybeans, bananas, and nuts could thus explain why patients associate their consumption with the onset of histamine intolerance symptoms. However, it should be noted that some foods with similar or even much higher putrescine contents, such as green pepper, peas, or corn, are permitted in low-histamine diets (Table 1).

The polyamines spermidine and spermine can also be metabolized by DAO, albeit with a lower affinity [68,69], and therefore their presence should also be taken into account in this type of diet (Table 4). Thus, the exclusion of foods such as soybeans, mushrooms, lentils, chickpeas, peanuts, nuts, and pears may be justified by their high polyamine content.

Finally, the levels of biogenic amines and polyamines found in kiwi, papaya, strawberry, pineapple, and plum are too low to justify their exclusion. Some authors consider these foods, along with others such as milk, shellfish, and eggs, as endogenous histamine releasers, although by mechanisms still not well understood [1,11,70].

6. Conclusions

Biogenic amine data in nonfermented plant-origin foods from the different reviewed studies showed a great variability both within the same food item and among them. Putrescine was the most frequent biogenic amine found in fresh vegetables, legumes, fruits, and cereals, and only a limited number of products contained relevant levels of histamine (eggplant, spinach, tomato, and avocado). Tyramine and cadaverine were usually more scarcely found in plant-origin foods. Generally, low levels of histamine and putrescine may have a physiological origin. However, undesirable microbial enzymatic activity during production or storage may lead to the accumulation of high levels of these amines.

No single trend has emerged in the evolution of amine contents during refrigerated storage, which might be at least partly due to the different experimental designs of the studies. In some cases, refrigeration seems to have prevented the formation of certain amines, but this remains a hypothesis, as no study performed a comparative analysis of samples stored under refrigeration and at room temperature. The increase in the biogenic amine content during refrigerated storage reported by other authors may be attributed to bacterial activity. Additionally, some studies have observed an influence of culinary process on the biogenic amine content, mainly derived from the transfer of these compounds to the boiling water or by dilution or concentration effects of the applied treatment.

The exclusion of a high number of plant-origin foods from low-histamine diets cannot be accounted for by their histamine contents, but is more likely due to high levels of putrescine or spermidine. The plant-origin foods consumed by people with histamine intolerance should be of maximum freshness, since histamine and other amines may continue to form during refrigerated storage. The cooking of vegetables in water (boiling) is another relevant strategy for this population, since it can reduce the contents of histamine and other amines in the food.

Author Contributions: Conceptualization, M.T.V.-N., M.L.L.-M., and M.C.V.-C.; Investigation, S.S.-P., O.C.-B., and J.R.-G.; Writing—Original draft preparation, S.S.-P., O.C.-B., and M.L.L.-M.; Writing—Review and editing, M.T.V.-N., M.L.L.-M., and M.C.V.-C.; Supervision, M.C.V.-C.

Acknowledgments: Sònia Sánchez-Pérez is a recipient of a doctoral fellowship from the University of Barcelona (APIF2018).

References

1.   Maintz, L.; Novak, N. Histamine and histamine intolerance. *Am. J. Clin. Nutr.* **2007**, *85*, 1185–1196. [CrossRef] [PubMed]

2.   Böhn, L.; Störsrud, S.; Törnblom, H.; Bengtsson, U.; Simrén, M. Self-reported food-related gastrointestinal

symptoms in IBS are common and associated with more severe symptoms and reduced quality of life. *Am. J. Gastroenterol.* **2013**, *108*, 634–641. [CrossRef] [PubMed]

3. Rosell-Camps, A.; Zibetti, S.; Pérez-Esteban, G.; Vila-Vidal, M.; Ferrés Ramis, L.; García-Teresa-García, E. Intolerancia a la histamina como causa de síntomas digestivos crónicos en pacientes pediátricos. *Rev. Esp. Enferm. Dig.* **2013**, *105*, 201–207. [CrossRef] [PubMed]

4. Veciana-Nogués, M.T.; Vidal-Carou, M.C. Dieta baja en histamina. In *Nutrición y dietética clínica*, 3rd ed.; Salas-Salvadó, J., Bonada-Sanjaume, A., Trallero-Casaña, R., Saló-Solà, M., Burgos-Peláez, R., Eds.; Ediciones Elsevier España: Barcelona, Spain, 2014; pp. 443–448.

5. Lefèvre, S.; Astier, C.; Kanny, G. Histamine intolerance or false food allergy with histamine mechanism. *Rev. Fr. Allergol.* **2016**, *57*, 24–34. [CrossRef]

6. Wagner, N.; Dirk, D.; Peveling-Oberhag, A.; Reese, I.; Rady-Pizarro, U.; Mitzel, H.; Staubach, P.A. Popular myth—Low-histamine diet improves chronic spontaneous urticaria—Fact or fiction? *J. Eur. Acad. Dermatol. Venereol.* **2017**, *31*, 650–655. [CrossRef] [PubMed]

7. Ede, G. Histamine intolerance: Why freshness matters? *J. Evol. Health* **2017**, *2*, 11. [CrossRef]

8. Swiss Interest Group Histamine Intolerance (SIGHI)—Leaflet Histamine Elimination Diet. Available online: http://www.histaminintoleranz.ch/downloads/SIGHI-Leaflet_HistamineEliminationDiet.pdf (accessed on 25 October 2018).

9. EFSA Panel on Biological Hazards (BIOHAZ). Scientific opinion on risk based control of biogenic amines formation in fermented foods. *EFSA J.* **2011**, *9*, 2393. [CrossRef]

10. World Health Organization and Food and Agriculture Organization of the United Nations. *Joint FAO/WHO Expert Meeting on the Public Health Risks of Histamine and other Biogenic Amines from Fish and Fishery Products: Meeting Report*; World Health Organization: Geneva, Switzerland, 2013.

11. Kovacova-Hanuskova, E.; Buday, T.; Gavliakova, S.; Plevkova, J. Histamine, histamine intoxication and intolerance. *Allergol. Immunopathol.* **2015**, *43*, 498–506. [CrossRef]

12. Jarisch, R.; Wantke, F.; Raithel, M.; Hemmer, W. Histamine and biogenic amines. In *Histamine Intolerance. Histamine and Seasickness*; Jarisch, R., Ed.; Springer: Stuttgart, Germany, 2014; pp. 3–44.

13. Comas-Basté, O.; Latorre-Moratalla, M.L.; Bernacchia, R.; Veciana-Nogués, M.T.; Vidal-Carou, M.C. New approach for the diagnosis of histamine intolerance based on the determination of histamine and methylhistamine in urine. *J. Pharm. Biomed. Anal.* **2017**, *14*, 379–385. [CrossRef]

14. Maintz, L.; Yu, C.F.; Rodríguez, E.; Baurecht, H.; Bieber, T.; Illig, T.; Weidinger, S.; Novak, N. Association of single nucleotide polymorphisms in the diamine oxidase gene with diamine oxidase serum activities. *Allergy* **2011**, *66*, 893–902. [CrossRef]

15. García-Martín, E.; Martínez, C.; Serrador, M.; Alonso-Navarro, H.; Ayuso, P.; Navacerrada, F.; Agúndez, J.A.G.; Jiménez-Jiménez, F.J. Diamine oxidase rs10156191 and rs2052129 variants are associated with the risk for migraine. *Headache* **2015**, *55*, 276–286. [CrossRef] [PubMed]

16. Meza-Velázquez, R.; López-Márquez, F.; Espinosa-Padilla, S.; Rivera-Guillen, M.; Ávila-Hernández, J.; Rosales-González, M. Association of diamine oxidase and histamine N-methyltransferase polymorphisms with presence of migraine in a group of Mexican mothers of children with allergies. *Neulología* **2017**, *32*, 500–507. [CrossRef]

17. Sattler, J.; Häfner, D.; Klotter, H.J.; Lorenz, W.; Wagner, P.K. Food-induced histaminosis as an epidemiological problem: Plasma histamine elevation and haemodynamic alterations after oral histamine administration and blockade of diamine oxidase (DAO). *Agent Action* **1988**, *23*, 361–365. [CrossRef]

18. Steinbrecher, I.; Jarisch, R. Histamine and headache. *Allergologie* **2005**, *28*, 85–91. [CrossRef]

19. Maintz, L.; Benfadal, S.; Allam, J.P.; Hagemann, T.; Fimmers, R.; Novak, N. Evidence for a reduced histamine degradation capacity in a subgroup of patients with atopic eczema. *J. Allergy Clin. Immunol.* **2006**, *117*, 1106–1112. [CrossRef] [PubMed]

20. Worm, M.; Fielder, E.; Döle, A.S.; Schink, T.; Hemmer, W.; Jarisch, R.; Suberbier, T. Exogenous histamine aggravates eczema in a subgroup of patients with atopic dermatitis. *Acta Derm. Venereol.* **2009**, *89*, 52–56. [CrossRef]

21. Honzawa, Y.; Nakase, H.; Matsuura, M.; Chiba, T. Clinical significance of serum diamine oxidase activity in inflammatory bowel disease: Importance of evaluation of small intestinal permeability. *Inflamm. Bowel Dis.* **2011**, *17*, E23–E25. [CrossRef] [PubMed]

22. Mušič, E.; Korošec, P.; Šilar, M.; Adamič, K.; Košnik, M.; Rijavec, M. Serum diamine oxidase activity as a diagnostic test for histamine intolerance. *Wien. Klin. Wochenschr.* **2013**, *12*, 239–243. [CrossRef]

23. Manzotti, G.; Breda, D.; Gioacchino, M.; Burastero, S.E. Serum diamine oxidase activity in patients with histamine intolerance. *Int. J. Immunopath. Pharmacol.* **2016**, *29*, 105–111. [CrossRef]

24. Hoffmann, M.; Gruber, E.; Deutschmann, A.; Jahnel, J.; Hauer, A. Histamine intolerance in children with chronic abdominal pain. *Arch. Dis. Child.* **2016**, *98*, 832–833. [CrossRef]

25. Pinzer, T.C.; Tietz, E.; Waldmann, E.; Schink, M.; Neurath, M.F.; Zopf, Y. Circadian profiling reveals higher histamine plasma levels and lower diamine oxidase serum activities in 24% of patients with suspected histamine intolerance compared to food allergy and controls. *Allergy* **2018**, *73*, 949–957. [CrossRef] [PubMed]

26. Kacik, J.; Wróblewska, B.; Lewicki, S.; Zdanowski, R.; Kalicki, B. Serum diamine oxidase in pseudoallergy in the pediatric population. *Adv. Exp. Med. Biol.* **2018**, *1039*, 35–44. [CrossRef] [PubMed]

27. Son, J.H.; Chung, B.Y.; Kim, H.O.; Park, C.W. A histamine-free diet is helpful for treatment of adult patients with chronic spontaneous urticaria. *Ann. Dermatol.* **2018**, *30*, 164–172. [CrossRef] [PubMed]

28. Izquierdo-Casas, J.; Comas-Basté, O.; Latorre-Moratalla, M.L.; Lorente-Gascón, M.; Duelo, A.; Vidal-Carou, M.C.; Soler-Singla, L. Low serum diamine oxidase (DAO) activity levels in patients with migraine. *J. Physiol. Biochem.* **2018**, *74*, 93–99. [CrossRef] [PubMed]

29. Bover-Cid, S.; Latorre-Moratalla, M.L.; Veciana-Nogués, M.T.; Vidal-Carou, M.C. Processing contaminants: Biogenic amines. In *Encyclopedia of Food Safety*; Motarjemi, Y., Moy, G.G., Todd, E.C.D., Eds.; Elsevier Inc.: Burlington, MA, USA, 2014; Volume 2, pp. 381–391.

30. Gardini, F.; Özogul, Y.; Suzzi, G.; Tabanelli, G.; Özogul, F. Technological factors affecting biogenic amine content in foods: A review. *Front. Microbiol.* **2016**, *7*, 1218. [CrossRef] [PubMed]

31. Vidal-Carou, M.C.; Veciana-Nogués, M.T.; Latorre-Moratalla, M.L.; Bover-Cid, S. Biogenic amines: Risks and control. In *Handbook of Fermented Meat and Poultry*, 2nd ed.; Toldrá, F., Hui, Y.H., Astiasarán, I., Sebranek, J.G., Talon, R., Eds.; Wiley-Blackwell: Hoboken, NJ, USA, 2014; pp. 413–428.

32. Lavizzari, T.; Veciana-Nogués, M.T.; Weingart, O.; Bover-Cid, S.; Mariné-Font, A.; Vidal-Carou, M.C. Occurrence of biogenic amines and polyamines in spinach and changes during storage under refrigeration. *J. Agric. Food Chem.* **2007**, *55*, 9514–9519. [CrossRef] [PubMed]

33. Kumar, M.N.K.; Bbu, B.N.H.; Venkayesh, Y.P. Higher histamine sensitivity in non-atopic subjects by skin prick test may result in misdiagnosis of eggplant allergy. *Immunol. Investig.* **2009**, *38*, 93–103. [CrossRef]

34. Moret, S.; Smela, D.; Populin, T.; Conte, L. A survey on free biogenic amine content of fresh and preserved vegetables. *Food Chem.* **2005**, *89*, 355–361. [CrossRef]

35. Trivedi, K.; Borkovcová, I.; Karpíšková, R. Tyramine production by enterococci from various foodstuffs: A threat to the consumers. *Czech J. Food Sci.* **2009**, *27*, S357–S360. [CrossRef]

36. Kalač, P. Health effects and occurrence of dietary polyamines: A review for the period 2005–mid 2013. *Food Chem.* **2014**, *161*, 27–39. [CrossRef]

37. Bouchereau, A.; Aziz, A.; Larher, F.; Martin-Tanguy, J. Polyamines and environmental challenges: Recent development. *Plant Sci.* **1999**, *140*, 103–125. [CrossRef]

38. Kalač, P.; Svecová, S.; Pelikánová, T. Levels of biogenic amines in typical vegetable products. *Food Chem.* **2002**, *77*, 349–351. [CrossRef]

39. Toro-Funes, N.; Bosch-Fuste, J.; Latorre-Moratalla, M.L.; Veciana-Nogués, M.T.; Vidal-Carou, M.C. Biologically active amines in fermented and non-fermented commercial soybean products from the Spanish market. *Food Chem.* **2015**, *173*, 1119–1124. [CrossRef] [PubMed]

40. Nishimura, K.; Shiina, R.; Kashiwagi, K.; Igarashi, K. Decrease in polyamines with aging and their ingestion from food and drink. *J. Biochem.* **2006**, *139*, 81–90. [CrossRef] [PubMed]

41. Preti, R.; Bernacchia, R.; Vinci, G. Chemometric evaluation of biogenic amines in commercial fruit juices. *Eur. Food Res. Technol.* **2016**, *242*, 2031–2039. [CrossRef]

42. Eliassen, K.A.; Reistad, R.; Risøen, U.; Rønning, H.F. Dietary polyamines. *Food Chem.* **2002**, *78*, 273–280. [CrossRef]

43. Esti, M.; Volpe, G.; Masignan, D.; Compagnone, E.; La Notte, E.; Palleschi, G. Determination of amines in fresh and modified atmosphere packaged fruits using electrochemical biosensors. *J. Agric. Food Chem.* **1998**, *46*, 4233–4237. [CrossRef]

44. Kalač, P.; Křížek, M.; Pelikánová, T.; Langová, M.; Veškrna, O. Contents of polyamines in selected foods. *Food Chem.* **2005**, *90*, 561–564. [CrossRef]

45. González-Aguilar, G.A.; Zacarias, L.; Perez-Amador, M.A.; Carbonell, J.; Lafuente, M.T. Polyamine content and chilling susceptibility are affected by seasonal changes in temperature and by conditioning temperature in cold-stored "Fortune" mandarin fruit. *Physiol. Plant* **2000**, *108*, 140–146. [CrossRef]

46. Preti, R.; Rapa, M.; Vinci, G. Effect of Steaming and boiling on the antioxidant properties and biogenic amines content in Green Bean (*Phaeseolus vulgaris*) varieties of different colours. *J. Food Quality* **2017**. [CrossRef]

47. Ziegler, W.; Hahn, M.; Wallnöfer, P.R. Changes in biogenic amine contents during processing of several plant foods. *Deut. Lebensm. Rundsch.* **1994**, *90*, 108–112.

48. Cipolla, B.G.; Havouis, R.; Moulinoux, J.P. Polyamine contents in current foods: A basis for polyamine reduced diet and a study of its long-term observance and tolerance in prostate carcinoma patients. *Amino Acids* **2007**, *33*, 203–212. [CrossRef] [PubMed]

49. Nishibori, N.; Fujihara, S.; Akatuki, T. Amounts of polyamines in foods in Japan and intake by Japanese. *Food Chem.* **2007**, *100*, 491–497. [CrossRef]

50. Bardócz, S.; Grant, G.; Brown, D.S.; Ralph, A.; Pusztai, A. Polyamines in food—Implications for growth and health. *J. Nutr. Biochem.* **1993**, *4*, 66–71. [CrossRef]

51. Okamoto, A.; Sugi, E.; Koizumi, Y.; Yanadiga, F.; Udaka, S. Polyamine content of ordinary foodstuffs and various fermented foods. *Biosci. Biotechnol. Biochem.* **1997**, *61*, 1582–1584. [CrossRef] [PubMed]

52. Kalač, P.; Krausová, P. A review of dietary polyamines: Formation, implications for growth and health and occurrence in foods. *Food Chem.* **2005**, *90*, 219–230. [CrossRef]

53. Dadáková, E.; Pelikánová, T.; Kalač, P. Content of biogenic amines and polyamines in some species of European wild-growing edible mushrooms. *Eur. Food Res. Technol.* **2009**, *230*, 163–171. [CrossRef]

54. Dionex. Determination of biogenic amines in fruit, vegetables, and chocolate using ion chromatography with suppressed, conductivity and integrated pulsed amperometric detections. *Appl. Update* **2016**, *162*, 1–8.

55. Glória, M.B.A.; Tavares-Neto, J.; Labanca, R.A.; Carvalho, M.S. Influence of cultivar and germination on bioactive amines in soybeans (*Glycine max* L. Merril). *J. Agric. Food Chem.* **2005**, *53*, 7480–7485. [CrossRef]

56. Byun, B.Y.; Bai, X.; Mah, J.H. Occurrence of biogenic amines in Doubanjiang and Tofu. *Food Sci. Biotech.* **2013**, *22*, 55–62. [CrossRef]

57. Santiago-Silva, P.; Labanca, R.; Gloria, B. Functional potential of tropical fruits with respect to free bioactive amines. *Food Res. Int.* **2011**, *44*, 1264–1268. [CrossRef]

58. Naila, A.; Flint, S.; Fletcher, G.; Bremer, P.; Meerdink, G. Control of biogenic amines in food-existing and emerging approaches. *J. Food Sci.* **2010**, *75*, 139–150. [CrossRef] [PubMed]

59. Simon-Sarkadi, L.; Holazapfel, W.H.; Halasz, A. Biogenic amine content and microbial contamination of leafy vegetables during storage at 5C. *J. Food Biochem.* **1994**, *17*, 407–418. [CrossRef]

60. Chong, C.Y.; Abu Bakar, F.; Russly, A.R.; Jamilah, B.; Mahyudin, N.A. The effects of food processing on biogenic amines formation. *Int. Food Res. J.* **2011**, *18*, 867–876.

61. Latorre-Moratalla, M.L.; Comas-Basté, O.; Veciana-Nogués, M.T.; Vidal-Carou, M.C. La cocción reduce el contenido de histamina de las espinacas. In *11a Reunión anual de la Sociedad Española de Seguridad Alimentaria*; Sociedad Española de Seguridad Alimentaria: Pamplona, Spain, 2015.

62. Veciana-Nogués, M.T.; Latorre-Moratalla, M.L.; Toro-Funes, N.; Bosh-Fusté, J.; Vidal-Carou, M.C. Efecto de la cocción con y sin sal en el contenido de poliaminas de las verduras. *Nutr. Hosp.* **2014**, *30*, 53.

63. Lo Scalzo, R.; Fibiani, M.; Francese, G.; D'Alessandro, A.; Rotino, G.L.; Conte, P.; Mennella, G. Cooking influence on physico-chemical fruit characteristics of eggplant (*Solanum melongena* L.). *Food Chem.* **2016**, *194*, 835–842. [CrossRef] [PubMed]

64. Chung, B.Y.; Park, S.Y.; Byun, Y.S.; Son, J.H.; Choi, Y.W.; Cho, Y.S.; Kim, H.O.; Park, C.W. Effect of different cooking methods on histamine levels in selected foods. *Ann. Dermatol.* **2017**, *29*, 706–714. [CrossRef]

65. Wantke, F.; Gotz, M.; Jarisch, R. Histamine-free diet: Treatment of choice for histamine-induced food intolerance and supporting treatment for chronic headaches. *Clin. Exp. Allergy* **1993**, *23*, 982–985. [CrossRef]

66. Guida, B.; De Martino, C.; De Martino, S.; Tritto, G.; Patella, V.; Trio, R.; D'Agostino, C.; Pecoraro, P.; D'Agostino, L. Histamine plasma levels and elimination diet in chronic idiopathic urticaria. *Eur. J. Clin. Nutr.* **2000**, *54*, 155–158. [CrossRef]

67. Siebenhaar, L.; Melde, A.; Magerl, T.; Zuberier, T.; Church, M.K.; Maurer, M. Histamine intolerance in patients with chronic spontaneous urticaria. *J. Eur. Acad. Dermatol. Venereol.* **2016**, *30*, 1774–1777. [CrossRef]

68. Schwelberger, H.G.; Bodner, E. Purification and characterization of diamine oxidase from porcine kidney and intestine. *Biochim. Biophys. Acta* **1997**, *1340*, 152–164. [CrossRef]

69. Finney, J.; Moon, H.J.; Ronnebaum, T.; Lantz, M.; Mure, M. Human copper-dependent amine oxidases.
     Arch Biochem. Biophys. **2014**, *546*, 19–32. [CrossRef] [PubMed]
70. Vlieg-Boerstra, B.J.; Van der Heide, S.; Oude, J.N.G.; Kluin-Nelemans, J.C.; Dubois, A.E. Mastocytosis and
     adverse reactions to biogènic amines and histamine-releasing foods. What is the evidence? *Neth. J. Med.*
     2005, *63*, 244–249. [PubMed]

7

Biogenic Amine Contents and Microbial Characteristics of Cambodian Fermented Foods

Dalin Ly [1,2,]*, Sigrid Mayrhofer [1], Julia-Maria Schmidt [1], Ulrike Zitz [1] and Konrad J. Domig [1]

[1]   Institute of Food Science, Department of Food Science and Technology, BOKU - University of Natural Resources and Life Sciences Vienna, Muthgasse 18, A-1190 Vienna, Austria; sigrid.mayrhofer@boku.ac.at (S.M.); jm_schmidt@gmx.net (J.-M.S.); ulrike.zitz@boku.ac.at (U.Z.); konrad.domig@boku.ac.at (K.J.D.)

[2]   Faculty of Agro-Industry, Department of Food Biotechnology, RUA - Royal University of Agriculture, Dangkor District, P.O. BOX 2696 Phnom Penh, Cambodia

*   Correspondence: dalin.ly@boku.ac.at or dalinely@rua.edu.kh

Abstract: Naturally fermented foods are an important part of the typical diet in Cambodia. However, the food safety status of these products has not been widely studied. The aim of this study was, therefore, to provide an overview of the quality of these foods in relation to microbiology and biogenic amines. Additionally, the obtained results were compared to the habits and practices of Cambodians in handling this type of food. A total of 57 fermented foods (42 fishery and 15 vegetable products) were collected from different retail markets in the capital of Cambodia. Pathogenic *Salmonella* spp., *Listeria* spp., and *Listeria monocytogenes* were not detected in 25 g samples. Generally, less than 10^2 cfu/g of *Staphylococcus aureus*, *Escherichia coli*, *Pseudomonas* spp., Enterobacteriaceae, and molds were present in the fermented foods. *Bacillus cereus* group members ($<10^2$ to 2.3×10^4 cfu/g), lactic acid bacteria ($<10^2$ to 1.1×10^7 cfu/g), halophilic and halotolerant bacteria ($<10^2$ to 8.9×10^6 cfu/g), sulfite-reducing *Clostridium* spp. ($<10^2$ to 3.5×10^6 cfu/g), and yeasts ($<10^2$ to 1.1×10^6 cfu/g) were detected in this study. Still, the presence of pathogenic and spoilage microorganisms in these fermented foods was within the acceptable ranges. Putrescine, cadaverine, tyramine, and histamine were detected in 100%, 89%, 81%, and 75% of the tested products, respectively. The concentrations of histamine (>500 ppm) and tyramine (>600 ppm) were higher than the recommended maximum levels in respectively four and one of 57 fermented foods, which represents a potential health risk. The results suggest that the production process, distribution, and domestic handling of fermented foods should be re-evaluated. Further research is needed for the establishment of applicable preservation techniques in Cambodia.

Keywords: Cambodian fermented foods; microbial characteristics; biogenic amines; food quality; food safety

1. Introduction

Cambodia is an agricultural country that has a tropical climate with two distinct monsoon seasons (dry and rainy seasons). Thus, the availability of certain products is not stable through the year, and food preservation and storage are required to maintain the food supply [1]. Since food spoilage is mainly caused by microorganisms, preventing their access to susceptible foods is one method of food preservation. Another one is the inhibition of microbial growth through fermentation, salting, drying, or smoking, as it is common in Cambodia [2,3]. The storage time of food depends on factors that affect the growth of spoilage microorganisms like intrinsic food characteristics (e.g., pH, a_w, composition) and extrinsic parameters (e.g., temperature, relative humidity, atmospheric gases). Due to higher ambient temperatures and moisture, food spoils faster in the tropics. As a result, it is not surprising that food security issues are reported in densely populated tropical cities [2]. However, the majority of the Cambodian population lives in rural areas where poverty is high and access to drinking water, electricity, and sanitation is limited [4].

Fermented foods are an important part of the typical diet in Cambodia [1]. Since Cambodia has an extensive network of waterways, freshwater fish, along with marine, fermented and preserved fish, is a major component of the diet of most Cambodians [5]. Fermented fishery products are consumed daily as main dishes, side dishes, or condiments/seasonings [5]. Additionally, they are applied as flavor enhancers due to their delicacy and high nutritional properties [6,7]. Vegetables also play an essential role in daily dishes for their nutrient content. The availability of certain fresh vegetables, however, does not last throughout the year. Depending on the varieties of domestic raw vegetables, many types of fermented vegetables have emerged in Cambodia with the popularity of traditional fermentation [8]. In the meantime, these foods have become a common part of the Cambodian diet [1,8].

Cambodian fermented foods are produced through knowledge that is passed on from generation to generation and from person to person [1]. The great majority of fermented products are locally produced by smallholders, many of them women, and sold in traditional wet markets where women also predominate as retailers [9]. Most of them are illiterate and have a poor knowledge about hygiene practices. Additionally, the awareness of food safety is limited in Cambodia. The quality of these foods is influenced by raw materials, processing methods, and climate, but there is no quality control of these determinants as well as of the finished products in Cambodia [1]. Fermented foods are generally not labelled with an appropriate shelf-life and usually stored at room temperature until they are completely consumed [1]. Fermented fishery products are usually cooked before consumption. However, fish paste and sauce can also be eaten raw and are often mixed with chili or lemon juice [1]. In contrast, fermented vegetables are normally considered as ready-to-eat (RTE) foods. As a result, it is not surprising that foodborne outbreaks are common in Cambodia [8]. But there is no coordinated food surveillance program and little analytical data regarding microbiological or chemical contamination of food are present [10]. Nevertheless, food safety is a key priority of the Cambodian government [11], and efforts to improve foodborne disease surveillance and food safety are being undertaken [10].

Escherichia coli, Cronobacter sakazakii, Enterobacter spp., opportunistic non-Enterobacteriaceae, *Staphylococcus* spp., and *Listeria* spp. have already been detected in fermented vegetables in Cambodia [8,12]. Furthermore, potentially pathogenic bacteria such as *Bacillus, Clostridium,* and *Staphylococcus* were found in traditional Cambodian fermented fish products [5]. Next to microbiological contamination, chemical substances can lead to acute poisoning or even long-term diseases such as cancer [13]. The most prevalent ones are biogenic amines (BAs) and biotoxins [14]. BAs are low molecular weight organic molecules, formed by microbial decarboxylation of their precursor amino acids or by transamination of aldehydes and ketones by amino acid transaminases [15]. Beside spoilage, preservative technological processes such as fermentation, salting, and ripening may increase BA formation in food. As BAs are thermostable, they cannot be inactivated by thermal treatment [13]. The most common BAs found in foods and beverages are histamine (HIS), tyramine (TYR), putrescine (PUT), and cadaverine (CAD) [16,17]. Low levels of BAs in food are not considered as a serious risk. However, when high amounts of BAs are consumed, various physiological effects may occur, namely, hypotension (in the case of HIS, PUT, and CAD) or hypertension (in the case of TYR), nausea, headache, rash, dizziness, cardiac palpitation, and even intracerebral hemorrhage and death in very severe cases [18]. BAs with more severe acute effects for human health are HIS and TYR [19]. PUT and CAD have low toxicological properties on their own, but they can act as precursor of carcinogenic N-nitrosamines when nitrite is present. These two BAs also potentiate the effects of HIS and TYR by inhibiting their metabolizing enzymes [19]. HIS is the only BA with regulatory limits [20]. In addition to their potential toxicity, BAs are also used to evaluate the hygienic quality of foods, as their levels in food can be an indirect indicator of excessive microbial proliferation [19].

Baseline surveillance data are essential to monitor the disease burden of fermented foods in Cambodia. To obtain such data, the physicochemical properties (pH, a_w, and salt content) as well as the presence of certain microorganisms (spoilage and pathogenic bacteria) and the concentrations of the BAs HIS, TYR, PUT, and CAD were determined in 57 Cambodian fermented food samples within this study. The main purpose of this manuscript is to give an overview of the quality of Cambodian fermented foods, to correlate physicochemical parameters with BA contents, and to describe the prevailing habits and practices of Cambodians in dealing with this type of food.

2. Materials and Methods

2.1. Sample Collection

Fifty-seven samples of naturally fermented foods (42 raw fermented fish and 15 RTE fermented vegetable products) were randomly purchased from wet markets in Phnom Penh, the capital city of Cambodia. These products originated from various provinces of the country. Fermented fishery samples included fish sauce (teuktrey; $n = 7$), fish paste (prahok; $n = 12$), shrimp paste (kapi; $n = 6$), fermented fish (paork chav; $n = 7$; mam trey; $n = 3$), sour fermented fish (paork chou; $n = 3$), and salted fish (trey proheum; $n = 4$). Fermented vegetables such as salty fermented radish (chaipov brey; $n = 3$), sweet fermented radish (chaipov paem; $n = 3$), fermented melon (trasork chav; $n = 3$), fermented mustard (spey chrourk; $n = 3$), and fermented papaya (mam lahong; $n = 3$) were bought on the next day. Detailed information about each fermented product is provided in Table 1. After purchasing, the samples were immediately packed into plastic boxes and stored at their storage temperature. One day later, the samples were transported to Vienna by airplane. At the Department of Food Science and Technology, Vienna, the samples were checked shortly after arrival and kept in their original containers at 4 °C until analysis. All samples were analyzed within the usual shelf-life of the products [21].

2.2. Physicochemical Properties Analysis

2.2.1. Determination of pH and Water Activity (a_w)

The pH value was determined by penetrating the spear tip of the Blueline 21 pH electrode (Schott AG, Mainz, Germany) into the samples. The pH values were then measured using a digital pH meter (Schott Lab 870, Mainz, Germany).

The a_w value was measured using the digital water activity meter Rotronic Hygropalm HP23-AW-A (Rotronic, Zurich, Switzerland) after equilibration at room temperature (~25 °C).

2.2.2. Determination of Salt Content (NaCl)

The salt content was analyzed by potentiometric precipitation titration of chloride-ions with the 877 Titrino plus-Titrator equipped with a calomel electrode (Metrohm AG, Herisau, Switzerland). The protocol was performed according to the producer with minor modifications. Depending on the expected salt content, 1 to 10 g ± 0.01 g (a) of the samples was weighed into a glass beaker and filled with distilled water to 200 g ± 0.01 g (b). Subsequently, the samples were homogenized for 2 min at 9500 rpm using an Ultra Turrax T25 (IKA, Germany). Fifty grams ± 0.01 g (c) of the homogenized samples was weighed into a new glass beaker, and 50 mL of distilled water was added. Afterwards, 2 mL (2 M) HNO_3 was added. The samples were then titrated with 0.1 M $AgNO_3$. With the obtained results, the salt content of the original samples was calculated as % (w/w) NaCl = $V \times M \times 0.0584 \times 100/m$; where V and M are the volume and molarity of the $AgNO_3$ standard solution used. The initial sample weight is m, which was calculated considering the sample preparation: $m = a \times c/b$. The test was conducted in duplicate.

Table 1. Detailed information of fermented food products.

Product Type	English Name	Local Name of Fish/Vegetable Species [a]	Scientific Family Name of Fish/Vegetable Species	Major Ingredients	Usage	Market Origin
Fermented Fishery Products						
Teuktrey (n = 7)	Fish sauce	Trey Kakeum, Trey Kamong, Trey Riel, Trey Kanchanhchras, Trey Linh, Trey Kralong	*Engraulidae, Scombridae, Cyprinidae, Ambassis, Cyprinidae, Cyprinidae*	Freshwater/sea fish, salt	Side dish, Condiment, Seasoning	Chamkadaung, Oreusey
		Trey Achkok	*Cyprinidae*			
Prahok (n = 12)	Fish paste	Trey Riel, Trey Chhkork, Trey Ptourk, Trey Kampleanh	*Cyprinidae, Cyprinidae, Channidae, Osphronemidae*	Freshwater fish, salt	Main dish, Side dish, Condiment, Seasoning	Deumkor, Oreusey, Chamkadaung
Kapi (n = 6)	Shrimp paste	-	-	Tiny marine shrimp, salt	Side dish, Condiment, Seasoning	Kandal, Chas, Oreusey
Paork chav (n = 7)	Fermented fish	Trey Por, Trey Pra, Trey Chhkork	*Pangasiidae, Pangasiidae, Cyprinidae*	Freshwater fish, brown glutinous rice, salt	Main dish, Side dish	Thmey, Oreusey
Paork chou (n = 3)	Sour fermented fish	Trey Sleuk Reusey	*Engraulidae*	Freshwater fish, rice, salt	Main dish, Side dish	Chamkadaung, Oreusey
Mam trey (n = 3)	Fermented fish	Trey Bondol Ampov	*Clupeidae*	Freshwater fish, palm sugar, salt	Main dish, Side dish	Thmey
Trey proheum (n = 4)	Salted fish	Trey Pra, Trey Proma	*Pangasiidae, Sciaenidae,*	Freshwater fish, salt	Main dish, Seasoning	Thmey
Fermented Vegetables						
Chaipov brey (n = 3)	Salty fermented radish	Chaitav	*Brassicaceae*	Chinese white radish, salt	Side dish, Seasoning, Appetizer	Kandal, Chas, Oreusey
Chaipov paem (n = 3)	Sweet fermented radish	Chaitav	*Brassicaceae*	Chinese white radish, sugar	Main dish, Side dish, Appetizer	Kandal, Chas, Oreusey
Trasork chav (n = 3)	Fermented melon	Trasork	*Cucurbitaceae*	Baby melon, salt, purple sticky rice	Side dish	Thmey
Spey chrourk (n = 3)	Fermented mustard	Speythom	*Brassicaceae*	Chinese mustard, salt	Side dish	Phumreusey Limcheanghak
Mam lahong (n = 3)	Fermented papaya	Lahong	*Caricaceae*	Green papaya, tiny fermented fish, salt, herb	Side dish	Phumreusey Limcheanghak

[a] Name in Cambodian.

2.3. Microbiological Analysis

From each fermented product, a 10 g sample was aseptically collected, transferred to a stomacher bag, and homogenized (Stomacher 400 Circulator, Seward, UK) with 90 mL of buffered peptone water for 45 s at 230 rpm. Appropriate dilutions of the samples were prepared using the same diluent, and 0.1 or 1 mL aliquots of each dilution were applied on various selective media using the spread plate method or pour plate method. Lactic acid bacteria (LAB) were anaerobically grown on DeMan Rogosa Sharpe agar (MRS, Merck, Darmstadt, Germany) at 30 °C for 72 h according to ISO 15214 [22]. Enterobacteriaceae were enumerated using the pour plate method with an additional overlay on Violet Red Bile Dextrose agar (VRBD, Merck, Darmstadt, Germany) and an incubation at 37 °C for 24 h according to ISO 21528-2 [23]. *Pseudomonas* spp. were detected by plating appropriate dilutions on Cephalothin-Sodium Fusidate-Cetrimide agar (CFC, Oxoid, Hampshire, UK) and incubation at 25 °C for 44 h based on ISO 13720 [24]. Yeasts and molds were determined according to ISO 21527-2 [25] using the spread plate method on Dichloran Glycerol agar (DG18, Merck, Darmstadt, Germany). The plates were incubated at 25 °C for 5–7 d and counted on the 5th and 7th day of incubation. Selected yeast colonies were confirmed by methylene blue staining and microscopy [26]. Halophilic and halotolerant bacteria were counted after an incubation at 30 °C for 2–4 d on Tryptone Soya agar (TS, Oxoid, Hampshire, UK) supplemented with 10% NaCl (Roth, Karlsruhe, Germany) [27]. *Staphylococcus aureus* was enumerated on Baird Parker agar (BP, Merck, Darmstadt, Germany), which was incubated at 37 °C for 24 h based on ISO 6888-1/AMD 1 [28]. The plates were evaluated again after an additional 24 h incubation. The confirmation of colonies was performed using Gram-stain and DNase agar (Oxoid, Hampshire, UK) according to Kateete et al. [29]. The number of presumptive *Bacillus cereus* group members was investigated by spreading dilutions on Mannitol Yolk Polymyxin agar (MYP, Merck, Darmstadt, Germany) and incubating plates at 30 °C for 18–48 h. The evaluation was also performed after 18 h and 48 h of incubation. Colonies were confirmed by endospore staining [30]. *E. coli* was enumerated by pour plating on Tryptone Bile Glucuronic medium (TBX, Oxoid, Hampshire, UK) with an incubation at 44 °C for 18–24 h based on ISO 16649-2 [31]. The presence of sulfite-reducing *Clostridium* spp. (SRC) was analyzed by pour plating with an additional overlay on Sulfite-Cycloserin agar (SC, Oxoid, Hampshire, UK). Plates were anaerobically incubated at 37 °C for 20 h. Confirmation tests were performed using Lactose-Gelatine medium (Conda, Madrid, Spain) and Motility-Nitrate medium (Conda, Madrid, Spain) according to ISO 7937 [32]. The presence of *Salmonella* spp. was investigated using the VIDAS UP (BioMerieux, Crappone, France) *Salmonella* (SPT) system, whereas that of *Listeria* spp. and *L. monocytogenes* was tested by the VIDAS LDUO (BioMerieux, Crappone, France) system. VIDAS SPT and VIDAS LDUO are based on an enzyme-linked fluorescent immunoassay. The preparation of the samples was similar to the previous method, but 25 g of the sample was weighted in instead of 10 g. After pre-enrichment (*Listeria* spp.) and enrichment (*L. monocytogenes*, *Salmonella* spp.) steps, which were carried out according to the manufacturer's manual, the assay steps were performed automatically by the instrument.

2.4. Determination of Biogenic Amines (BAs)

The concentrations of BAs in the supernatant were analyzed by reverse-phase HPLC (Waters 2695 Separations Module, Waters, MA, USA) according to the method of Šimat et al. and Saarinen et al. [33,34]. Briefly, 1 g of the homogenized sample was extracted overnight with 5 mL of 0.4 M perchloric acid (Merk, Darmstadt, Germany). Then, the sample was centrifuged at 5000 rpm for 10 min and the supernatant was kept for further analysis. For derivatization, 80 μL of 2 M NaOH (Roth, Karlsruhe, Germany), 120 μL of saturated sodium bicarbonate solution (Merck, Darmstadt, Germany), and 400 μL of derivatization reagent (1% dansyl chloride in acetone; prepared daily; Fluka, Seelze, Germany) were added to 400 μL of sample solution. The sample was mixed and incubated for 45 min at 40 °C. Afterwards, 60 μL of 1 M ammonia solution (Roth, Karlsruhe, Germany) was added, mixed, and incubated in the dark for 60 min at room temperature. Finally, 940 μL acetonitrile (Roth, Karlsruhe, Germany) was added. The sample was mixed and centrifuged for 10 min at 13,400 rpm. A RP-18 column (Li Chro CART 250-4, 5 μm,

Merck, Darmstadt, Germany) with a LiChroCART 4-4 Guard Column (RP-18, 5 µm) (Merk, Darmstadt, Germany) and a manu-CART NT cartridge holder (Merck, Darmstadt, Germany) was used for separation. The flow rate was 1 mL/min, the column temperature was 40 °C, and the injection volume was 20 µL. The mobile phase A consisted of 0.1 M ammonium acetate (Roth, Karlsruhe, Germany) and the mobile phase B was 100% acetonitrile. The following gradient was used for the separation: time = 0 min, 50% A and 50% B; time = 19 min, 10% A and 90% B; time = 20 min, 50% A and 50% B; time = 28 min, 50% A and 50% B. The detection was performed by UV–vis (Waters 2489 UV-visible detector, Waters, MA, USA) at a wavelength of 254 nm. Heptylamine (Fluka, Seelze, Germany) was used as an internal standard that was well separated from other compounds. The specificity of the method was checked using standard mixtures of 12 BA chemicals, which included spermine tetrahydrochloride, spermidine trihydrochloride, ethanolamine, isopropylamine, histamine dihydrochloride, putrescine (1,4-diaminobutan dihydrochloride), methylamine hydrochloride, agmatine sulfate, cadaverine (1,5-diaminopentan dihydrochloride), tyramine hydrochloride, dimethylamine hydrochloride, and pyrrolidine. Except for the last four chemicals, which were from Sigma-Aldrich (St. Louis, MO, USA), all chemicals were from Fluka. Standard stock solutions of BAs were prepared at 500 mg/L in 0.01 M perchloric acid. The stock solutions were diluted with 0.4 M perchloric acid to obtain series of working standard solutions (0.25, 1, 5, 10, and 15 mg/L). The derivatization procedure was the same as for the samples. All compounds were separated and could be identified by their retention times. The linearity of the method was tested by analyzing the series of working standard solutions. The correlation coefficients for the linear regression lines were better than 0.99 for all compounds. The limit of detection (LOD = 3x standard deviation of y-residuals of low concentrations/slope of calibration curve) of all BAs ranged between 0.5 and 1.5 ppm, and the limit of quantification (LOQ = 10x standard deviation of y-residuals of low concentrations/slope of calibration curve) ranged between 1.5 and 4.8 ppm. PUT, CAD, HIS, and TYR were analyzed in duplicate.

2.5. Statistical Analysis

Result units of quantitative microbiological analyses were cfu/g. The physicochemical parameters and BA concentrations results were analyzed with statistical analyses using the Statistical Package for the Social Sciences (SPSS, Version 20.0.0 for Windows, 2011; IBM Co., Somers, NY, USA). Data were analyzed for the degree of variation by calculating the mean and standard deviations (SDs) of the results. The significance of differences was evaluated using analysis of variance (ANOVA). A p value of less than 0.05 was considered statistically significant. The least-squares difference (LSD) test was used to determine the significance of differences in the physicochemical parameters and BA contents among the samples. The relationship value was determined using the Pearson correlation coefficient.

3. Results

3.1. Physicochemical Characteristics in Fermented Foods

Physicochemical parameters such as pH, a_w, and salt content were measured and compared to discuss possible causes for the different BA levels. Table 2 shows the physicochemical parameters of the tested fermented products. The pH values in fermented fishery samples were in the range of 4.4 to 7.6. Lower pH values were found in paork chav (fermented fish), while higher values were detected in kapi (shrimp paste) and trey proheum (salted fish). The a_w values of the fermented fishery products ranged from 0.69 to 0.84. The lowest a_w value (0.69) was detected in kapi (shrimp paste) and the highest (0.84) in paork chav (fermented fish). The salt contents were in the range of 6% to 34%, with the lowest value (6%) found in trey proheum (salted fish) and the highest (34%) in prahok (fish paste). In the fermented vegetables, the pH values were between 3.6 and 5.5. Lower pH values were found in spey chrourk (fermented mustard, 3.6–3.9) and mam lahong (fermented papaya, 3.7–3.8), while the highest value was detected in chaipov brey (salty fermented radish) (4.6). The a_w values in these fermented products were between 0.75 and 0.97. The highest salt concentration (25%) was found in chaipov brey (salty fermented radish), while the lowest (2%) was detected in spey chrourk (fermented mustard) (Table 2).

Table 2. Physicochemical characteristics (pH, a_w, % NaCl) of fermented food products.

Product Types	English Name	Physicochemical Characteristics (Mean ± SD)		
		pH	a_w	Salt Content (%)
Fermented Fishery Products				
Teuktrey ($n = 7$)	Fish sauce	4.8–6.3 [≠] (5.5 ± 0.6 [¥]) [a]	0.72–0.82 (0.76 ± 0.04) [a]	19–25 (22 ± 2) [a]
Prahok ($n = 12$)	Fish paste	5.3–5.7 (5.5 ± 0.1) [a]	0.71–0.80 (0.73 ± 0.02) [ab]	15–34 (21 ± 5) [a]
Kapi ($n = 6$)	Shrimp paste	6.6–7.3 (7.0 ± 0.2) [b]	0.69–0.72 (0.71 ± 0.01) [b]	13–28 (19 ± 7) [a]
Paork chav ($n = 7$)	Fermented fish	4.4–5.1 (4.8 ± 0.3) [c]	0.74–0.84 (0.80 ± 0.03) [c]	9–14 (11 ± 2) [b]
Paork chou ($n = 3$)	Sour fermented fish	4.9–5.1 (5.0 ± 0.1) [c]	0.74–0.76 (0.75 ± 0.01) [a]	13–15 (14 ± 1) [b]
Mam trey ($n = 3$)	Fermented fish	4.9–5.6 (5.3 ± 0.4) [ac]	0.76–0.78 (0.77 ± 0.01) [ac]	8–14 (10 ± 3) [b]
Trey proheum ($n = 4$)	Salted fish	6.3–7.6 (6.8 ± 0.6) [b]	0.78–0.83 (0.80 ± 0.02) [c]	6–17 (12 ± 5) [b]
Fermented Vegetables				
Chaipov brey ($^n = 3$)	Salty fermented radish	4.4–4.8 (4.6 ± 0.2) [g]	0.75–0.76 (0.75 ± 0.01) [g]	24–25 (24 ± 0.4) [g]
Chaipov paem ($n = 3$)	Sweet fermented radish	3.9–5.5 (4.5 ± 0.9) [gh]	0.78–0.87 (0.82 ± 0.04) [h]	10–18 (13 ± 4) [h]
Trasork chav ($n = 3$)	Fermented melon	4.1–4.4 (4.3 ± 0.2) [gh]	0.76–0.83 (0.80 ± 0.03) [h]	9–13 (10 ± 2) [h]
Spey chrourk ($n = 3$)	Fermented mustard	3.6–3.9 (3.7 ± 0.2) [h]	0.96–0.97 (0.97 ± 0.01) [i]	2–5 (4 ± 1) [i]
Mam lahong ($n = 3$)	Fermented papaya	3.7–3.8 (3.7 ± 0.1) [h]	0.91–0.94 (0.92 ± 0.01) [i]	3–4 (3 ± 1) [i]

[≠] Ranged values (minimum to maximum). [¥] Mean ± SD (standard deviation). Values with different superscript letters in the same column indicate significant differences at $p < 0.05$ by LSD test. Statistical analysis of fermented fish and vegetable samples was conducted separately.

Based on the statistical analysis (ANOVA) of fermented fishery products, there were significant differences ($p < 0.05$) between the physicochemical parameters of teuktrey (fish sauce) and those of paork chav (fermented fish) and trey proheum (salted fish). There was no statistically significant difference ($p > 0.05$) among the samples of teuktrey (fish sauce) and prahok (fish paste), and of paork chav (fermented fish), paork chou (sour fermented fish), and mam trey (fermented fish). Statistical analysis of fermented fish and vegetables samples was conducted separately. Regarding the fermented vegetable products, the physicochemical values of chaipov brey (salty fermented radish) were significantly different from that of spey chrourk (fermented mustard) and mam lahong (fermented papaya) ($p < 0.05$), while no significant difference was found among chaipov paem (sweet fermented radish) and trasork chav (fermented melon) ($p > 0.05$) (Table 2).

3.2. Presence of Microorganisms

Counts of LAB ($<10^2$ to 1.1×10^6 cfu/g), halophilic and halotolerant bacteria ($<10^2$ to 8.9×10^6 cfu/g), Enterobacteriaceae ($<10^2$ cfu/g), *Pseudomonas* spp. ($<10^2$ cfu/g), yeasts ($<10^2$ to 1.1×10^6 cfu/g), and molds ($< 10^2$ to 2.3×10^2 cfu/g) from the different types of fermented fish tested are indicated in Table 3. Table 3 also presents the results regarding the *B. cereus* group members ($<10^2$ to 2.3×10^4 cfu/g), SRC ($<10^2$ to 3.5×10^6 cfu/g), *S. aureus*, and *E. coli* ($<10^2$ cfu/g, respectively).

The microbial profiles found in fermented vegetables are displayed in Table 3 as well. The LAB counts were in the range of $<10^2$ to 1.1×10^7 cfu/g. The highest LAB counts were detected in spey chrourk (fermented mustard) and mam lahong (fermented papaya). Halophilic and halotolerant bacteria were found in numbers ranging from 2×10^2 to 5.5×10^4 cfu/g. The counts of *B. cereus* group members ranged from $<10^2$ to 1.2×10^4 cfu/g. SRC and yeasts were detected in the range of $< 10^2$ to 1.5×10^3 cfu/g and $<10^2$ to 2.6×10^5 cfu/g in the tested vegetable samples, respectively (Table 3). The counts of all other microorganisms were $<10^2$ cfu/g.

Table 3. Microbial profiles found in fermented food products.

Product Type	English Name	Microorganisms (cfu/g)									
		LAB [a]	Halophilic & Halotolerant Bacteria	Entero-Bacteriaceae	Pseudomonas spp.	Yeasts	Molds	B. cereus Group Members	SRC [b]	S. aureus	E. coli
					Fermented Fishery Products						
Teuktrey (n = 7)	Fish sauce	$<10^2$	$<10^2$	$<10^2$	$<10^2$	$<10^2$	$<10^2$	$<10^2$	$<10^2$	$<10^2$	$<10^2$
Prahok (n = 12)	Fish paste	$<10^2$–1.5×10^2	10^2–9.5×10^3	$<10^2$	$<10^2$	$<10^2$	$<10^2$	$<10^2$–1.6×10^3	2.4×10^2–2.8×10^6	$<10^2$	$<10^2$
Kapi (n = 6)	Shrimp paste	$<10^2$–2.8×10^3	5.2×10^3–5×10^5	$<10^2$	$<10^2$	$<10^2$	$<10^2$	$<10^2$–6.8×10^3	2×10^2–1.3×10^5	$<10^2$	$<10^2$
Paork chav (n = 7)	Fermented fish	$<10^2$–1.6×10^3	1.2×10^3–2.6×10^5	$<10^2$	$<10^2$	$<10^2$–1.1×10^6	$<10^2$	10^2–2.3×10^4	$<10^2$–9×10^4	$<10^2$	$<10^2$
Paork chou (n = 3)	Sour fermented fish	$<10^2$	2×10^2–9×10^3	$<10^2$	$<10^2$	$<10^2$–2.7×10^4	$<10^2$	$<10^2$–3.5×10^3	4×10^2–1.2×10^4	$<10^2$	$<10^2$
Mam trey (n = 3)	Fermented fish	$<10^2$–1.2×10^3	10^3–5.1×10^4	$<10^2$	$<10^2$	$<10^2$	$<10^2$	1.1×10^2–1.2×10^4	1.9×10^3–3.5×10^6	$<10^2$	$<10^2$
Trey proheum (n = 4)	Salted fish	$<10^2$–1.1×10^6	2.3×10^4–8.9×10^6	$<10^2$	$<10^2$	$<10^2$	$<10^2$	$<10^2$–2.6×10^2	$<10^2$–1.6×10^5	$<10^2$	$<10^2$
					Fermented Vegetables						
Chaipov brey (n = 3)	Salty fermented radish	$<10^2$	3×10^3–1.6×10^4	$<10^2$	$<10^2$	$<10^2$	$<10^2$	$<10^2$	$<10^2$	$<10^2$	$<10^2$
Chaipov paem (n = 3)	Sweet fermented radish	$<10^2$	1.7×10^4–5.5×10^4	$<10^2$	$<10^2$	$<10^2$	$<10^2$	$<10^2$–2.2×10^2	$<10^2$–3×10^2	$<10^2$	$<10^2$
Trasork chav (n = 3)	Fermented melon	$<10^2$	2×10^2–8×10^3	$<10^2$	$<10^2$	$<10^2$	$<10^2$	$<10^2$	$<10^2$	$<10^2$	$<10^2$
Spey chrourk (n = 3)	Fermented mustard	$<10^2$–1.1×10^7	7.3×10^2–2×10^4	$<10^2$	$<10^2$	$<10^2$–2.6×10^3	$<10^2$	2×10^2–1.2×10^4		$<10^2$	$<10^2$
Mam lahong (n = 3)	Fermented papaya	$<10^2$–5.8×10^6	2×10^2–1.1×10^3	$<10^2$	$<10^2$	1.7×10^2–2.6×10^5	$<10^2$	$<10^2$–1.6×10^2	3.2×10^2–1.5×10^3	$<10^2$	$<10^2$

[a] Lactic acid bacteria; [b] sulfite-reducing clostridia.

3.3. Quantification of Biogenic Amines (BAs) in Fermented Foods

Table 4 shows the BA contents of 57 fermented product samples. The detection limits in this study were <0.5 ppm (PUT, CAD, and TYR) and <2 ppm for HIS. The results indicate that PUT was detected in quantifiable amounts in all tested fishery samples (100%), while CAD, TYR, and HIS concentrations were quantified in approximately 95%, 88%, and 86% of these products, respectively. PUT and CAD were the most frequently detected BAs in the tested samples. The highest concentrations of PUT (830 ppm), CAD (2035 ppm), HIS (840 ppm), and TYR (691 ppm) were detected in paork chav (fermented fish). PUT concentrations in 42 fishery samples were in the range between 23 to 830 ppm, with the lowest (23 ppm) presented in paork chou (sour fermented fish) and the highest (830 ppm) found in paork chav (fermented fish). The concentrations of HIS in the quantifiable fishery products (86%) ranged from 32 to 840 ppm (Table 4). The current results show that, overall, less than 50 ppm HIS was determined in all kapi (shrimp paste) samples. The concentrations of TYR in the quantifiable fishery samples (88%) ranged from 10 to 691 ppm (Table 4). In general, lower levels of TYR were detected in kapi (shrimp paste) and paork chou (sour fermented fish) than in other fermented fishery products in this study.

Table 4. Contents of biogenic amines in fermented food products.

Product Type	English Name	Biogenic Amines (ppm)			
		PUT *	CAD	HIS	TYR
Fermented Fishery Products					
Teuktrey (n = 7)	Fish sauce	75–404 [#] (233 ± 126 [¥]) [ab]	99–766 (368 ± 231) [abc]	40–253 (155 ± 74) [ab]	39–342 (144 ± 110) [ab]
Prahok (n = 12)	Fish paste	191–649 (360 ± 150) [b]	119–899 (522 ± 231) [abc]	35–408 (179 ± 115) [abc]	76–594 (218 ± 159) [ab]
Kapi (n = 6)	Shrimp paste	29–294 (112 ± 99) [ac]	ND [#]–270	ND–46	ND–57
Paork chav (n = 7)	Fermented fish	33–830 (386 ± 337) [b]	38–2035 (672 ± 782) [bc]	32–840 (422 ± 264) [c]	10–691 (299 ± 294) [b]
Paork chou (n = 3)	Sour fermented fish	23–92 (49 ± 37) [a]	26–69 (43 ± 24) [a]	46–559 (260 ± 267) [bc]	ND–82
Mam trey (n = 3)	Fermented fish	37–569 (378 ± 296) [bc]	23–1470 (930 ± 788) [c]	33–732 (320 ± 366) [bc]	ND–196
Trey proheum (n = 4)	Salted fish	72–278 (153 ± 90) [ab]	187–485 (297 ± 130) [ab]	ND–183	53–118 (79 ± 30) [a]
Fermented Vegetables					
Chaipov brey (n = 3)	Salty fermented radish	12–18 (15 ± 3) [g]	ND–12	ND	ND
Chaipov paem (n = 3)	Sweet fermented radish	11–16 (14 ± 3) [g]	ND–10	ND	ND
Trasork chav (n = 3)	Fermented melon	28–107 (70 ± 40) [g]	ND–23	ND–18	7–30 (15 ± 13) [g]
Spey chrourk (n = 3)	Fermented mustard	33–197 (95 ± 89) [g]	16–51 (29 ± 19) [g]	34–103 (66 ± 35) [g]	22–86 (44 ± 36) [gh]
Mam lahong (n = 3)	Fermented papaya	72–184 (119 ± 58) [g]	22–118 (68 ± 48) [g]	33–72 (49 ± 20) [g]	38–63 (53 ± 13) [h]

* PUT, putrescine; CAD, cadaverine; HIS, histamine; TYR, tyramine; [#] Ranged values (minimum to maximum) [¥] Mean ± SD; [#] ND, not detected (Limit of detection < 0.5 ppm for PUT, CAD, and TYR; <2 ppm for HIS). Values with different superscript letters in the same column indicate significant differences at $p < 0.05$ by LSD test. Statistical analysis of fermented fish and vegetable samples was conducted separately.

The four types of BAs were also analyzed for the safety evaluation of fermented vegetables from Cambodia. The BA levels varied among the collected RTE fermented vegetables (Table 4). PUT, CAD, TYR, and HIS were detected in 100%, 73%, 60%, and 47% of the fermented vegetables, respectively. The ranges of the quantifiable BAs were from 11 to 197 ppm for PUT, 10 to 118 ppm for CAD, 18 to 103 ppm for HIS, and 7 to 86 ppm for TYR (Table 4). The results clearly show that most BA concentrations in the five types of fermented vegetables were less than 100 ppm. Even no HIS and TYR could be detected in chaipov brey (salty fermented radish) and chaipov paem (sweet fermented radish) samples (Table 4).

According to one-way ANOVA and LSD tests of 42 fermented fisheries samples, statistically significant differences ($p < 0.05$) were found among the detected concentrations of PUT, CAD, HIS, and TYR in each product type. The statistical analysis of 15 fermented vegetable samples showed no statistical difference ($p > 0.05$) among concentrations of PUT, CAD, and HIS, while TYR concentrations were statistically different ($p < 0.05$) (Table 4).

Analyzing the correlation between total BA contents and the physicochemical parameters pH, a_w, and salt content (%) in 42 fermented fishery products, a weak positive relationship between total BAs and a_w values ($r = 0.22$, $p > 0.05$; $n = 42$), and a weak negative with pH values ($r = -0.22$, $p > 0.05$; $n = 42$) were found. There was no correlation among total BAs and salt content ($r = 0.00$, $p > 0.05$; $n = 42$) (Figure S1A–C). Furthermore, the linear functions between total BA contents and parameters of pH ($r = -0.57$, $p < 0.05$; $n = 15$) and salt content ($r = -0.81$, $p < 0.05$; $n = 15$) were characterized by a moderate and strong negative correlation coefficient, respectively, while a strong positive correlation between total BAs and a_w value ($r = 0.79$, $p < 0.05$; $n = 15$) were determined in fermented vegetable products (Figure S2A–C).

4. Discussion

4.1. Physicochemical Characteristics in Fermented Foods

Based on the physicochemical results, types of fermented fishery products were more different than those of fermented vegetables. Nevertheless, the pH values of this study are in good agreement with those of fermented fish products in Thailand, Vietnam, Laos, Myanmar, China, Korea, Japan, Malaysia, and Taiwan [5,35–38]. The results of the salt concentration analysis are also consistent with previous data for fermented fish products [5,38], shrimp paste [6], and fish sauce [37]. The a_w values of fermented fish products were comparable to fermented fish products from other countries, for example, Thai shrimp paste (0.65–0.72) and Indonesian fermented fish (0.75–0.93) [39,40].

The pH values found in the fermented vegetables were between 3.6 and 5.5 (Table 2). This is in agreement with a previous study, which reported that the pH of Cambodian fermented vegetables ranged from 3.6 to 6.5, depending on the raw materials and processing techniques [8]. Chaipov brey (salty fermented radish) was found to have the highest salt value (25%) of the fermented vegetables. Salty fermented radish with high salt concentrations (20–25%) has also been reported elsewhere [8]. As salt reduces a_w, the lowest a_w values were also determined in these samples (0.75–0.76) (Table 2).

Growth of microorganisms in foods are mainly influenced by the a_w and pH [41]. The addition of salt, in turn, has an inhibitory effect on the growth of microorganisms due to its impacts on the a_w value [42].

4.2. Microbiological Parameters in Fermented Foods

Microorganisms associated with fermented foods are commonly present on the external surface and in the pre- and post-harvest environment of raw materials. Additionally, they exist in the gill and gut of seafood [14].

Regarding *Bacillus* spp. and *Clostridium* spp., our results are comparable to those of Chuon et al., who also analyzed Cambodian traditional fermented fish sauce, fish paste, and shrimp paste [5]. Such traditionally home-prepared salted or fermented products are often associated with foodborne botulism [43]. However, routine testing for *C. botulinum* to ensure food safety is not recommended [43]. Instead, SRC have been proposed to identify risks from *C. botulinum* [43]. In addition to *C. botulinum*, *C. perfringens*—the most important of the SRC—poses a frequent problem and challenge in fish industry [44]. It is estimated that 10^5 to 10^8 cfu/g *C. perfringens* are capable of generating toxinfection [44]. Foodborne diseases that have *C. perfringens* as causative agent are related to inadequate storage, processing, and food service operations [44]. Nevertheless, no *C. perfringens* could be confirmed within this study. It is known that $>10^5$ cfu/g *B. cereus* group members are potentially harmful for human consumption [45]. None of the fermented products exceeded this limit (Table 3). The survival of *B. cereus* in low numbers in several fermented products, including those based on fish and vegetables, has already been described [46]. The inactivation of this pathogen could be attributed to the presence of organic acids or higher salt concentrations [46]. Moreover, no *S. aureus* could be quantified ($<10^2$ cfu/g) in the tested samples, which is in contradiction to a previous study [5]. In addition, it is reported that *S. aureus* is uniquely resistant to adverse conditions such as low a_w values (0.83), high salt contents, and pH stress. Thus, most strains can grow over an a_w and pH range of 0.83 to >0.99 and 4.5–9.3, respectively [47]. Although 11 of all 57 food products tested (19.3%) had an a_w value in the range specified above, 10 of them (e.g., all trasork chav (fermented melon), spey chrourk (fermented mustard), and mam lahong (fermented papaya), and one chaipov paem (sweet fermented radish) product had a pH < 4.5 (Table 2). Overall, *S. aureus* could only have grown in one sample. Although *L. monocytogenes* appears to be relatively tolerant to acidic conditions, no representatives of this species as well as of other *Listeria* species were verified, which may be due to the low a_w (<0.9) of most food samples tested (90%). Also, less than 10^2 cfu/g of *Pseudomonas* spp., Enterobacteriaceae, and *E. coli* were detected in all products examined (Table 3). Furthermore, no *Salmonella* spp. could be determined using the VIDAS system. These Gram-negative bacteria are often inhibited by a salt concentration >10%, an a_w value <0.95, a pH value <3.8 or >9.0 (depending on the acidulant), and the fermentation process itself [48]. LAB are not only responsible for the fermentation, they also significantly contribute to the flavor, texture, and nutritional value of fermented products [48], produce effective antimicrobial agents, and are the primary preservation factor in fermented fish products [49]. However, LAB are generally only tolerant to moderate salt concentrations (10%–18%). Consequently, their counts decrease as the salt concentration increases [50]. Forty-three samples (~75%) of all fermented products tested in this study contained more than 10% salt (Table 3). LAB were only present in high numbers in samples with less than 10% salt (Table 2 and 3).

Since typical spoilage bacteria are generally non- or only slightly halotolerant (e.g., pseudomonads, enterobacteria), the extensive use of salt is another technological process for food preservation besides fermentation [51]. Up to 25% and 34% salinity was respectively determined for fermented vegetable and fishery products in this study. Classifying the various products according to their salt content (e.g., 0–10%, 11–20%, >20%, data not shown), the numbers of halophilic and halotolerant bacteria generally decreased with increasing salinity.

The unfavorable conditions for bacterial growth (high salt content, a low pH or a_w) may result in higher yeasts and mold numbers. These microorganisms are quite salt-tolerant [51]. As recommended by the European Food Safety Authority (EFSA), the accepted limit for molds in foods is $<10^3$ cfu/g [52]. As shown in Table 3, all 57 fermented food products were acceptable regarding molds. It has been reported that $<10^6$ cfu/g of yeasts are acceptable in RTE foods placed on the market [53]. An unsatisfactorily higher yeasts count ($>10^6$ cfu/g) was only found in one paork chav (fermented fish) sample, which could lead to spoilage by acid and gas production [45,53]. However, the limit was just exceeded marginally (Table 3).

According to different organizations and previous studies [45,52⁻56], the detected counts of the investigated microorganisms in this study are satisfactory. Thus, the fermented foods tested are suitable for human consumption regarding the microbiological quality.

4.3. Formation of BAs in Fermented Foods

A deviation in BA concentrations within a specific food category is probably due to intrinsic food characteristics such as pH and a_w values, nutrients, and microbiota, as well as extrinsic factors including storage time, temperature, and manufacturing processes [57–59]. This may explain the wide variation of BA concentrations between the fermented fishery products and even within the same tested product type. Shalaby (1996) stated that BA levels differ not only between different food varieties but also within the same variety [57]. However, no significant difference was observed in fermented vegetables within this study.

Fish species associated with a high amount of histidine belong to the families *Scombridae, Clupeidae, Engraulidae, Coryphenidae, Pomatomidae,* and *Scombreresosidae* [60]. As seen in Table 1, the fish species of some fermented fish products belong to the families *Engraulidae* and *Scombridae* for teuktrey (fish sauce), *Engraulidae* for paork chou (fermented fish) and *Clupeidae* for mam trey (fermented fish). Hence, these products contained higher HIS amounts (Table 4). In contrast, low HIS and TYR contents are reported in crustaceans such as shrimp [61]. Corresponding values were determined for six kapi (shrimp paste) samples within this study. Fresh fruits and vegetables such as melon, cabbage, radishes, and cucumber contain lower HIS levels; however, papaya is considered as a HIS liberator [62]. Mustard is generally an allergen, and sometimes listed as moderately high in HIS [63]. Accordingly, HIS has been found in all fermented spey chrourk (fermented mustard) and mam lahong (fermented papaya) samples, but only in one trasork chav (fermented melon) and in no chaipov (fermented radish) sample within this study. TYR has been detected in more fermented vegetable samples than HIS, although in lower concentrations. TYR and CAD have been described in few vegetables in relatively low concentrations [64]. In contrast, it has been reported that PUT is the most common BA found in food of plant origin. It is particularly abundant in vegetables [64,65] and fermented products [60]. As seen in Table 4, this BA was the only one that was verified in all fermented vegetable samples with relatively high values.

The possible involvement of molds and yeasts in BA (especially CAD and PUT) accumulation is still discussed [19]. However, it is known that different genera, species, and strains of Gram-positive and Gram-negative bacteria are able to produce BAs by the action of microbial decarboxylases [66]. In particular, Enterobacteriaceae were identified as HIS-producing bacteria, but also halophilic and halotolerant bacteria (among other representatives of the families Enterobacteriaceae, Pseudomonadaceae, and the genera *Photobacterium*, *Vibrio*, and *Staphylococcus*), LAB, *Bacillus* spp., and *Clostridium* spp. were said to be capable of HIS formation [14,57,67]. According to Rodriguez-Jerez et al., microbial species with the capacity to form HIS and those with the capacity to form other BAs are similar [68]. Thus, Enterobacteriaceae were also reported to produce PUT, CAD, and to a lesser extent TYR. These BAs have also been detected when testing various *Bacillus* strains [67]. However, TYR should be mainly formed by LAB (*Lactobacillus*, *Enterococcus*) during fermentation [16]. Next to strains of the genera *Clostridium*, *Pseudomonas*, and *Staphylococcus*, LAB (*Enterococcus*, *Lactococcus*) are also involved in the production of PUT. It should be kept in mind that decarboxylase activities are often related to strains rather than to species or genera [69]. The capabilities of such strains, in turn, vary depending on the type and even batch of food product from which the strains are isolated [67].

The main factors affecting microbial activities in food are temperature, salt concentration, and pH [19]. Most fermented foods in this study had a pH value within the range of 3.0 to 6.0 (79%, Table 2), providing an acidic environment. The transcription of many decarboxylase genes is induced by a low pH value, which improves the fitness of the microbial cells subjected to acidic stress [19]. As the decarboxylation of amino acids is a mechanism of BA-forming bacteria to counteract acidic stress and to adapt to environmental conditions, their decarboxylase activity increases, resulting in higher BA concentrations [7,58]. Hence, it contributed to higher BA contents (Table 4). This effect could be confirmed within this study, as weak and moderate negative correlations ($r = -0.22$ and -0.57) were respectively found between the total BA contents and pH values for fermented fishery and fermented vegetable products. A strong negative linear fit ($r = -0.81$) could be detected between total BAs and salt content in fermented vegetables, whereas there was no correlation between these parameters in fermented fishery products (Figures S1 and S2). In general, increasing salt concentrations contribute to the reduction of BA accumulation in foods, mainly reducing the metabolic activities of decarboxylase-positive microorganisms [19] as it may have been the case for the fermented vegetables. However, a possible enhancing effect of NaCl on the BA production has also been described [19]. Thus, stressed cells seem to activate the decarboxylating pathways in the framework of more complex response systems [19] being probably more present in fermented fishery than vegetable products. In contrast, the rate of BA accumulation decreases with the decrease of a_w values due to the water loss [19]. Correspondingly, a positive correlation should be observed between total BA contents and a_w values. In fact, weak and strong positive relationships were determined for fermented fishery ($r = 0.22$) and vegetable ($r = 0.79$) products, respectively.

The ability of microorganisms to produce BAs is limited by low temperature [19]. Within this study, samples were stored at 4°C until investigation. However, fermented fishery and vegetable products are usually stored at room temperature in Cambodia due to the given conditions. Thus, even higher BA amounts could be expected for these products in this country. Paork chav (fermented fish) and other fermented fish products are stored at room temperature up to a year [21]. In the case of fermented vegetables, the salt content seems to be particularly relevant for the storage time. Thus, vegetables with 5–6% salt should be sold as soon as possible, while chaipov brey (salty fermented radish) samples with high salt concentrations (20–25%) have a longer storage time [8]. In this regard, higher BA values were detected in fermented vegetables with lower salt contents (2–5%).

4.4. BAs and Food Safety

Table 5 shows the distribution of the tested fermented food products according to the different allowable limits. Several organizations have set legal maximum limits on HIS concentrations in fermented foods that should ensure safe human consumption if the limits are not exceeded. Such organizations are the US Food and Drug Administration (FDA) with 50 ppm, FAO/WHO with 200 ppm for fish and fishery products, respectively, and EFSA with 400 ppm for fishery products that have undergone enzyme maturation treatment in brine [60,70,71]. The HIS level in fish sauce has been regulated in particular by the Codex Alimentarius Commission (CAC) and EFSA, with a maximum allowable limit of 400 ppm [55,72]. Correspondingly, the contents of HIS in all teuktrey (fish sauce) products did not exceed 400 ppm (Table 5). Thus, the levels of HIS in teuktrey (fish sauce) products in the current study can be regarded as safe for human consumption according to EFSA and CAC. Due to numerous outbreaks with toxic HIS concentrations ≥500 ppm [60], one paork chou (sour fermented fish), one mam trey (fermented fish), and two paork chav (fermented fish) products, representing about 7% (4/57) of the fermented products (Table 5), could pose a health risk. Although a food safety criterion is only set for HIS, HIS is not the only BA responsible for health hazards. Healthy individuals should also not be exposed to TYR values of 600 ppm or more by meal as recommended by EFSA [60]. The concentrations of TYR in all tested products were less than 600 ppm (Table 4), except for one paork chav (fermented fish) sample, which may constitute a health hazard [60]. Nevertheless, this sample is still fine according to Prester et al., who suggested a dietary value of up to 800 ppm of TYR as acceptable. More than 1080 ppm are toxic for adults [61].

Table 5. Distribution of fermented foods with quantifiable histamine contents.

Product Type	English Name	Histamine Contents (ppm)					
		≤50	>50 to 200	>200 to 400	>400 to <500	≥500	Total
Fermented Fishery Products—Number of Quantifiable Samples							
Teuktrey (n = 7)	Fish sauce	1 (14%)	3 (43%)	3 (43%)			7
Prahok (n = 12)	Fish paste	1 (9%)	7 (58%)	3 (25%)	1 (8%)		12
Kapi (n = 6)	Shrimp paste	6 (100%)					6
Paork chav (n = 7)	Fermented fish	1 (14%)		2 (29%)	2 (29%)	2 (29%)	7
Paork chou (n = 3)	Sour fermented fish	1 (33%)	1 (33%)			1 (33%)	3
Mam trey (n = 3)	Fermented fish	1 (33%)	1 (33%)			1 (33%)	3
Trey proheum (n = 4)	Salted fish	2 (50%)	2 (50%)				4
Total		13 (31%)	14 (33%)	8 (19%)	3 (7%)	4 (10%)	42 (100%)
Fermented Vegetables—Number of Quantifiable Samples							
Chaipov brey (n = 3)	Salty fermented radish	3 (100%)					3
Chaipov paem (n = 3)	Sweet fermented radish	3 (100%)					3
Trasork chav (n = 3)	Fermented melon	3 (100%)					3
Spey chrourk (n = 3)	Fermented mustard	1 (33%)	2 (67%)				3
Mam lahong (n = 3)	Fermented papaya	2 (67%)	1 (33%)				3
Total		12 (80%)	3 (20%)				15 (100%)

TYR concentrations from <0.4 to 270.6 ppm in commercially Chinese fish sauces [73], from 77.5 to 381.1 ppm in Korean anchovy sauces [74], and from 0 to 1178 ppm in commercial fish sauces the Far East sold at German markets [75] were reported. Hence, the concentrations of TYR in teuktrey (fish sauce) from retail markets in Cambodia were generally within or even below these concentrations (Table 4).

It has also been reported that the acute toxicity levels of TYR and CAD are respectively greater than 2000 ppm and the oral toxicity level of PUT is 2000 ppm [15]. It is known that TYR has a stronger and more rapid cytotoxic effect than HIS [76]. Unlike HIS and TYR, the pharmacological activities of PUT and CAD seem to be less potent. Nonetheless, both amines show in vitro cytotoxicity at concentrations easily reached in inherently BA-rich foods [77] and enhance the toxicity of HIS and TYR [19]. In the tested teuktrey (fish sauce) products, the highest levels of PUT (404 ppm) and CAD (766 ppm) were higher than those in fish sauce sold at Malaysian markets (242.8 ppm PUT and 704.7 ppm CAD) [78] and Chinese markets (276.6 ppm PUT and 606.3 ppm CAD) [73] but much lower than the levels in imported fish sauce products sold at German markets (1257 ppm PUT and 1429 ppm CAD) [75], Austrian markets (510 ppm PUT and 1540 ppm CAD) [65], and other European markets (1220 ppm PUT and 1150 ppm CAD) [60]. Extremely high PUT and CAD contents characterize inferior fish sauces, which may be due to the minor production hygiene, less salt content (<20%), the type of fish species, and storage condition [19,61]. Nevertheless, a health risk from consuming such a fish sauce is likely to be excluded due to the relatively small average intake [75]. Interestingly, the PUT concentrations of fish sauce samples were generally lower than the associated CAD concentrations (Table 4). The complexity of fish protein, which releases more lysine (precursor of CAD) during the fermentation of fish sauce, resulting in increased CAD concentrations could be the reason [78]. The current results also show that the highest PUT and CAD values in fish and shrimp pastes were higher than those in paste products in Taiwan [35] and in the Maldives [79]. Generally, BAs were detected in low levels in the tested fermented vegetables, which should not cause any risk when consumed. These results were consistent with previous studies [60,80], which reported that fermented vegetables should be considered as low-risk products in terms of BAs.

4.5. BAs and Food Quality

The HIS content alone may be a reliable indicator of food safety, but not of food quality. TYR and CAD are used as spoilage index [81]. Other authors have considered PUT and CAD as spoilage indicators [82]. Furthermore, PUT and CAD increase with longer storage times [19] and give strong unpleasant decaying odors at very low concentrations [75]. Therefore, the PUT and CAD concentration could be used as quality indicator [19], and their accumulation should be avoided [15,77]. These BAs are also included in the biogenic amine index (BAI) [83]. The BAI, the sum of HIS, TYR, PUT, and CAD, is more indicative of food quality, as these BAs are mostly produced at the end of shelf-life, indicating spoilage [83,84]. This index was also established to facilitate the evaluation and comparison of the BA concentrations in food. However, the usefulness of the BAI as quality index depends on many factors, mainly concerning the nature of the product (e.g., fresh or fermented food). Owing to the number of different factors (e.g., fermentation, maturation, starters), BA amounts vary much more in fermented products [13]. Thus, the BAI has proven to be more satisfactory for fresh products, and there is a BAI for freshwater fish of 50 ppm [85], while it is missing for fermented fishery products. The only BAI for a fermented food product was given by Wortberg and Woller [84] for Bologna sausages at 500 ppm. The higher BAI mainly results from the fermentation process and/or ripening. Using this limit, about one-third (31%) of the fermented fishery products in this study had a BAI of less than 500 ppm, while two-thirds (69%) had a higher BAI (>500 ppm), indicating a poor hygienic quality (Figure S3). Of the fermented vegetables, about 13% (2/15) had a BAI higher than 300 ppm (Figure S4). This value corresponds to the sum of HIS, TYR, PUT, and CAD, which should not be exceeded by acceptable sauerkraut [57].

In view of these results, the production process, distribution, and domestic handling of fermented products should be re-evaluated under strict hygienic practices together with the hazard analysis critical control point (HACCP) approach in order to minimize the content of BAs and microbiological contamination. The storage of food by cooling or freezing requires electricity that is not available to all Cambodians. Therefore, preservation techniques, such as the use of antimicrobial substances and/or autochthonous starter cultures, which are characterized by the absence of any BA formation ability or the presence of a BA detoxification activity, should be tested for their possible application on an industrial and small scale.

5. Conclusions

The presence of microorganisms in the examined fermented samples presented no health risk since pathogenic and spoilage microorganisms were in acceptable ranges. Nevertheless, one paork chou (sour fermented fish), one mam trey (fermented fish), and two paork chav (fermented fish) products represent a health risk because of the high level of HIS (>500 ppm). One of the paork chav (fermented fish) samples additionally exceeded the recommended TYR maximum (>600 ppm) per meal. The totals of all BAs tested were higher than the recommended corresponding BAI values in about 69% of the tested fermented fishery and 13% of the vegetable products. This may indicate a poor hygienic quality for these products. Hence, the production process, distribution, and domestic handling of fermented products in Cambodia should be re-evaluated in order to minimize the content of BAs and microbiological contamination. Further research is required to establish preservation techniques that could be applied on an industrial and small-scale in Cambodia.

Supplementary Materials
Figure S1: Linear fitting between total biogenic amines and physicochemical parameters, including pH (A), water activity (B), and salt content (C), in fermented fishery products (n = 42). Each dot indicates a data set obtained from a single sample, Figure S2. Linear fitting between total biogenic amines and physicochemical parameters, including pH (A), water activity (B), and salt content (C), in fermented vegetable products (n = 15). Each dot indicates a data set obtained from a single sample, Figure S3. Biogenic amine index (BAI) evaluated for 42 samples of fermented fishery products. The bold horizontal line describes the limit value of 500 ppm, which is used to distinguish between fermented fishery products of good and poor hygienic quality, Figure S4. Biogenic amine index (BAI) evaluated for 15 samples of fermented vegetable products. The bold horizontal line describes the limit value of 300 ppm, which is used to distinguish between fermented vegetable products of good and poor hygienic quality.

Author Contributions: Conceptualization, D.L., K.J.D., and S.M.; Analysis, D.L.; Investigation, J.-M.S.; Resources, U.Z. and K.J.D.; Writing—original draft, D.L.; Writing—Review and Editing, D.L., S.M., and K.J.D.; Supervision, K.J.D. and S.M. All authors have read and agreed to the published version of the manuscript.

Acknowledgments: Special thanks is given to Vibol San for his valuable contribution. This work was supported by the European Commission for the Erasmus Mundus scholarship under the ALFABET project Reference number: 552071 and partially funded by the Schlumberger Foundation, Faculty for the Future Program for supporting the first author to pursue a Ph.D.

References

1. Ly, D.; Mayrhofer, S.; Domig, K.J. Significance of traditional fermented foods in the lower Mekong subregion: A focus on lactic acid bacteria. *Food Biosci.* **2018**, *26*, 113–125. [CrossRef]
2. Hammond, S.T.; Brown, J.H.; Burger, J.R.; Flanagan, T.P.; Fristoe, T.S.; Mercado-Silva, N.; Nekola, J.C.; Okie, J.G. Food Spoilage, Storage, and Transport: Implications for a Sustainable Future. *BioScience* **2015**, *65*, 758–768. [CrossRef]
3. Hubackova, A.; Kucerova, I.; Chrun, R.; Chaloupkova, P.; Banout, J. Development of solar drying model for selected Cambodian fish species. *Sci World J.* **2014**, *2014*, 439431. [CrossRef]
4. Collignon, B.; Gallegos, M.; Kith, R. *Global Evaluation of UNICEF's Drinking Water Supply Programming in Rural Areas and Small Towns 2006–2016: Country Case Study Report—Cambodia*; Unicef: Cambodia, 2017; p. 71.
5. Chuon, M.R.; Shiomoto, M.; Koyanagi, T.; Sasaki, T.; Michihata, T.; Chan, S.; Mao, S.; Enomoto, T. Microbial and chemical properties of Cambodian traditional fermented fish products. *J. Sci. Food Agric.* **2014**, *94*, 1124–1131. [CrossRef] [PubMed]
6. Faithong, N.; Benjakul, S.; Phatcharat, S.; Binsan, W. Chemical composition and antioxidative activity of Thai traditional fermented shrimp and krill products. *Food Chem.* **2010**, *119*, 133–140. [CrossRef]
7. Waché, Y.; Do, T.-L.; Do, T.-B.-H.; Do, T.-Y.; Haure, M.; Ho, P.-H.; Kumar Anal, A.; Le, V.-V.-M.; Li, W.-J.; Licandro, H.; et al. Prospects for food fermentation in South-East Asia, topics from the tropical fermentation and biotechnology network at the end of the AsiFood Erasmus+project. *Front. Microbiol.* **2018**, *9*. [CrossRef]

8.   Chrun, R.; Hosotani, Y.; Kawasaki, S.; Inatsu, Y. Microbioligical hazard contamination in fermented vegetables sold in local markets in Cambodia. *Biocontrol Sci.* **2017**, *22*, 181–185. [CrossRef]

9.   Grace, D. Food Safety in Low and Middle Income Countries. *Int. J. Environ. Res. Public Health* **2015**, *12*, 10490–10507. [CrossRef]

10.   FAO/WHO. Cambodia Country Report on Food Safety. In Proceedings of the FAO/WHO Regional Conference on Food Safety for Asia and the Pacific, Seremban, Malaysia, 24–27 May 2004.

11.   Cheng, M.; Spengler, M. *How (Un)Healthy and (Un)Safe is Food in Cambodia?* Konrad-Adenauer-Foundation Cambodia: Phnom Penh, Cambodia, 2016; p. 11.

12.   Soeung, R.; Phen, V.; Buntong, B.; Chrun, R.; LeGrand, K.; Young, G.; Acedo, A.L. Detection of coliforms, *Enterococcus* spp. and *Staphylococcus* spp. in fermented vegetables in major markets in Cambodia. *Acta Hortic.* **2017**, 139–142. [CrossRef]

13.   Ruiz-Capillas, C.; Herrero, A.M. Impact of Biogenic Amines on Food Quality and Safety. *Foods* **2019**, *8*, 62. [CrossRef]

14.   Visciano, P.; Schirone, M.; Tofalo, R.; Suzzi, G. Biogenic amines in raw and processed seafood. *Front. Microbiol.* **2012**, *3*, 188. [CrossRef] [PubMed]

15.   Biji, K.B.; Ravishankar, C.N.; Venkateswarlu, R.; Mohan, C.O.; Gopal, T.K. Biogenic amines in seafood: A review. *J. Food Sci. Technol.* **2016**, *53*, 2210–2218. [CrossRef]

16.   Spano, G.; Russo, P.; Lonvaud-Funel, A.; Lucas, P.; Alexandre, H.; Grandvalet, C.; Coton, E.; Coton, M.; Barnavon, L.; Bach, B.; et al. Biogenic amines in fermented foods. *Eur. J. Clin. Nutr.* **2010**, *64*, S95–S100. [CrossRef] [PubMed]

17.   Prester, L.; Orct, T.; Macan, J.; Vukusic, J.; Kipcic, D. Determination of biogenic amines and endotoxin in squid, musky octopus, Norway lobster, and mussel stored at room temperature. *Arh. Hig. Rada Toksikol.* **2010**, *61*, 389–397. [CrossRef]

18.   Rawles, D.D.; Flick, G.J.; Martin, R.E. Biogenic amines in fish and shellfish. *Adv. Food Nutr. Res.* **1996**, *39*, 329–365. [CrossRef]

19.   Gardini, F.; Özogul, Y.; Suzzi, G.; Tabanelli, G.; Özogul, F. Technological factors affecting biogenic amine content in foods: A review. *Front. Microbiol.* **2016**, *7*, 1–18. [CrossRef]

20.   Suzzi, G.; Torriani, S. Biogenic amines in fermented foods. *Front. Microbiol.* **2015**, *6*, 472. [CrossRef]

21.   Ly, D.; Mayrhofer, S.; Agung Yogeswara, I.B.; Nguyen, T.-H.; Domig, K.J. Identification, Classification and Screening for γ-Amino-butyric Acid Production in Lactic Acid Bacteria from Cambodian Fermented Foods. *Biomolecules* **2019**, *9*, 768. [CrossRef]

22.   ISO 15214. *Microbiology of Food and Animal Feeding Stuffs—Horizontal Method for the Enumeration of Mesophilic Lactic Acid Bacteria—Colony-Count Technique at 30 °C*; International Organization for Standardization: Geneva, Switzerland, 1998; p. 7.

23.   ISO 21528–2. *Microbiology of Food and Animal Feeding Stuffs—Horizontal Methods for the Detection and Enumeration of Enterobacteriaceas—Part. 2: Colony-Count Method*; International Organization for Standardization: Geneva, Switzerland, 2004; p. 15.

24.   ISO 13720. *Meat and Meat Products—Enumeration of Presumptive Pseudomonas spp.*; International Organization for Standardization: Geneva, Switzerland, 2010; p. 7.

25.   ISO 21527–2. *Microbiology of Food and Animal Feeding Stuffs—Horizontal Method for the Enumeration of Yeasts and Moulds—Part. 2: Colony Count Technique in Products with Water Activity Less Than or Equal to 0,95*; International Organization for Standardization: Geneva, Switzerland, 2008; p. 9.

26.   Kwolek-Mirek, M.; Zadrag-Tecza, R. Comparison of methods used for assessing the viability and vitality of yeast cells. *Fems Yeast Res.* **2014**, *14*, 1068–1079. [CrossRef]

27.   Essghaier, B.; Fardeau, M.L.; Cayol, J.L.; Hajlaoui, M.R.; Boudabous, A.; Jijakli, H.; Sadfi-Zouaoui, N. Biological control of grey mould in strawberry fruits by halophilic bacteria. *J. Appl. Microbiol.* **2009**, *106*, 833–846. [CrossRef]

28.   ISO 6888–1/AMD 1. *Microbiology of Food and Animal Feeding Stuffs—Horizontal Method for the Enumeration of Coagulase-Positive Staphylococci (Staphylococcus Aureus and Other Species)—Part. 1: Technique Using Baird-Parker Agar Medium—Amendment 1: Inclusion of Precision Data*; International Organization for Standardization: Geneva, Switzerland, 2003; p. 9.

29.   Kateete, D.P.; Kimani, C.N.; Katabazi, F.A.; Okeng, A.; Okee, M.S.; Nanteza, A.; Joloba, M.L.; Najjuka, F.C. Identification of *Staphylococcus aureus*: DNase and Mannitol salt agar improve the efficiency of the tube coagulase test. *Ann. Clin. Microbiol. Antimicrob.* **2010**, *9*, 23. [CrossRef] [PubMed]

30. Robinow, C.F. Observations on the structure of *Bacillus* spores. *J. Gen. Microbiol.* **1951**, *5*, 439–457. [CrossRef] [PubMed]

31. ISO 16649–2. *Microbiology of Food and Animal Feeding Stuffs—Horizontal Method for the Enumeration of Beta-Glucuronidase-Positive Escherichia Coli—Part. 2: Colony-Count Technique at 44 Degrees C Using 5-Bromo-4-Chloro-3-Indolyl Beta-D-Glucuronide*; International Organization for Standardization: Geneva, Switzerland, 2001; p. 8.

32. ISO 7937. *Microbiology of Food and Animal Feeding Stuffs—Horizontal Method for the Enumeration of Clostridium Perfringens—Colony-Count Technique*; International Organization for Standardization: Geneva, Switzerland, 2004; p. 16.

33. Šimat, V.; Dalgaard, P. Use of small diameter column particles to enhance HPLC determination of histamine and other biogenic amines in seafood. *LWT Food Sci. Technol.* **2011**, *44*, 399–406. [CrossRef]

34. Saarinen, M.T. Determination of biogenic amines as dansyl derivatives in intestinal digesta and feces by reversed phase HPLC. *Chromatographia* **2002**, *55*, 297–300. [CrossRef]

35. Tsai, Y.-H.; Lin, C.-Y.; Chien, L.-T.; Lee, T.-M.; Wei, C.-I.; Hwang, D.-F. Histamine contents of fermented fish products in Taiwan and isolation of histamine-forming bacteria. *Food Chem.* **2006**, *98*, 64–70. [CrossRef]

36. Lopetcharat, K.; Park, J.W. Characteristics of fish sauce made from pacific whiting and surimi by-products during fermentation storage. *J. Food Sci.* **2002**, *67*, 511–516. [CrossRef]

37. Park, J.-N.; Fukumoto, Y.; Fujita, E.; Tanaka, T.; Washio, T.; Otsuka, S.; Shimizu, T.; Watanabe, K.; Abe, H. Chemical composition of fish sauces produced in southeast and east Asian countries. *J. Food Compos. Anal.* **2001**, *14*, 113–125. [CrossRef]

38. Kobayashi, T.; Taguchi, C.; Kida, K.; Matsuda, H.; Terahara, T.; Imada, C.; Moe, N.K.; Thwe, S.M. Diversity of the bacterial community in Myanmar traditional salted fish *yegyo ngapi*. *World J. Microbiol. Biotechnol.* **2016**, *32*, 166. [CrossRef]

39. Daroonpunt, R.; Uchino, M.; Tsujii, Y.; Kazami, M.; Oka, D.; Tanasupawat, S. Chemical and physical properties of Thai traditional shrimp paste (Ka-pi). *J. Appl. Pharm. Sci.* **2016**, 58–62. [CrossRef]

40. Purnomo, H.; Suprayitno, E. Physicochemical characteristics, sensory acceptability and microbial quality of *Wadi Betok* a traditional fermented fish from South Kalimantan, Indonesia. *Int. Food Res. J.* **2013**, *20*, 933–939.

41. Lee, D.S. Packaging and the microbial shelf life of food. In *Food Packaging and Shelf Life: A Practical Guide*; Robertson, G.L., Ed.; CRC Press-Taylor and Francis Group: Boca Raton, FL, USA, 2010; pp. 55–79.

42. Doyle, M.E.; Glass, K.A. Sodium Reduction and Its Effect on Food Safety, Food Quality, and Human Health. *Compr. Rev. Food Sci. Food Saf.* **2010**, *9*, 44–56. [CrossRef]

43. ICMSF. Fish and Seafood Products. In *Microorganisms in Foods 8: Use of Data for Assessing Process Control and Product Acceptance*, 2nd ed.; Swanson, K.M., Ed.; Springer: New York, NY, USA, 2011; pp. 107–133. [CrossRef]

44. Cortés-Sánchez, A.D.J. *Clostridium perfringens* in foods and fish. *Regul. Mech. Biosyst.* **2018**, *9*, 112–117. [CrossRef]

45. Health Protection Agency. *Guidelines for Assessing the Microbiological Safety of Ready-To-Eat Foods*; Health Protection Agency: London, UK, 2009; pp. 1–33.

46. Panagou, E.Z.; Tassou, C.C.; Vamvakoula, P.; Saravanos, E.K.; Nychas, G.J. Survival of Bacillus cereus vegetative cells during Spanish-style fermentation of conservolea green olives. *J. Food Prot.* **2008**, *71*, 1393–1400. [CrossRef] [PubMed]

47. Bennett, R.W. Staphylococcal enterotoxin and its rapid identification in foods by enzyme-linked immunosorbent assay-based methodology. *J. Food Prot.* **2005**, *68*, 1264–1270. [CrossRef]

48. Ijong, G.G.; Ohta, Y. Physicochemical and microbiological changes associated with Bakasang processing—A traditional Indonesian fermented fish sauce. *J. Sci. Food Agric.* **1996**, *71*, 69–74. [CrossRef]

49. Aarti, C.; Khusro, A.; Arasu, M.V.; Agastian, P.; Al-Dhabi, N.A. Biological potency and characterization of antibacterial substances produced by *Lactobacillus pentosus* isolated from Hentak, a fermented fish product of North-East India. *Springerplus* **2016**, *5*, 1743. [CrossRef]

50. Besas, J.R.; Dizon, E.I. Influence of salt concentration on histamine formation in fermented Tuna Viscera (*Dayok*). *Food Nutr. Sci.* **2012**, *3*, 201–206. [CrossRef]

51. Larsen, H. Halophilic and halotolerant microorganisms-an overview and historical perspective. *Fems Microbiol. Lett.* **1986**, *39*, 3–7. [CrossRef]

52. EFSA. *Working Document on Microbial Contaminant Limits for Microbial PEST control Products*; OECD Environment, Health and Safety Publications: Paris, France, 2012; pp. 1–53.

53.   Food Safety Authority of Ireland. *Guidance Note No. 3: Guidelines for the Interpretation of Results of Microbiological Testing of Ready-To-Eat Foods Placed on the Market (Revision 2)*; FSAI: Dublin, Ireland, 2016; pp. 1–41.

54.   Centre for Food Safety. *Microbiological Guidelines for Food for Ready-To-Eat Food in General and Specific Food Items*; Centre for Food Safety, Food and Environmental Hygiene Department: Hong Kong, China, 2014; pp. 1–38.

55.   EFSA. Commission regulation (EC) No 2073/2005 of 15 November 2005 on microbiological criteria for foodstuffs. *Off. J. Eur. Union* **2005**, *338*, 1–26.

56.   Gilbert, R.; Louvois, J.D.; Donovan, T.; Little, C.; Nye, K.; Ribeiro, C.; Richards, J.; Roberts, D.; Bolton, F. Guidelines for the microbiological quality of some ready-to-eat foods sampled at the point of sale. *Commun. Dis. Public Health* **2000**, *3*, 163–167.

57.   Shalaby, A.R. Significance of biogenic amines to food safety and human health. *Food Res. Int.* **1996**, *29*, 675–690. [CrossRef]

58.   Silla Santos, M.H. Biogenic amines: Their importance in foods. *Int. J. Food Microbiol.* **1996**, *29*, 213–231. [CrossRef]

59.   Fardiaz, D.; Markakis, P. Amines in fermented fish paste. *J. Food Sci.* **1979**, *44*, 1562–1563. [CrossRef]

60.   EFSA. Scientific opinion on risk based control of biogenic amine formation in fermented foods. *EFSA J.* **2011**, *9*, 2393. [CrossRef]

61.   Prester, L. Biogenic amines in fish, fish products and shellfish: A review. *Food Addit. Contam. Part A Chem. Anal. Control Expo. Risk Assess.* **2011**, *28*, 1547–1560. [CrossRef]

62.   Food Intolerance Network. Histamine Levels in Foods. Available online: https://www.food-intolerance-network.com/food-intolerances/histamine-intolerance/histamine-levels-in-food.html (accessed on 25 November 2019).

63.   Healing Histamine. High Histamine Foods I Still Eat. Available online: https://healinghistamine.com/supposedly-high-histamine-foods-i-still-eat/ (accessed on 25 November 2019).

64.   Sanchez-Perez, S.; Comas-Baste, O.; Rabell-Gonzalez, J.; Veciana-Nogues, M.T.; Latorre-Moratalla, M.L.; Vidal-Carou, M.C. Biogenic Amines in Plant-Origin Foods: Are They Frequently Underestimated in Low-Histamine Diets? *Foods* **2018**, *7*, 205. [CrossRef]

65.   Rauscher-Gabernig, E.; Gabernig, R.; Brueller, W.; Grossgut, R.; Bauer, F.; Paulsen, P. Dietary exposure assessment of putrescine and cadaverine and derivation of tolerable levels in selected foods consumed in Austria. *Eur. Food Res. Technol.* **2012**, *235*, 209–220. [CrossRef]

66.   Ladero, V.; Calles-Enríquez, M.; Fernández, M.; Alvarez, A.M. Toxicological effects of dietary biogenic amines. *Curr. Nutr. Food Sci.* **2010**, *6*, 145–156. [CrossRef]

67.   Mah, J.H.; Park, Y.K.; Jin, Y.H.; Lee, J.H.; Hwang, H.J. Bacterial Production and Control of Biogenic Amines in Asian Fermented Soybean Foods. *Foods* **2019**, *8*, 85. [CrossRef]

68.   Rodriguez-Jerez, J.J.; Lopez-Sabater, E.I.; Roig-Sagues, A.X.; Mora-Ventura, M.T. Histamine, Cadaverine and Putrescine Forming Bacteria from Ripened Spanish Semipreserved Anchovies. *J. Food Sci.* **1994**, *59*, 998–1001. [CrossRef]

69.   Bover-Cid, S.; Hugas, M.; Izquierdo-Pulido, M.; Vidal-Carou, M.C. Amino acid-decarboxylase activity of bacteria isolated from fermented pork sausages. *Int. J. Food Microbiol.* **2001**, *66*, 185–189. [CrossRef]

70.   FDA. *Fish and Fishery Products Hazards and Controls Guidance*, 4th ed.; Health and Human Services, Public Health Service, Food and Drug Administration, Center for Food Safety and Applied Nutrition, Office of Food Safety: College Park, MD, USA, 2011; pp. 113–152.

71.   Food and Agriculture Organization of the United Nations/World Health Organization (FAO/WHO). *Joint FAO/WHO Expert Meeting on the Public health Risks of Histamine and other Biogenic Amines from Fish and Fishery Products*; Joint FAO/WHO Expert Meeting Report; FAO/WHO: Rome, Italy, 2012; pp. 1–111.

72.   Codex Alimentarius Commission (CAC). *Standard for Fish Sauce, CXS 302-2011 (Amended in 2012, 2013, 2018)*; FAO/WHO: Rome, Italy, 2011.

73.   Jiang, W.; Xu, Y.; Li, C.; Dong, X.; Wang, D. Biogenic amines in commercially produced Yulu, a Chinese fermented fish sauce. *Food Addit. Contam. Part B Surveill.* **2014**, *7*, 25–29. [CrossRef] [PubMed]

74.   Moon, J.S.; Kim, Y.; Jang, K.I.; Cho, K.-J.; Yang, S.-J.; Yoon, G.-M.; Kim, S.-Y.; Han, N.S. Analysis of biogenic amines in fermented fish products consumed in Korea. *Food Sci. Biotechnol.* **2010**, *19*, 1689–1692. [CrossRef]

75.   Stute, R.; Petridis, K.; Steinhart, H.; Biernoth, G. Biogenic amines in fish and soy sauces. *Eur. Food Res. Technol.* **2002**, *215*, 101–107. [CrossRef]

76.   Linares, D.M.; del Rio, B.; Redruello, B.; Ladero, V.; Martin, M.C.; Fernandez, M.; Ruas-Madiedo, P.;

Alvarez, M.A. Comparative analysis of the *in vitro* cytotoxicity of the dietary biogenic amines tyramine and histamine. *Food Chem.* **2016**, *197*, 658–663. [CrossRef]

77. Del Rio, B.; Redruello, B.; Linares, D.M.; Ladero, V.; Ruas-Madiedo, P.; Fernandez, M.; Martin, M.C.; Alvarez, M.A. The biogenic amines putrescine and cadaverine show in vitro cytotoxicity at concentrations that can be found in foods. *Sci. Rep.* **2019**, *9*, 120. [CrossRef]

78. Zaman, M.Z.; Bakar, F.A.; Selamat, J.; Bakar, J. Occurrence of biogenic amines and amines degrading bacteria in fish sauce. *Czech J. Food Sci.* **2010**, *28*, 440–449. [CrossRef]

79. Naila, A.; Flint, S.; Fletcher, G.C.; Bremer, P.J.; Meerdink, G. Biogenic amines and potential histamine—Forming bacteria in Rihaakuru (a cooked fish paste). *Food Chem.* **2011**, *128*, 479–484. [CrossRef]

80. Andersson, R.E. Biogenic amines in lactic acid-fermented vegetables. *Lebensm. Wiss. Technol.* **1988**, *21*, 68–69.

81. Galgano, F.; Favati, F.; Bonadio, M.; Lorusso, V.; Romano, P. Role of biogenic amines as index of freshness in beef meat packed with different biopolymeric materials. *Food Res. Int.* **2009**, *42*, 1147–1152. [CrossRef]

82. Li, M.; Tian, L.; Zhao, G.; Zhang, Q.; Gao, X.; Huang, X.; Sun, L. Formation of biogenic amines and growth of spoilage-related microorganisms in pork stored under different packaging conditions applying PCA. *Meat Sci.* **2014**, *96*, 843–848. [CrossRef] [PubMed]

83. Jairath, G.; Singh, P.K.; Dabur, R.S.; Rani, M.; Chaudhari, M. Biogenic amines in meat and meat products and its public health significance: A review. *J. Food Sci. Technol.* **2015**, *52*, 6835–6846. [CrossRef]

84. Wortberg, W.; Woller, R. Quality and freshness of meat and meat products as related to their content of biogenic amines. *Fleischwirtschaft* **1982**, *62*, 1457–1463.

85. Venugopal, V. Postharvest quality changes and safety hazards. In *Seafood Processing: Adding Value Through Quick Freezing, Retortable Packaging and Cook-Chilling*; Venugopal, V., Ed.; CRC Press, Taylor and Francis Group: Boca Raton, FL, USA, 2006; pp. 23–60.

8

What we Know and what we Need to Know about Aromatic and Cationic Biogenic Amines in the Gastrointestinal Tract

Alberto Fernández-Reina [1], José Luis Urdiales [1,2,*] and Francisca Sánchez-Jiménez [1,2]

[1]   Departamento de Biología Molecular y Bioquímica, Facultad de Ciencias, Universidad de Málaga, 29071 Málaga, Spain; afernandezreina4@gmail.com (A.F.-R.); kika@uma.es (F.S.-J.)

[2]   CIBER de Enfermedades Raras & IBIMA, Instituto de Salud Carlos III, 29010 Málaga, Spain

*   Correspondence: jlurdial@uma.es

Abstract: Biogenic amines derived from basic and aromatic amino acids (B/A-BAs), polyamines, histamine, serotonin, and catecholamines are a group of molecules playing essential roles in many relevant physiological processes, including cell proliferation, immune response, nutrition and reproduction. All these physiological effects involve a variety of tissue-specific cellular receptors and signalling pathways, which conforms to a very complex network that is not yet well-characterized. Strong evidence has proved the importance of this group of molecules in the gastrointestinal context, also playing roles in several pathologies. This work is based on the hypothesis that integration of biomedical information helps to reach new translational actions. Thus, the major aim of this work is to combine scientific knowledge on biomolecules, metabolism and physiology of the main B/A-BAs involved in the pathophysiology of the gastrointestinal tract, in order to point out important gaps in information and other facts deserving further research efforts in order to connect molecular information with pathophysiological observations.

Keywords: histamine; serotonin; catecholamines; polyamines; gastrointestinal tract; nutrition; inflammation; gastric cancer; bowel diseases; colon cancer

---

1. Introduction

Biogenic amines (BAs) are low molecular weight organic compounds synthetized in vivo by decarboxylation of L-amino acids or their derivatives, thus containing one or more amino groups [1]. BAs can be derived from L-basic amino acids, as for instance, histamine (HIS) derived from L-histidine, as well as putrescine (Put), agmatine (Agm), spermidine (Spd) and spermine (Spm) derived from L-arginine or L-ornithine, depending on the organism. Other BAs can also be synthetized from L-aromatic amino acids or derivatives in mammalian tissues, as is the case for serotonin (5-HT) and catecholamines (CAs) that have L-aromatic amino acids such as L-tryptophan and L-tyrosine as their precursors, respectively. Figure 1 shows chemical structures of BAs. Throughout this work, this set of biogenic amines derived from basic or aromatic L-amino acids are abbreviated as BA. We will focus our attention on the role of B/A-BAs, as they are the most important ones in the gastrointestinal context. Another important BA for the central nervous system (CNS), the gamma-aminobutyric acid (GABA), is derived from the amino acid L-glutamate. Many other BAs, outside the scope of this review, can also be synthetized in nature playing different roles along the phylogenetic scale (for instance, tyramine from L-tyrosine and cadaverine from L-lysine, among others) [2,3].

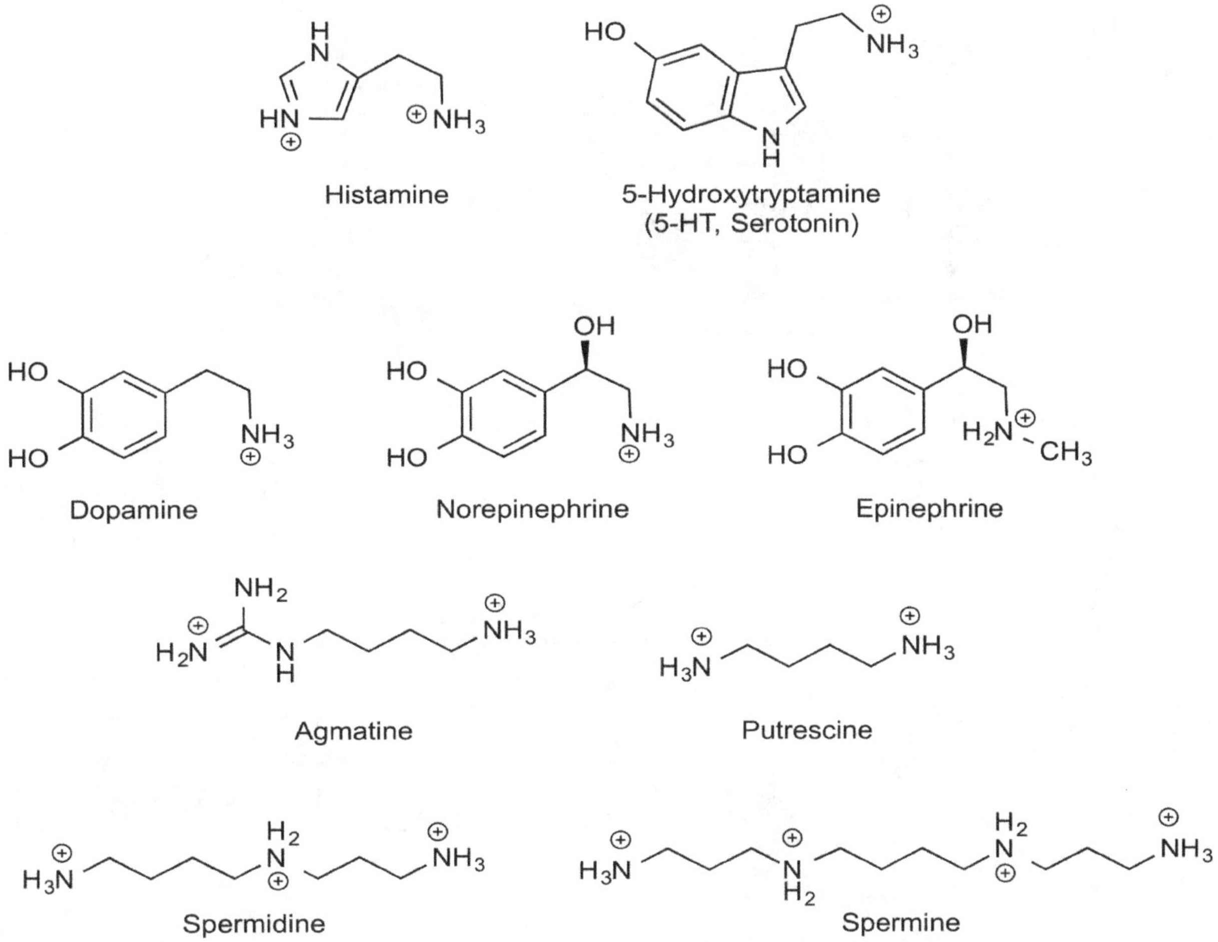

Figure 1. Chemical structures of B/A-BAs in their major forms at physiological pH. Histamine imidazole group is only partially protonated at pH 7 (pI ≈ 6).

All B/A-BA synthetic pathways include the alpha-decarboxylation of L-amino acids with cationic or aromatic side chains, or methylated or hydroxylated amino acid derivatives, as in the cases of 5-HT and CAs, respectively (Figure 2). In mammalian cells, B/A-BA synthesis involves the action of three pyridoxal 5'-phosphate (PLP)-dependent enzymes: ornithine decarboxylase (ODC, EC 4.1.1.17), histidine decarboxylase (HDC, EC 4.1.1.22) and aromatic L-amino acid decarboxylase (or DOPA decarboxylase, DDC, EC 4.1.1.28) [4–6]. In some cases, their common names used to derive from the precursor amino acid, as for HIS, that is synthetized from L-histidine, or 5-HT synthetized from L-tryptophan, but it is not a general rule. The metabolic origins of these BAs are shown in Figure 2 and further explained in the following sections.

Expressions of the involved PLP-decarboxylases—ODC, HDC and DDC—are cell-specific events, therefore linked to cell-specific developmental programs, for which we still ignore many involved factors. Both mammalian HDC and DDC share a high degree of homology; however mammalian ODC has a different evolutionary origin [7,8]. Nevertheless, all of them could compete for the cofactor PLP, in cases of vitamin B6 deficit or altered hepatic PLP metabolism (for instance, during aging [9]).

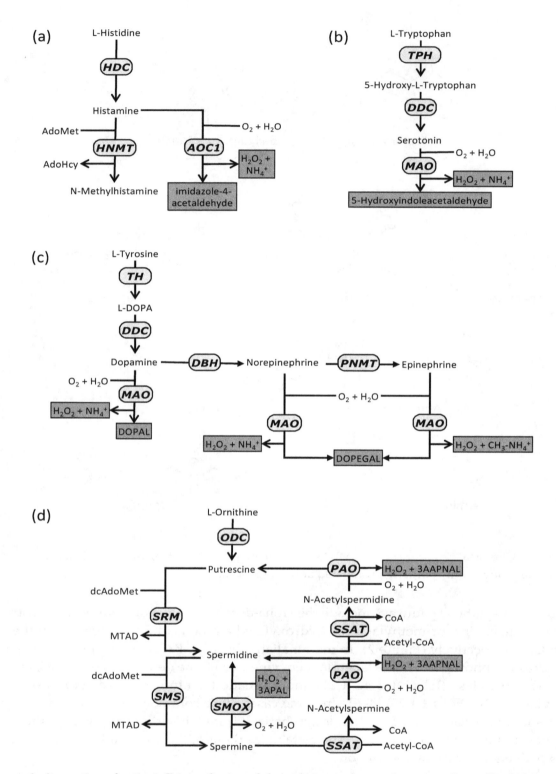

Figure 2. Aromatic and cationic BA synthesis and degradation pathways in mammalian cells. (**a**) histamine; (**b**) serotonin; (**c**) cathecolamines; (**d**) polyamines. Degradation products are depicted in orange boxes. Abbreviations (by alphabetical order): AdoMet, adenosylmethionine; AdoHcy, adenosylhomocysteine; AOC1; CoA, coenzyme A; diamine oxidase; DBH, dopamine β-hydroxylase; dcAdoMet, decarboxylated adenosylmethionine; ADDC, DOPA decarboxylase; DOPA, dihydroxyphenylalanine; DOPAL, 3,4-dihydroxyphenylacetaldehyde; DOPEGAL, 3,4-dihydroxyphenylglycoaldehyde; HDC, histidine decarboxylase; HNMT, histamine N-methyltransferase; MAO, monoamine oxidase; MTAD, methylthioadenosine; PAO, polyamine oxidase; PNMT, phenylethanolamine N-methyltransferase; SMOX, spermine oxidase; SMS, spermine synthase; SRM, spermidine synthase; SSAT, spermidine/spermine N^1-acetyltransferase; TH, tyrosine hydroxylase; TPH, tryptophan hydroxylase.

Another common fact is that BA degradation in vivo involves the action of amino oxidases. These reactions produce aldehydes (sometimes very toxic ones) and H_2O_2. A high oxidase activity could therefore cause local ROS and/or toxic aldehyde elevations. Amine oxidase specificities for each B/A-BA will be mentioned below. There are two families of amine oxidases, copper- or flavine-dependent oxidases [10,11]. These enzymes can be extra- or intracellular located and they also differ in the amine substrate specificities. For instance, the copper-dependent diamine oxidase (AOC1 or DAO, EC 1.4.3.22) can accept both HIS and Put as a substrate; these BAs come from different synthesis pathways. DDC products also share MAO activities (EC 1.4.3.4) [12]. Thus, degradation is also a process in which different BA metabolic pathways can eventually be confluent in the same physiological context.

In the following subsections, we will focus on descriptive overviews of the different B/A-BA specificities of their respective metabolic pathways and physiological functions in the gastrointestinal tract (GIT) system. It is a very complex physiological scenario still unveiled or confusing in many aspects. However, it is well known that BA metabolism in GIT can be highly decisive for health and quality of life, as occurring in the other physiological contexts mentioned above, therefore also deserving further biochemical and cellular research efforts to reach more efficient translational actions.

From a biochemical point of view, BAs were considered to be a part of secondary metabolism for many years, and consequently neglected in many general biochemistry textbooks. However, evidence has accumulated revealing important roles of these metabolites in mammalian pathophysiology. For instance, it is well known that HIS is an important mediator of the immune system, as well as a key biomolecule for correct gastric function [13–15] (Table 1). Nowadays, we can say that they are very important for human homeostasis, as they play important roles in the most important human physiological functions (neurotransmission, defence, digestion and nutrition, growth, apoptosis, and reproduction). Consequently, impairments in their metabolic (including signalling) pathways are related to many different pathological phenotypes and diseases.

Table 1. Nomenclature, precursors and main functions of the basic and aromatic amines involved in the gastrointestinal pathophysiology.

Common Names (Abbreviations) *	IUPAC Names	Precursor L-Amino Acids	Physiological Roles
Histamine (HIS)	2-(1H-Imidazol-4-yl)ethanamine	L-Histidine	Neurotransmitter. Immune mediator. Gastric acid secretion inducer.
Serotonin (5-HT)	3-(2-Aminoethyl)-1H-indol-5-ol	L-Tryptophan	Neurotransmitter related to reward motivated behaviour. Modulator of vessel constriction and intestinal motility.
Catecholamines (CAs):			
Dopamine (DA)	4-(2-Aminoethyl)benzene-1,2-diol	L-Tyrosine	Blood pressure regulators. Modulators of nutrient absorption and intestinal motility.
Epinephrine	(R)-4-(1-Hydroxy-2-(methyl amino)ethyl)benzene-1,2-diol		
Norepinephrine	(R)-4-(2-amino-1-hydroxy ethyl)benzene-1,2-diol		
Polyamines (PAs):			
Putrescine (Put)	Butane-1,4-diamine	L-Ornithine	Essential for cell viability, proliferation and correct differentiation.
Spermidine (Spd)	N'-(3-aminopropyl)butane-1,4-diamine	L-Ornithine + L-Methionine	
Spermine (Spm)	N,N'-bis(3-aminopropyl)butane-1,4-diamine		
Agmatine (Agm)	2-(4-aminobutyl)guanidine	L-Arginine	Anti-apoptotic effects. Positive effects on brain, hepatic and renal functions.

Data from references [16–24]. *, Abbreviations used in the text.

BAs can be synthesized de novo by specific mammalian cell types, but can also have an exogenous origin [19]. Microbiota, as well as microorganisms taking part in food processing or contamination, can produce biogenic amines from dietary amino acids at different rates and with different structures to those synthetized by human cells, which can have physiological effects; for instance, the decarboxylation product of L-arginine, Agm [20]. Its endogenous synthesis is, at least, controversial [21]. BAs are also present in a huge variety of drinks and foods, especially those in which microbial activity takes place during storage or preparation, sometimes with negative consequences for human health; for instance, toxicity due to HIS overproduction in contaminated seafood (i.e., by *Morganella morganii* sp.) [22] or high levels of amines in cold cuts and fermented foods (lactic products, fermented vegetables, wine, beer, etc.) [23,24]. In addition, BAs could form carcinogenic nitrosamines in the presence of nitrites during food processing [25].

A full characterization of the physiological effects of exogenous BAs has always faced two big handicaps: the multiple difficulties to evaluate the degree in which dietary amines are absorbed by gut epithelium, and the complexity of the characterization of the BA metabolism capacities of our particular microbiota, as this factor can induce important changes in the BA concentrations available to gut epithelium.

The importance of B/A-BAs in our digestive systems has been observed throughout the 20th century. In spite of all these valuable pathophysiological data (thousands of indexed publications) available, many gaps in molecular information still exist with regard to the mechanisms involved in each case, delaying the progress towards more personalized and accurate solutions for digestive-related pathologies [26]. As a research group working on several Systems Biology initiatives [27,28] and BAs [1,29,30], our hypothesis is based on the concept that integration of information can reveal emergent information, offering light to new hypothesis and translational actions. As far as we know, there is no recent similar review devoted to gathering biochemical and pathophysiological information on B/A-BAs in the GIT. Thus, the major aim of this work is to present an overview of the known facts of biochemical and pathophysiological information on B/A-BAs in the GIT context. The objective is to point out interesting facts deserving further research in order to eliminate gaps of molecular information currently blocking or delaying translational possibilities for prevention, diagnosis, and/or intervention of gastrointestinal diseases.

2. Histamine Biochemistry and Physiology

2.1. Histamine Synthesis

HIS was one of the first discovered low molecular weight (111 Da) immune mediators at the beginning of the last century [17,31,32]. Its precursor, the semi essential amino acid L-histidine, can be at least partially endogenous or derived from dietary proteins [33]. In mammalian cells, HIS synthesis occurs by decarboxylation of L-histidine, which is catalysed by the enzyme named L-histidine decarboxylase (HDC) (Figure 2a). This activity is carried out by PLP-dependent enzymes in both Gram-negative bacteria and Metazoa [34]. However, in Gram positive bacteria potentially present in intestines (for instance, *Lactobacillus* sp.), the reaction is catalysed by a non-homologous pyruvoil-dependent enzyme [5,34,35]. In human tissue, HDC is only expressed in a short list of cell types. Among them, several immune differentiated cells (mast cells and basophils) [36–38], histaminergic neurons [18,39], and gastric enterochromaffin-like cells (ECL cells) [40] are able to both synthetize and store HIS. Transformed HIS producing cells can preserve or even increase their HIS-producing capacity, as in the case of malignant mastocytosis and several types of gastric cancer cells [41,42]. Other cells (for instance, macrophages, eosinophils, and platelets) can also synthetize HIS to any extent, being unable to store it in specialized vesicles [31,35].

A big gap of information still exists on mechanisms controlling cell type-specificities with respect to HDC expression, but it seems to be clear that epigenetic events play important roles (i.e., methylation/demethylation of CpG islands present in mammalian HDC gene promoter) [43,44].

It has been observed that HDC gene expression can increase in response to several stimuli such as gastrin, estrogens, or several interleukins (ILs) as IL-1, IL-3, IL-12 or IL-18. It depends on the specific receptors expressed by the target HIS-producing cells [35,45].

Alternative splicing events have been observed during mammalian (and human) HDC expression in HIS-producing cells [46,47]. The meaning of these aberrant messengers is still unknown. As the active protein is a dimer and taking into account that dimerization involves interaction of both N-terminus [34], some of the truncated sequences could act as natural HDC inhibitors.

In addition, the protein needs to be processed to reach the active conformation and is a very unstable enzyme [48–53]. Regulation of the enzyme processing and turnover can be important as a determinant of active HDC levels. However, HDC processing and maturation is a process not fully characterized. It seems to be clear that the monomer mature form must correspond to a 53–63 KDa fragment of the N-terminus of the primary translation product [49]. Nevertheless, the precise sequence of this fragment in vivo is not yet known.

The action mechanism of mammalian PLP-dependent L-amino acid decarboxylases has been previously described [54,55]. Briefly, it involves two transaldamination reactions from the PLP-enzyme complex to the L-amino acid-enzyme complex, which is decarboxylated in the substrate α-carboxylic group to form a covalent amine product-enzyme complex. This last complex suffers a second transaldimination reaction with PLP to recover the initial PLP-enzyme complex, thus releasing the amine product. Important changes in the global decarboxylase conformation have been observed for both mammalian HDC and DDC during catalysis [56,57]. The quaternary structure of HDC and DDC only differs in tautomeric forms of intermediates along the reaction, most probably due to slight differences in the active dimer conformation [54,55]. In fact, both enzymes can share substrates (i.e., L-histidine, but with different affinities) and inhibitors (for instance, epigallocathechine-3-gallate). This fact needs to be taken into account for design of specific inhibitors of any of these activities with pharmacological purposes.

2.2. Exogenous Histamine Synthesis

Important quantities of HIS can be present in some natural products, such as oranges or tomatoes. Fermented products (cheese, alcoholic drinks, fermented vegetables and fish) can also contain high quantities of HIS (and other BAs), as a result of the metabolic properties of the living organism involved in each case. Contamination during food processing or storage can also allow undesirable growth of HIS-producing microorganisms. Many efforts of EU COST actions (i.e., COST Action 917, 922 and BN0806) have been devoted to the study and control of BA levels in foods (see for instance, [58,59]) However, it is a very complex subject with many variables and a lot of uncertainty with regard to amine absorption mechanisms and traceability. This issue requires applying more holistic approaches, and the use of high-throughput technologies, in order to efficiently translate the knowledge to both new general nutritional recommendations and personalized diets.

In addition, GIT microbiota species can synthetize HIS (and many other BAs) by using PLP- or pyruvoyl-dependent enzymes. Thus, microbiota characterization should also be considered during personalized medicine initiatives of patients affected by BA-related diseases [26].

2.3. Histamine Degradation

In human tissues, HIS can be degraded by two different pathways (Figure 2a). The first one involves the intracellular N-methylation of HIS in its imidazole group catalysed by histamine N-methyl transferase (HNMT, EC 2.1.1.8) [60]. It is an ubiquitous enzyme expressed in liver and also in intestinal mucosa in a minor extent [61,62]. Its product, N-tele methyl-histamine, is a substrate of monoamine oxidase (MAO), which produces N-methylimidazole acetaldehyde. Finally, the enzyme aldehyde dehydrogenase (AD, EC 1.2.1.5) reduces this metabolite to N-methylimidazole acetic acid. This seems to be the major pathway in the brain [63]. However in GIT, the main pathway for HIS degradation involves the direct HIS oxidation by the action of human DAO producing imidazole acetaldehyde [35]. It is a copper-containing glycoprotein associated with cytosolic membrane and expressed in the stomach, duodenum, small intestine, and colon [61,62]. It can be released from membranes of their producing cells and is active in human serum [64]. It is also known as amyloride-binding protein-1 (AOC1) and histaminase.

2.4. Histamine Transport and Storage Mechanisms

In mammalian cells, HIS can be transported into epithelial cells throughout organic cations transporters (OCT 2 and 3), and the plasmatic membrane monoamine transporter (PMAT). OCT 2 and 3, and PMAT are located in the basolateral plasmatic membranes. OCT 3 has also been located in the luminal membranes of bronchial and small intestine epithelial cells [35,65].

Inside the cell, HIS can be transported through endosomal membranes by using the vesicular amine/proton antiporter systems named vesicular monoamine transporter 2 (VMAT2 or SLC18A2), which is also able to transport other monoamines such as DA, norepinephrine and 5-HT, and can be modulated by drugs such as amphetamines and cocaine [66].

With respect to storage, as mentioned previously, only a few cell types are able to store HIS. The major HIS-storing cells in a human body are mature mast cells, which can accumulate HIS, as well as 5-HT and even PAs into secretory granules, most probably derived from *trans*-Golgi vesicles [67]. Mast cells are infiltrated into mammalian epithelia, including GIT epithelia. Other important HIS storing cells in GIT are gastric ECL cells [68].

Three different mechanisms have been described for HIS secretion [35]:

1. Mast cell degranulation by immune stimuli. The presence of specific antigens induces IgE synthesis, inducing a high affinity binding between the specific IgE and IgE receptor known as FcεRI. This high affinity complex induces degranulation after further expositions to the antigen.

2. Cytokines can also induce degranulation. It is mediated by vesicular trafficking events involving fusion and/or content interchange between secretory granules and vesicles driven to exocytosis.

3. Constitutive HIS leakage due to non-active transport through cytosolic membranes or *trans*-Golgi vesicles driven to exocytosis.

2.5. Histamine Signalling and Physiological Functions

HIS could be considered the most pleiotropic amine, as it is involved in a wide spectrum of physiological processes concerning the most important function for a human being. It is a well-known immune mediator, as well as a neurotransmitter, thus involved in the most complex and still not fully characterized physiological capabilities. It also plays a key role in gastric acid secretion, and has also been described as a cell proliferation modulator, nutrition and cell proliferation being two essential functions for life [13].

HIS effects in different physiological scenarios are elicited by different HIS receptors. Four specific HIS receptors have been detected so far, namely H_1R, H_2R, H_3R and H_4R. All of them are members of the G-protein coupled receptor family. Their expressions are cell-type dependent and the elicited signals sometimes contradictory. Nevertheless, all of them somehow participate in GIT functions and homeostasis. Table 2 summarizes their specific characteristics.

Table 2. Molecular and functional properties of human histamine receptor types.

Properties	HIS Receptor 1 (H$_1$R)	HIS Receptor 2 (H$_2$R)	HIS Receptor 3 (H$_3$R)	HIS Receptor 4 (H$_4$R)
Chromosome	3	5	20	18
Molecular weight (KDa)	56	40	49	44
G protein signalling	Gα_q	Gα_s	G$_{i/o}$	G$_{i/o}$
Elicited signalling	PLC activation Increase of Ca^{2+} Production of NOS and cGMP	PKA activation Increase of cAMP PLC activation Increase of Ca^{2+}	Decrease of cAMP Inhibition of Ca^{2+} channels	Inhibition of cAMP Stimulation of MAP kinase phosphorylation
Expression	Brain, smooth muscle, skin, gastrointestinal and genitourinary tract, adrenal medulla, immune system and heart	Brain, smooth muscle, skin, gastrointestinal and genitourinary tract, adrenal medulla, immune system and heart	Widely found in brain and gastric mucosa	Inflammatory cells, dendritic cells and peripheral nerves
Physiological effects	Smooth muscle contraction Vasodilation and increase of vascular permeability	Inhibition of chemotaxis in basophils, gastric secretion of HCl and duodenal bicarbonate secretion	Release regulation of HIS (and other neurotransmitters) release from neurons Inhibition the secretion of gastric acid	Inflammatory processes such as allergies and asthma

Data from references [16–18]. Abbreviations; cAMP, 3′-5′-cyclic adenosine monophosphate; cGMP, 3′-5′-cyclic guanosyl monophosphate; HIS, histamine; MAP, mitogen-activated protein; NOS, nitric oxide synthase; PLC, phospholipase C.

H$_1$R and H$_2$R are the most ubiquitous receptors along the GIT. H$_1$R, H$_2$R and H$_3$R are located in gastric mucosa and their affinities for HIS are dependent on the expressed isoforms. H$_4$R can be expressed by inflammatory cells, as well as peripheral neurons associated with GIT. Specific HIS actions throughout the GIT are summarized below. H$_4$R was the most recent HIS receptor discovered, and is still not fully characterized, in spite of the multiple efforts made by international research groups [69]. Nevertheless, insights point out to its importance in GIT physiology, thus encouraging new actions to decipher this still veiled but important information for GIT pathophysiology. In fact, the receptor is proposed to be involved in gastric acid secretion, gastric mucosa defence, intestinal motility and secretion, visceral sensitivity, inflammation, immunity and gastric and colorectal carcinogenesis [70].

2.5.1. Histamine and Acid Gastric Secretion

Gastric acid secretion is regulated by different positive stimuli, such as acetylcholine, HIS and gastrin, and inhibitors such as somastostatin. Acetylcholine is a neurotransmitter coming from enteric neurons. HIS, gastrin and somatostatin are secreted by different endocrine cells infiltrated in GIT mucosa; they include ECL cells, G cells and D cells, respectively [71].

Figure 3 is a scheme of the balance between stimuli and inhibitors of gastric secretion. Briefly, on the one hand, binding of acetylcholine (from enteric neurons) to specific receptors stimulates parietal cells to secrete HCl; as well as gastrin (from gastric epithelium G cells), which binds to the cholecystokinin receptor 2 (CCK2 receptor) of ECL cells, thus inducing HIS secretion. As mentioned before, HIS is a stimulus for HCl secretion by parietal cells through the signalling pathway elicited by H$_2$R. On the other hand, circulating cholecystokinin (CCK) binds to CCK1 receptors of gastric D cells, thus stimulating somatostatin secretion [71]. Somatostatin directly inhibits acid secretion by parietal cells, as well as both HIS and gastrin secretion [72].

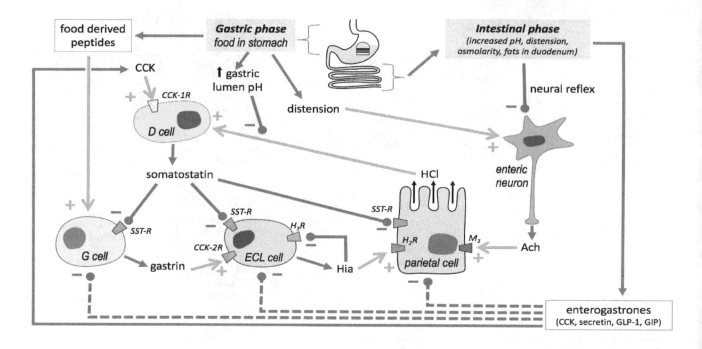

Figure 3. Balance between stimuli and inhibitors of gastric secretion. CCK, cholecystokinin; CCK_nR, different types of cholecystokinin receptors (1 or 2); D cell, somatostatin-releasing cell; ECL cell, enterochromaffin-like cell; G cell, gastrin-producing cell. GLP-1, glucagon-like peptide; GIP, gastric inhibitory polypeptide; H_nR, different types of histamine receptors; M_3, muscarinic acetylcholine receptor type 3; SST-R, somatostatin receptors. Products are represented by grey arrows, activations by green plus symbols and arrows, and inhibitions by minus symbols and red bars.

The balance between stimuli and inhibitors change throughout different phases involved in the process, including an intracranial phase, a gastric phase, and an intestinal phase. In the next paragraphs, we will focus on phases directly related to GIT.

In the gastric phase, the presence of food in the stomach induces acid gastric secretion by three different ways (Figure 3): the stomach distention caused by the food is detected by mechanoreceptors, which in turn induces neuronal reflexes for acetylcholine production; food derived-peptides and amino acids stimulate gastrin secretion by G cells; food increases gastric lumen pH, which is an inhibitory signal for somatostatin secretion [72].

When chyme reaches the duodenum, negative feedback mechanisms operate to reduce acid secretion (Figure 3). On the one hand, neuronal reflexes are activated, therefore blocking acetylcholine induced HCl secretion. On the other hand, in enteroendocrine cells, the synthesis of somatostatin synthesis activators (i.e., CCK, secretin, glucagon-like peptide and gastric inhibitory polypeptide) [73] are also promoted in different enteroendocrine cell types, which finally lead to gastrin, HIS and HCl secretion inhibition (Figure 3).

It has been proposed recently that HIS could also inhibit its own secretion through binding to H_3R present in ECL cells membranes [17]. Acting through other receptors, HIS has also been proposed as involved in the gastric vasodilatation and reactive hyperaemia produced in response to acid challenge (through H_1R), and the modulation of the gastric mucosal defence, the enteric neurotransmission and the feedback regulation of HIS release (through H_3R, and maybe also through H_4R) [73,74]. Nevertheless, the precise roles of H_4R in gastric physiology are still controversial [75].

2.5.2. Histamine and Immune Response in Gastrointestinal Tract

As mentioned before, mast cells (and basophyls) are the major producers of HIS. In these cells, HIS release can be induced by IgE, but also by cytokines, neuropeptides, growth factors, free radicals and anaphylotoxins [17], many of them potentially present in the GIT. Other eventually HIS-producing, but not HIS-storing cells, like lymphocytes, fibroblasts and macrophages are also present and interact with GIT epithelia [76,77].

HIS modulates immune response mainly through H_1R, H_2R and H_4R, depending on the receptor type expressed in each immune cell type. HIS elicited immune actions include the capability to modulate expression and/or activity of many cytokines and the complement system [78–80]. Again in turn, a cross regulation between cytokines and HIS seems to control GIT functions, as it has been proven that several cytokines such as TNF-α, IL-1 and IL-6 modulate HIS synthesis and secretion. Several components of complement systems like C3a, C4a and C5a (anaphylotoxins) have the capability to induce HIS release from mast cells and basophils [17]. During the last decade, interest of the role of H_4R in the GIT context has increased [81].

3. Serotonin Biochemistry and Physiology

3.1. Serotonin Synthesisn

5-HT is an L-tryptophan-derivative (Table 1). Meat, milk and fruit are the major sources of the essential amino acid precursor [82]. About 95% of the total 5-HT content in a human body is synthetized by GIT-associated cells (approximately 9/10 by intestinal enterochromaffin-cells (EC), and 1/10 by serotoninergic neurons located in the myenteric plexus. Only 5% of the 5-HT content in a human body is estimated to be synthetized in CNS [83].

The biosynthetic 5-HT pathway begins with the hydroxylation of the indole moiety C5 (Figure 2b) catalysed by the enzyme tryptophan hydroxylase (TPH, EC 1.14.16.4). It is a tetrameric non-heme iron-dependent monooxygenase that uses L-Trp and oxygen as substrates and tetrahydrobiopterin (BH_4) as the cofactor. The reaction occurs as two sequential half reactions: a reaction between the active site iron, oxygen, and the tetrahydropterin to produce a reactive $Fe^{IV}O$ intermediate and the hydroxylation of the amino acid by $Fe^{IV}O$ [84]. Two isoforms have been detected for this enzyme. TPH-2 expression is almost exclusive for neurons, and TPH-1 is expressed in other 5-HT-producing cell types [85].

The TPH product, 5-hydroxy-L-tryptophan (5-HTP), is the substrate of DDC that produces the amine 5-HT by decarboxylation of the 5-HTP α-carbon. This enzyme also decarboxylates other aromatic L-amino acids or derivatives; for instance, L-dihydroxyphenylalanine (L-DOPA) to produce DA [86]. In addition, it is also able to catalyse other reactions under different environmental circumstances or specific mutations (for instance, half-transaminations and oxidative deaminations) [87,88]. The mammalian enzyme is also a PLP-dependent enzyme, highly homologous to mammalian HDC, as mentioned above [26,89]. In fact, it is able to accept L-His as a substrate but with a much lower affinity than for 5-HTP or DOPA; however, the human DDC gene lacks the sequence encoding the carboxy-terminal fraction present in mammalian HDC, which is involved in mammalian HDC sorting to endoplasmic reticulum and activation [49]. This could suggest a different intracellular location for both enzymes. Mammalian HDC and DDC share the catalytic mechanism explained above for HDC [54,90]. However, at least in the case of the purified recombinant wild proteins, mammalian DDC seems to be a more efficient enzyme according to their respective catalytic constant values obtained in silico and in vitro [54,55]. In the case of DDC, slight modifications of the catalytic site environment seem to induce important changes in catalytic constant (k_{cat}) values [90].

Both enzymes (DDC and HDC) also share other structural properties related to enzyme stability and catalysis. For instance, the presence of PEST regions in the N-terminus of the monomers and a highly labile flexible loop, which is essential for conformation changes of the enzymes during PLP binding and for catalysis itself [6,53,88]. In the case of 5-HT biosynthesis, the limiting step is not decarboxylation but TPH activity.

It is noteworthy that HIS and 5-HT synthesis exhibit antagonist time-course patterns during differentiation of mouse bone marrow derived cells to mast cells in vitro, as well as opposite responses to PA inhibitors [67]. These results suggest a sort of regulatory coordination among all of these amine biosynthetic processes, which are not fully characterized yet, but should be taken into account in all of the pathophysiological scenarios where synthesis of these amines can be confluent, as GIT is.

3.2. Serotonin Degradation

5-HT degradation is catalysed mainly by any MAO activity (Figure 2b). MAO catalysis requires FAD as the cofactor to carry out an oxidative deamination of 5-HT, thus producing hydrogen peroxide and 5-hydroxy-3-indolacetaldehyde, which is rapidly processed to 5-hydroxy-3-indolacetic acid by the action of AD [91]. Human genome contains two different genes encoding MAO activities, namely MAO-A and MAO-B [12]. Both proteins are located in the external mitochondrial membrane [92]. MAO-A has a higher affinity by 5-HT as well as a wider expression spectrum. However, MAO-B is the only one detected in serotoninergic neurons [91]. Both MAO isozymes are expressed in GIT [61,62,93].

3.3. Serotonin Transport and Storage Mechanism

In GIT, 5-HT is mainly produced and secreted by the neuroendocrine enterochromaffin (EC) cells, located alongside the intestinal epithelium lining the lumen of the digestive tract. Recently, it was described that the sodium channel $Na_v 1.3$ plays an important role for EC excitability and 5-HT release [94]. Once 5-HT is secreted by EC cells and binds to the specific receptors of the surrounding cells, it is removed from the interstitial space by the sodium-dependent 5-HT transporter (SERT), also named as the solute carrier family 6 member 4 (SLC6A4), which is expressed by GIT epithelial cells. SERT is a protein with 12-transmembrane fragments, which is able to transport 5-HT by a Na^+/K^+- and Cl-dependent mechanism [83,95]. Inside the GIT epithelial cells, 5-HP is rapidly degraded by MAO activity, as explained above [83,95,96].

In addition, postprandial 5-HT can also enter systemic circulation and is absorbed by platelets. Actually, most of the circulating 5-HT is stored in platelets, as these cells also express 5-HT transporters. SERT is negatively regulated by activation of tool-like receptors and several pro-inflammatory cytokines. On the contrary, other anti-inflammatory cytokines, such as IL-10, increase transporter activity. Treatment with specific SERT inhibitors leads to an increase of free 5-HT content, thus empowering 5-HT effects, not only in GIT but also in CNS [83].

3.4. Serotonin Signalling and Physiological Functions

It is well known that 5-HT has also been involved in very complex physiological processes such as being an essential neurotransmitter and paracrine molecule for brain-intestine crosstalk, commonly known as gut-brain axis [97]. The amine is involved in modulation of body temperature and circadian rhythm [98,99], as well as in cardiovascular activity, morphogenesis and cell proliferation [100,101]. In the GIT context, it has been described as a gastric motility and secretion, a nutrient absorption regulator and an immunoregulatory compound. Consequently, dysfunctions in 5-HT metabolism usually have very important negative consequences on human physiology including gut-brain communication [82,102]. 5-HT also elicits both motor and sensitive responses in the intestine by binding to different receptors expressed by mesenteric and mucosal neurons (Table 3).

Table 3. Molecular and functional properties of the best-known human 5-HT receptor subtypes important for GIT functions.

Properties	5-HT$_{1A}$ Receptors (5-HT$_{1A}$R)	5-HT$_{1D}$ Receptors (5-HT$_{1D}$R)	5-HT$_2$ Receptors (5-HT$_2$R)	5-HT$_3$ Receptors (5-HT$_3$R)	5-HT$_4$ Receptors (5-HT$_4$R)	5-HT$_7$ Receptors (5-HT$_7$R)
Chromosome	5	6	13/2/X	11 (A, B and C) and 3 (D and E)	5	10
Molecular weight (KDa)	421	390	471/481/458	Pentameric 478 (A); 441 (B); 447 (C); 279 (D); 471 (E)	387	445
G protein signalling	G$_{i/o}$	G$_{i/o}$	G$_{q/11}$	Activated by ligand binding and opening channels	G$_s$	G$_s$
Expression	Enteric neurons, substantia nigra, hippocampus	Enteric neurons, substantia nigra, basal ganglia	Stomach, fundus, caudate nucleus, cerebellum	Enteric, sympathetic and vagus nerves, area postrema	Enteric neurons (myenteric plexus), hippocampus	Smooth muscle, thalamus, hypothalamus and hippocampus
Physiological effects	Neuronal inhibition	Neuronal inhibition	Muscle contraction	Neuronal depolarization Increased neurotransmitter release	Muscle contraction Positive effects on cholinergic transmission.	Muscle relaxation

Data from references [95,103].

3.4.1. Regulation of GIT Smooth Muscle Contraction and Relaxation

5-HT is a regulator of both intestinal smooth muscle contraction and relaxation through the activation of enteric excitatory motor neurons and intrinsic inhibitory neurons, respectively [83,104]. The amine can bind 5-HT$_3$R and 5-HT$_4$R of excitatory cholinergic motor neurons, thus inducing acetylcholine release and smooth muscle contraction. However, 5-HT binding to 5-HT$_4$R, 5-HT$_{1A}$R, and/or the badly characterized 5-HT$_{1D}$R, present in inhibitory nitrergic motor neurons induces nitric oxide (NO) synthesis and consequently smooth muscle relaxation (Figure 4). In addition, 5-HT also participates in gastric muscle motility regulation [83,105].

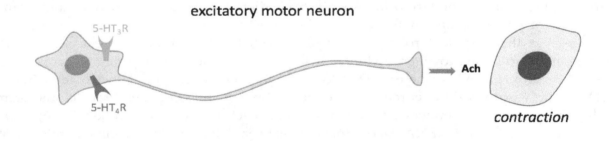

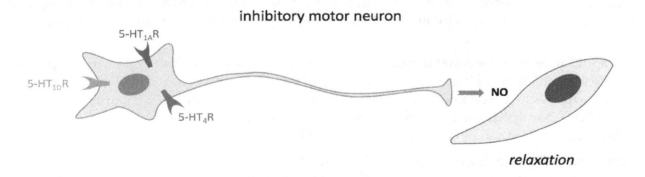

Figure 4. Regulation of GIT smooth muscle contraction and relaxation through 5-HT receptors. AcH, acetylcholine; 5-HT$_{1A}$R serotonin receptor type 1A; 5-HT$_{1D}$R; serotonin receptor type 1D, 5-HT$_3$R; serotonin receptor type 3, 5-HT$_4$R; serotonin receptor type 4.

3.4.2. Mucosal Sensory Transduction

EC cells secrete 5-HT in response to intraluminal pressure. The released amine can stimulate both the intrinsic primary afferent neurons (IPANs) located in submucosal and myenteric plexus and the extrinsic afferent neurons (vagal and spinal), through their binding to different receptors: 5-HT$_3$R, 5-HT$_4$R, 5-HT$_7$R; and 5-HT$_{1D}$R [83].

On the one hand, submucosal neurons release both acetylcholine and calcitonin gene-related peptide; however, myenteric neurons only release acetylcholine. Both neuron types are involved in the modulation of intestinal motility, secretion, and vasodilatation. Thus, submucosal neurons initiate peristaltic and secretory reflexes, and myenteric neurons start migratory contractions. On the other hand, spinal afferent neurons transmit signals related to digestive reflexes, satiety, and pain from the intestine to the CNS.

In addition, some authors claim an important role of neuronal 5-HT in promotion of development/survival of some classes of late-born enteric neurons, including dopaminergic neurons, which appear to innervate and activate the adult enteric nervous system [104].

3.4.3. Serotonin and Immune Response in GIT

It has been reported that 5-HT can also elicit pro-inflammatory responses in GIT that involve different transduction pathways most probably started by the amine binding to the 5-HT receptors expressed by dendritic cells located in *lamina propria*. Recently, this immuneregulatory role of 5-HT in GIT has been the subject of very relevant reviews on the topic [95,102].

4. Biochemistry and Physiology of Catecholamines

4.1. Synthesis of Catecholamines

DA, and their derivatives noradrenaline/norepinephrine and adrenaline/epinephrine, are the most important cathecolamines for human physiology (Table 1). All of them are synthetized from L-phenylalanine or L-tyrosine mainly from diet (Figure 2c). L-phenylalanine can be the substrate of phenylalanine hydroxylase (PAH) to produce L-tyrosine. The limiting step for DA synthesis is the enzyme tyrosine hydroxylase (TH, tyrosine 3-monooxygenase, EC 1.14.16.2), which introduces a hydroxyl group in the meta position of the cathecol ring of L-tyrosine to obtain L-3,4-dihydroxyphenylalanine or L-DOPA). This reaction requires Fe^{2+}, the cofactor BH$_4$ and O$_2$. TH is mainly expressed in neuroendocrine cells in both soluble and membrane bound forms. It is a highly stereo specific enzyme, although it can also act on L-phenylalanine [82,106]. Up to four different alternative TH mRNA spliced forms have been detected. The meaning of this variability is still uncertain [106].

L-DOPA is a substrate of the DDC mentioned in the previous section, producing DA. In fact, it has also been named as dopa decarboxylase in the literature [6,90]. La AADC or DDC is a ubiquitous enzyme expressed by cell types located in different organs like the adrenal medulla, kidney, liver, GIT and brain [61,106].

In addition to the above-mentioned pathway, there are two other alternative ways to produce DA (not shown in Figure 2). One way, L-tyrosine, can also be decarboxylated by AADC to produce tyramine, which is hydroxylated by a member of the cytochrome P450 family (family 2, subfamily D, or CYP2D). Nevertheless, L-phenylalanine can also be decarboxylated by AADC to produce phenyltyramine, which can be converted to DA by CYP2D [92].

In several peripheral tissues (mainly in adrenal medulla), two other further reactions can take place to produce norepinephrine and epinephrine. Firstly, the action of the enzyme dopamine β-hydroxylase (DBH, dopamine β-monooxygenase, EC 1.14.17.1) produces norepinephrine. This oxidase requires ascorbic acid as the electron donor. It is a highly antigenic homotetramer (Mr around 290 kD) with low substrate specificity [106].

Finally, the enzyme phenylethanolamine *N*-methyltransferase (PNMT, EC 2.1.1.28) catalyses the *N*-methylation of norepinephrine to produce epinephrine (Figure 1c). PNMT is a cytosolic enzyme that uses *S*-adenosylmethionine (SAM) as the methyl donor. It has a low substrate specificity that allows it to carry out the beta carbon methylation of a variety of amines. Its expression is mainly but not exclusively restricted to the suprarenal glands [106,107].

4.2. Degradation of Catecholamines

As occurring with other BAs, CAs can be the subject of oxidative deamination catalysed by MAO, thus producing H_2O_2 and aldehydes, 3,4-dihydroxyphenylacetaldehyde (DOPAL) from DA, and 3,4-dihydroxyphenylglycoaldehyde (DOPEGAL) from epinephrine and norepinephrine (Figure 2c). Both are instable compounds rapidly oxidized to dihydroxyphenylacetic acid and 3,4-dihydroxymandelic acid, respectively, by the action of AD [108,109]. Alternatively, the enzyme aldehyde reductase (AR, EC 1.1.1.21) can reduce DOPAL to 3,4-dihydroxyphenylethanol, and DOPEGAL to dihydroxyphenylglycol. The lack of a beta-hydroxyl group in DOPAL favors its oxidation by AD. Conversely, the presence of the β-hydroxyl group in DOPEGAL makes it a better substrate of AR [108].

Nevertheless, the major product of norepinephrine degradation in humans seems to be vanillylmandelic acid (VMA), produced mainly by a pathway that requires the consecutive actions of MAO, AR, catechol O-methyltransferase (COMT, EC 2.1.1.6) alcohol dehydrogenase (ADH, EC 1.1.1.1) and AD [108]. COMT also uses SAM as the methyl donor. Two different isoforms are encoded by a unique gene, the cytosolic isoform being the one present in glia and peripheral organs (for instance, liver and kidney) (not shown in Figure 2).

DA and their derivatives can be converted into other molecules in CNS and peripheral tissues before being excreted. The activity phenolsulfotransferase (EC 2.8.2.1) is able to produce dopamine-3-*O*-sulfate and dopamine-4-*O*-sulfate by transferring a sulfate group of 3′-phosphoadenosine-5′-fosfosulfate to any hydroxyl groups of the catechol (preferably to dopamine-3-*O*-sulfate). Moreover, the enzyme uridine diphosphoglucuronosyltransferase (EC 2.4.1.17) transfers glucuronic acid from UDP-glucuronic acid to both hydroxyl groups of the dopamine catechol ring. This enzyme is linked to the reticulum endoplasmic membrane [109].

4.3. Signalling and Physiological Functions of Catecholamines

CAs act as both neurotransmitters and hormones, depending on their targets in different tissues/organs [110]. In CNS, they are mainly involved in motor and emotional control, cognition, and memory [111].

In peripheral tissues, they also play important roles as modulators of blood pressure and renal excretion, as well as in the immune system and GIT functions [112]. DA receptors are G protein-coupled receptors (GPCRs) classified into 5 types (D1–D5) [113]. Heterodimerization among the subtypes and with other receptor type monomers have been reported, which results in a very complex cell-dependent signalling network. In addition, DA can also elicit physiological signalling through G-protein independent mechanisms (i.e., ion channels, tyrosine kinases and even arrestins [113].

Epinephrine and norepinephrine preferably bind to adrenergic receptors or adrenoceptors (all GCPRs) that are classified as α (1a,1b,1d and 2a,2b,2c) and β (1–3) subtypes [82,114]. In the intestine, the main epinephrine receptors are α1 and β2. However, in the case of norepinephrine, they are α1 and α2 (as well as β2, but in a minor degree). These differences determine the specific effects of the different CAs on absorption, blood flux, and motility in the intestine [82].

Regulation of Intestinal Blood Flux, Immunity and Motility

Norepinephrine binding to α-adrenergic receptors induces vasoconstriction and increases vascular resistance, thus reducing the blood flux in the intestine. Low epinephrine levels stimulate β receptors, inducing vasodilation and consequently an increase in blood flux. However, at high levels, it induces similar effects to norepinephrine [82]. DA preferentially binds to D1 at low concentrations, and to adrenoceptors β1, and even to α1, as the DA concentrations increase. Thus, low DA levels induce vasodilation and increase blood flux, but high DA levels can induce vasoconstriction and consequently abdominal flux blood decrease.

Recently, in 2017, Mittal et al. [82], summarized the major effects of CAs in GIT:

- *Nutrient absorption.* Both epinephrine and norepinephrine play important roles in nutrient absorption regulation. Epinephrine is able to induce a hyperglycemic response acting through β-adrenergic receptors, and it increases absorption of oligopeptides when bound to α-adrenoceptors.
- *Intestinal motility.* CAs binding to β-adrenoreceptors induces smooth muscle relaxation that lead to a global food transit delay. On the contrary, their bindings to α-adrenoreceptors stimulate intestinal smooth muscle contraction, and consequently gut motility and food transit.
- *CAs, immune system and GIT.* Recently, CAs, as well as 5-HT, have been described as regulators of the innate immune system, which can be related to food intolerance. In addition, it is also reported that these amines can influence the intestinal microbiota [115].

These effects are very interesting in the context of nutrition, therefore deserving further research [116]. Fortunately, the advances of high-throughput technologies can help us to reveal the bases of these complex but important subjects for healthy and personalized nutrition.

5. Biochemistry and Physiology of Polyamines

5.1. Synthesis of Polyamines

PAs synthetized by mammalian cells are aliphatic low molecular weight polycations with 2–4 protonated amino groups at physiological pH: Put, Spd and Spm. They are essential for all living organisms from Archaea to humans, as they modulate the most basic mechanism for life, macromolecular synthesis and structure and membrane dynamics. The diamine Put is the precursor of the triamine Spd and the tetramine Spm. In mammalian cells, Put is synthetized directly from α-decarboxylation of the amino acid L-ornithine. The aminopropyl groups of Spd (1) and Spm (2) are added from decarboxylated SAM (dcAdoMet) (Figure 2d) [117]. Plants and microorganisms can synthetize other BAs with different lengths, positive charges and configurations; for instance, cadaverine (pentane-1,5-diamine), Agm, and other PAs synthetized by thermophilic microorganisms [118,119].

The first step for PA biosynthesis in mammalian cells is the hydrolysis of the guanidinium group of L-arginine catalysed by arginase (EC 3.5.3.1) activity (not shown in Figure 2d). Its product, L-ornihine, is the substrate of the PLP-dependent decarboxylase, ODC, a minor and instable protein, which is a limiting step of PA synthesis (Figure 2d). ODC product (1,4-butanodiamine) is commonly known as Put (Figure 2d). Eukaryotic ODC structure follows a model that belongs to the group IV of the PLP-dependent L-amino acid decarboxylases [7]. It needs to be a dimer to be active (≈a 102 kDa homodimer in mammals). The enzyme has one of the shortest half-lives known so far for mammalian proteins (10–50 min) and is located mainly in cytosol but it has also been detected in nucleus [120,121]. Its activity is highly regulated in response to different growth factors, oncogenes, trophic hormones, among other proliferative stimuli [122].

As mentioned before, dcAdoMet is required to synthetize higher PAs (Spd and Spm). Its synthesis involves the condensation of L-methionine and ATP to produce SAM, catalysed by any of the isoforms of S-adenosylmethionine synthetase or methionine adenosyltransferase (MAT, EC 2.5.1.6). SAM can then be decarboxylated by the action of S-adenosyl-L-methionine decarboxylase (SAMDC, EC 4.1.1.50), producing the nucleoside dcAdoMet, which acts as the aminopropyl donor for Spd and Spm synthesis (Figure 2d). SAMDC activity can also be a limiting step of the PA biosynthesis pathway. The mature enzyme suffers a post-translational maturation process, which renders the essential pyruvoyl prosthetic group [123]. Spd is synthetized by the transfer of the dcAdoMet aminopropyl moiety to the N^4 of Put through the action of spermidine synthase (SpdS, EC 2.5.1.16). Finally, the addition of a new dcAdoMet aminopropyl moiety to the N^8 of Spd gives rise to the tetramine Spm through the action of spermine synthase (SpmS, EC 2.5.1.22) [124].

The aminopropyl tranferases SpdS and SpmS are homologus homodimers but with high substrate specificity. Steric restrictions avoid binding of Put to SpdS, as well as binding of Spd to SpmS. Nevertheless, both human enzymes contain two key Asp residues (Asp^{104} and Asp^{173} in SpdS, and Asp^{201} and Asp^{276} in SpmS), which are essential for the catalytic mechanism [119].

5.2. Degradation and Recycling of Polyamines

PA metabolism can be considered to be a very robust bicycle involving both metabolic branches (synthesis and degradation) (Figure 2d) [117]. Degradation involves a series of different amine oxidases. Spermine oxidase (SMO, EC 1.5.3.16) is a FAD-dependent oxidase able to directly transform Spm to Spd, 3-aminopropanal and H_2O_2 in the presence of H_2O and O_2. Several alternative splicing variants have been observed. It is highly inducible by PAs and their analogues, among other stimuli [124].

Al alternative pathway to convert Spm into Spd (as well as Spd into Put) requires the action of spermidine/spermine N^1-acetyltranferase (SSAT, EC 2.3.1.57). SSAT reaction using Spm and acetyl-CoA as the substrates transforms Spm into N^1-acetylspermine. This product may, in turn, follow two alternative pathways. It can be a substrate of the peroxisomal polyamine oxidase (PAO, EC 1.5.3.13) that produces the lower PA (Spd), 3-acetamidopropanal and H_2O_2 (Figure 2d). N^1-acetylspermine can even be acetylated again in its other terminal producing N^1,N^{12}-diacetylspermine. This metabolite can be a substrate for peroxisomal PAO to produce 3-acetamidopropanal, H_2O_2 and N^1-acetylspermidine. In a second reaction on N^1-acetylspermidine, PAO produces 3-acetamidopropanal, H_2O_2 and Put [125,126].

SSAT also can act on Spd to form N^1-acetylspermidine (and CoA). N^1-acetylspermidine, which is subsequently oxidized by PAO producing 3- the diamine Put, 3-acetamidopropanal and H_2O_2 [125]. The diamine Put can be further degraded by DAO [118], or alternatively recycled for higher PA synthesis. Acetylated PA are more easily excreted than their deacetylated counterparts, and their levels in urine have been used as biomarkers of elevated PA metabolism [127].

In summary, the net result of the action of SMO or the tandem SSAT plus PAO is the conversion of higher PAs or their acetylated versions into their respective lower poly- or diamine, which can be again recycled in the biosynthetic pathway (Figure 2d). It conforms two energy-consuming cycles being apparently futile. However, following the metabolic control theory, "futile cycles" confer high sensitivity for modulation of metabolic pathways that need to respond to regulatory stimuli in a coordinated way as a response (sometimes a compensation) to external alterations [128]. That indeed is the case for mammalian PA metabolism, as predicted by the mathematical model of mammalian PA metabolism and proven by its further validation [117,129]. This fact explains the difficulties experienced by many experimental groups when trying to deplete intracellular PA levels as an anticancer strategy [130,131].

5.3. Polyamine Transport Systems

PA import systems in mammalian cells are not yet fully characterized, in spite of the multiple efforts made by different group members of the PA research community; this fact is still one of the most important handicaps to control intracellular PA levels under pathological circumstances (for instance, cancer) [132,133].

Currently, three models have been proposed and reviewed by Poulin et al. [133]. A first model proposes the action of two permeases with different locations, one located in the cytosolic inner membrane (PMPP) and an H^+–coupled PA transporter located in vesicular membranes (VPA). A second mechanism (one step model) involves glypican-1, acting as a high affinity PA receptor. This binding could induce endocytosis leading to PA internalization, as described by Belting et al., for Spm transport [134]. The presence of NO and Ca^{++} in the endosomes would revert glypican-Spm binding. A third model proposes PA interaction with caveoline-1, which would also be reverted by NO. These mechanisms could explain the presence of higher PA in vesicles of several mammalian cell types. However, many doubts still remain concerning PA transport mechanisms through the different cellular compartments, as well as cell- and PA-specificities and regulation of each transport mechanism. Abdulhussein and Wallace recently reviewed this topic [135]. Specifically, Uemura et al., identified the amino acid transporter SLC3A2 as a Put export protein in colon cancer-derived cells [136] and also studied the specific characteristics of PA absorption by the intestinal tract, providing methods for PA transport analysis in the colon and the small intestine using membrane vesicles, culture cells, and mouse models [137].

5.4. Physiological Functions of Polyamines

PAs can be considered protonated amino groups kept together by aliphatic skeletons that impose specific distances among them. Thus, their positive charges and aliphatic chains can interact specifically with negative charges and hydrophobic residues/surfaces of other biomolecules (nucleic acid sequences, proteins and lipids) located at the correct distances, therefore modifying their conformations and consequently the functional properties of macromolecular structures. PAs are present and absolutely essential to keep cell viability in almost all living organisms. PA-DNA, PA-RNA and PA-membrane interactions and their conformational consequences can be reproduced and studied in vitro (in an abiotic environment) working with their purified components (PA, polynucleotides and/or lipids). This knowledge on specific molecular PA interactions gives rise to new hypotheses [138–140]. On the one hand, it is clear that some specific binding modes have been detected (among an immense quantity of possibilities for PA-biomolecules interactions) with physiological consequences or applications (for instance, nanotechnology applied to drug delivery) [141]. On the other hand, it is tempting to hypothesize that their interaction with both nucleic acids and membranes was of absolutely essential value from the beginning of life on Earth [142].

The interactive properties of PAs allow them to modulate a long list of processes involved in cell cycle progression and gene expression as DNA condensation, replication fidelity, RNA secondary and tertiary structure stabilization, translation initiation, elongation and fidelity, and posttranslational modification of proteins, among others [143]. The importance of the mentioned physiological functions modulated by PAs explains that their metabolism is strictly regulated, as well as very robust. Nevertheless, imbalance in PA levels is associated with a long list of human diseases that includes aberrant cell growth and differentiation and/or abnormal protein expression and folding; for instance, cancer, GIT and neurodegenerative diseases [144], as will be mentioned further on. Nevertheless, many molecular questions still remain unsolved on the molecular bases of the cellular functions of PAs [145], thus delaying the biotechnological applications of this yet unveiled knowledge.

As the other above mentioned BAs, PAs also play important roles in GIT physiology [146,147]. Synthetized de novo or uptaken from the intestinal lumen intestinal, they promote two different processes described for intestinal mucous reparation: DNA-independent cell migration and replacement of damaged cells by cell proliferation [148].

It is also proven that PAs are important for the correct biochemical and morphological maturation of intestine during the postnatal period. The benefits are dose and PA-specific (being Spm > Spd > Put). In addition, PAs have also been proposed as being involved in the correct immunological system development in neonatal intestine [146,149].

5.5. The Particular Case of Agmatine

Agm is a member of the PA family, as it is the product of L-arginine decarboxylation. It can be converted into Put through the action of a liver agmatinase activity, a hydrolase that removes the agmatine guanidinium moiety. It has finally been clarified that human cells express active agmatinase, but not an active arginine decarboxylase (ADC, EC 4.1.1.19) [21]. However, some bacteria present in human microbiota have, in fact, active ADC, so producing Agm that can be absorbed by intestinal epithelium. Agm protects mitochondrial functions and confers resistance to cellular apoptosis [150], being able to modulate several processes such as hepatic regeneration and renal function [151], among other proposed physiological functions. For instance, it is able to block N-methylaspartate receptor receptor in brain areas related to learning and memory [152], as well as modulate mental stress [153,154]. Thus, it seems to be a good candidate to participate in gut-brain axis. As HIS and Put, Agm is also a DAO substrate to produce γ-butiramide that is finally converted into γ-guanidinobutirate in CNS [39]. It can also compete with the other diamines for binding to OCT transporters [155].

6. Biogenic Amines and Microbiota-Intestine Crosstalk

In addition to BAs from diet and the endogenous BA metabolism of GIT cells, there is another important contributor to BA levels in the intestine: the intestinal microbiota metabolism.

On the one hand, intestinal microbiota species, like any other living organism, are able to synthesize PA [156]. These PAs can be uptaken by intestinal epithelia, thus contributing to cellular growth and tissue renewal, especially in the colon [137,156].

On the other hand, microbiota species can also synthetize any of the other BAs mentioned in this text. For instance, a PLP-dependent HDC homologous to the mammalian enzyme can be expressed by Gram negative Enterobacteria. On the contrary, Gram positive bacteria (for instance, *Lactobacillus sp.*) can express a non-homologous pyruvoil-dependent HDC [157]. The physiological effects of HIS produced by microbiota are controversial and need more research to be fully understood [15]. For instance, in spite of the general idea of a deleterious role of diet HIS on human health (see next section), it has been reported that HIS synthetized by the probiotic *Lactobacillus reuteri* acts as a positive immune regulator acting through H_2R [158].

In addition, some bacteria potentially taking part in human microbiota can synthetize other BAs, different for the mentioned endogenous ones, and can also be bioactive for human physiology (for instance, tyramine) [159]. In general, as all BAs are described as neurotransmitters, neuroendocrine factors or neuromodulators, amines synthetized by microbiota may interact with host signals establishing a microbiota-endocrinology system crosstalk that is a part of the larger microbiota-gut-brain axis, with consequences in health and diseases [160].

Moreover, other products of the microbiota metabolism different from amines can regulate endogenous BA metabolism. For instance, bile acids and short-chain fatty acids, affect 5-HT synthesis that, in turn, directly or indirectly regulates gut motility [161] and enteric neuroimmune mechanisms [102]. A protective role of probiotics against histamine-mediated colon carcinogenesis have been recently reported [162], as well as against gastric cancer by modulating PA metabolism [163].

Summarizing, the so called microbiota-gut-brain axis is a very interesting but extremely complex open subject of current biomedicine that absolutely requires the help of new approaches from system biology and high-throughput technologies.

7. What Is Known about Biogenic Amines Roles in Human Gastro Intestinal Pathologies?

It is clear that alterations of elements of the BA metabolism, including transport and signalling pathways, are involved in a wide diversity of GIT pathologies. However, further efforts are needed to fully characterize the specific aberrant element(s) and/or mechanism(s) responsible for each pathological consequence. In the next subsection, we will summarize the current state of the art for the better known relationships between BA-related elements and human GIT diseases.

7.1. Gastric Diseases

7.1.1. Peptic Ulcers

Peptic Ulcers are mucosal erosions induced by gastric secretions. In addition to those located in gastric mucosa (gastric ulcers), similar damages can appear at the entrance of duodenum (duodenal ulcers), or even in a minor percentage in the oesophagus or other intestinal segments [164]. Gastric ulcers are usually located along the minor curvature, particularly in the corpus-antrum transitional mucosa, their prevalence being higher in over 40 year-old humans. Duodenal ulcers are located between the lower part of the stomach and the start of the intestine (that is, the intestinal area exposed to gastric acid) with the highest frequency being between 20–50 year-old humans [70].

Abdominal pain is the most common symptom of peptic ulcer. Swelling, loss of appetite, nausea and/or indigestion are also usual symptoms. Associated complications include bleeding, perforation and stenosis. Bleeding is the most common complication among peptic ulcer patients (15–20%). Reciprocally, 40% of humans suffering from upper GIT bleeding are peptic ulcer patients [70].

Peptic ulcers are classified depending on their anatomical location [70,165,166]. When they are located in the stomach, they evolve as atrophic gastritis. Initially, inflammation induces parietal cell apoptosis, which in turn induces gastric acid hyposecretion and mucosal atrophy. Chronic inflammation and mucosal atrophy lead to gastric ulcers and eventually to gastric cancer.

Other individuals present gastritis located in the pylorus area. In this case, excessive quantities of gastric acid are usually produced, which lead to duodenal ulcers. The inflammatory response induces cytokine secretion that finally induces dysregulation of endocrine cells located in this area. Thus, G cells are stimulated to overproduce gastrin, while somatostatin secretion is inhibited. As a consequence, HIS synthesis and secretion increase, which in turn promotes proliferation and stimulation of parietal cells, leading to gastric acid hypersecretion (Figure 3). These facts explain the treatment of the ulcers with proton pump inhibitors and H_2R antagonists [167,168].

Infection by *Helicobacter pylori* is a very common origin of peptic ulcers. Nevertheless, other risk factors as alcohol and smoking abuse, as well as a continual use of several drugs (for instance, nonsteroidal anti-inflammatory drugs or NSAIDs) have been described. *H. pylori* is a microaerophilic flagellated bacteria able to colonize human gastric mucosa. It is a highly common infection that can take place for tens of years. This Gram-negative bacteria is considered an important pathogenic agent associated with several human pathologies like chronic gastritis, gastrointestinal ulcers, and neoplasms such as gastric adenocarcinomas and gastric mucosa-associated lymphomas (Table 4) [169].

H. pylori induces HDC expression, and consequently an increase in endogenous HIS synthesis, which leads to an inflammatory response including an increased presence of neutrophils and lymphocytes, which in turn produces different cytokines and chemokines (for instance, IL-1, IL-6, TNF-α and IFN-γ). Gene polymorphisms detected in some of these human cytokine genes are proposed as being involved in resistance or susceptibility to *H. pylori* [75,170].

Table 4. Features associated with *H. pylori* infections and subsequent inflammation located in different parts of the gastrointestinal tract.

Location	Acid Secretion	Gastric Features and Histology	Intestinal Features and Histology	Pathology
Stomach (pan-gastritis)	Hyposecretion	Chronic inflammation and parietal cell apoptosis Atrophy Intestinal metaplasia	Normal	Gastric ulcer Gastric cancer
Pylorus area	Hypersecretion	Chronic inflammation and increased gastrin released Inhibition of somatostatin Increase parietal cell stimulation	Gastric metaplasia Active chronic inflammation	Duodenal ulcer

Data from reference [16–18].

Different signalling mechanisms have been described to explain the pathological consequences of *H. pylori* infection. On the one hand, there is a CAG (cytotoxin-associated gene)-dependent pathway (involving signal transduction elements like Rho GTPases, PKA, MKK4 and JNK, and the transcription factors AP-1 and NF-κB), which result in synthesis and secretion of cytokines acting as an innate immune response. On the other hand, MAP kinase pathway (involving Raf-1, ERK, MEK) is activated by a CAG-independent mechanism, which finally results in activation of BP1 and BP2.

These transcription factors act as inducers on the HDC promoter, which lead to an increase in gastric acid synthesis [171].

7.1.2. Gastric Cancer

Gastric cancer presents a high morbidity and mortality, being described as one of the most common worldwide malignant neoplasms. Nevertheless, its prevalence has decreased in the last decades in most European countries probably due to changes in lifestyle such as smoking reduction and *H. pylori* control [172].

Chronic inflammation produced by the pathogen can result in changes in the normal architecture of gastric mucosa, destruction of gastric glands, parietal cell loss, decreased gastric acid secretion, intestinal metaplasia, and finally gastric cancer. Thus, World Health Organization estimates that chronic *H. pylori* infection increases the risk of gastric cancer by 10 times, considering it as a class I carcinogen [70]. In fact, a reduced gastric acid secretion is not only a predisposition to gastric ulcers, but also to gastric cancer. Low levels of gastric acid reduce vitamin C absorption and allow an excessive growth of salivary and intestinal bacteria in the stomach, which can promote carcinogenesis [165].

It has been observed that treatments with gastrin/CCK-2 receptor antagonists reduce parietal cell apoptosis and inhibit gastric atrophy. Similar results were obtained with an irreversible H_2R antagonist (i.e., cimetidine); suggesting that both receptors could be involved in gastric cell apoptosis and carcinogenesis [173]. However, a prospective, double blind trial carried out among hundreds of gastric cancer patients by the British stomach cancer group did not result in any increase of survival [174].

Recent research on H_4R physiological roles reveals interesting information on the involvement of the receptor in the relationship between the immune system and GIT carcinogenesis [70].

As PAs (mainly Spd and Spm) are also essential for gastric cancer progression, some authors claim that new probiotic-based anticancer strategies are able to reduce endogenous PA levels and gastric cancer growth as a result [163].

7.2. Intestinal Diseases

7.2.1. Irritable Bowel Syndrome

Formerly known as spastic colon, irritable bowel syndrome (IBS) is a prevalent disorder that is characterized by alterations in intestinal secretions and motility, mainly affecting the colon. The symptoms (cramps, abdominal pain, intestinal habit alterations, and food intolerances) can appear during childhood or in adults [175,176]. Different subtypes have been described. The IBS-D subtype is characterized by diarrhoea; IBS-C is characterized by constipation; and IBS-M presents both intestinal alterations [175].

Several research groups have proposed the involvement of 5-HT-related elements (enzymes, transporters and receptors) in the pathology. For instance, results of several studies point out genetic variants of the gene encoding the 5-HT transport SERT (chromosome 17) that could predispose to IBS [177]. However, other results are not conclusive and sometimes contradictory [178]. Thus, it is one of the complex BA-GIT relationships needing further investigation.

Different treatments are prescribed for IBS patients depending on the severity of the symptoms. Diet adjustments and behaioaral education are usually enough for mild and moderate symptoms. However, for most severe forms, multidisciplinary approaches including pharmacotherapy are required. For IBS-D, treatment with antidiarrhoeal drugs, such as loperamide, can be necessary. In the case of women with severe symptoms, the 5-HT$_3$R antagonist alosetron can be used while the 5-HT$_4$R agonist prucalopride is used to relieve constipation in IBS-patients [179].

Other drug discovery initiatives are trying to develop effective inhibitors of tryptophan hydroxylase 1 (TPH1, the isoform expressed in GIT), which are unable to pass the blood-brain barrier, as an alternative to reduce 5-HT synthesis by EC cells, and consequently reduce/avoid their deleterious effects induced by dysregulation of the gastrointestinal serotonergic system (for instance IBS and carcinoid syndrome) [180,181].

7.2.2. Inflammatory Bowel Diseases

Crohn disease (CD) and ulcerative colitis (UC) are inflammatory bowel diseases (IBD) with different characteristics. Nevertheless, both diseases occur with alternating periods of remission and relapse. UC is characterized by inflammation with the presence of superficial colon mucosa ulcers that usually originate in the colon and then progress towards upper colon sections. UC symptoms include diarrhoea, cramping, and rectal bleeding. CD is characterized by a discontinuous pattern along the intestinal tract and can present larger ulcerations and sometimes granulomas; abdominal pain, diarrhoea, weight loss and bleeding being its most common symptoms [45].

Different research groups have observed a HIS secretion increase in jejunum of CD patients and high HIS levels in the intestinal mucosa of UC patients. In addition, levels of the HIS degradation product N-methylhistamine are elevated in the urine of both disease patients, which suggest a more active HIS metabolism (synthesis and degradation) in the intestine with respect to control individuals. The results indicate that degranulation of mast cells infiltrated along the intestinal tract must be involved in these diseases. However, there is a lack of information about molecular details of the signalling mechanisms responsible for the HIS effects on IBD evolution, thus blocking the development of efficient intervention strategies [75,182]. It has been proven that the chronic use of H_2R receptor antagonists increases the risk of more severe CD, suggesting a protective role of HIS on the intestinal mucosa when acting through an H_2R receptor [45]. IBD intervention acting through H_4R receptor has also been recently proposed [183,184]. In fact, several studies highlight the potential of H_4 receptor targeted therapy in the treatment of various gastrointestinal disorders such as IBD, IBS and cancer [184].

DAO has been suggested as an IBD marker, but it is a controversial subject [64,185]. Currently, IBD treatment consists of diet adjustment, psychological support, and eventually surgery. Recently, it has been proven that microbiota is usually altered in IBD patients. Working with a mouse model of intestinal inflammation, it has been demonstrated that the probiotic *Lactobacillus reuteri* is able to reduce intestinal inflammation by a HIS signalling-dependent mechanism, thus acting as a preventive factor for cancer risk associated with chronic inflammation [186]. Microbiota can also have a positive role in IBD patients, as microbial-derived metabolites (for instance, bile acids and short-chain fatty acids) can regulate intestinal 5-HT synthesis, and consequently intestinal motility, which opens new perspectives for probiotic-based strategies [161].

7.2.3. Intestinal Neoplasias

Colon cancer is considered the most common GIT cancer. It is a multifactorial disease and its etiology can combine genetic inherited factors, exposition to environmental risk factors (including diet), as well as other endogenous circumstances such as chronic intestinal inflammation. At present, it is one of the most frequent human cancers, besides being one of the principal causes of death among human cancer patients [187]. As in other cancer types, colon cancer biopsies present increased activities of PA-synthesis key enzymes and elevated PA content up to 10-15-fold with respect to the levels observed in normal colon epithelium. Thus, together with inflammation, PAs are considered to be markers of colon carcinogenesis [188].

It is known that several oncogenes and suppressor genes that regulate specific phases of colon carcinogenesis are also involved in PA metabolism regulation. Under normal conditions, the suppressor gene adenomatous polyposis coli (APC) repress MYC transcription, a family of transcription factors required for cell proliferation. MYC overexpression is related to uncontrolled proliferation and progression of carcinogenic process in different cancer types, including colon cancer. Members of MYC family are inducers of ODC transcription [189]. In addition, APC up-regulate the expression of ODC antizymes, a protein family acting as ODC inhibitors, as they bind to ODC monomer blocking the active quaternary conformation of the enzyme, thus targeting the ODC monomer for an antizyme-dependent and ubiquitin-independent proteasomal degradation. When APC is deleted or inactivated by mutation during the early stages of carcinogenesis (as occurring in adenomatous polyposis patients), MYC APC-induced repression is lost, and consequently ODC and PA synthesis are upregulated. In addition, the upregulation of ODC antizyme expression is lost, which leads to an increase in ODC turnover [190].

Active KRAS oncogene downregulates the peroxisome proliferator-activated receptor gamma (PPARγ), which has been proposed as a marker for colorectal cancer survival [191]. Transcription of the key enzyme for PA degradation (SSAT) is upregulated by PPARγ response elements (PPREs) present in SSAT promoter [192]. In advanced stages of colon cancer, oncogenic mutations in KRAS lead to permanent KRAS activation, which in turn downregulates PPARγ and, consequently, SSAT expression and PA degradation [193]. These regulatory mechanisms explain the molecular bases of the relationship between oncogenic events and PA elevation in colon cancer patients. As mentioned in previous sections, elevation of PA levels helps replication and macromolecular synthesis of the transformed cells and confers other advantages for cancer progression.

In addition to the endogenous regulation of PA metabolism, it is also worth mentioning that diet and microbiota are also potential PA sources [194]. The lack of effective PA absorption inhibitors, as well as the robustness of PA metabolism, is blocking the success of antitumoural strategies based on PA depletion in different cancer types [190]. In colon cancer, treatment depends on the progression stage. Surgery can be enough during the first stages; then, chemotherapy or immunotherapy must be required. What about chemoprevention? The irreversible ODC inhibitor DFMO is able to act on both Enterobacteria and mammalian ODC activities. As PAs are essential for colon cancer progression, DFMO could be effective as a chemopreventive agent for putative familiar colon cancer patients. This was the hypothesis claimed by Gerner and Meyskens several years ago, and validated with positive results when administered as a combined therapy with NSAIDs, the latest acting as PPARγ inducers [188,190].

Working with HDC knocked out mice under treatment with probiotics (*Lactobacillus reuteri*), results obtained by Gao et al. [162] indicate that luminal HIS produced by gut microbiota could suppress inflammation-associated colon carcinogenesis.

As mentioned above (Table 1), HIS is an immune system regulator playing important roles in immune cell development, span lives, and functions. These effects can involve any of the four receptors (mainly H_1R and H_4R) [37,195]. HIS has also been described as a cell proliferation regulator of several cancer types (for instance, breast cancer, GIT cancer, leukaemia, lung cancer, lymphoma) [196]. These effects can be different depending on the HIS receptors expressed by the different cell types. In addition, communication between the immune system and cancer is a dynamic process involving different immune cells; for instance, macrophages, monocytes, mast cells and regulatory B cells and T cells [197,198]. Consequently, HIS effects on carcinogenesis and cancer progression is a very complex but interesting topic, which also deserves further research efforts. Recent results support the therapeutic potential of H_4R ligand in several cancer types, including colon cancer [198].

Evolution of Zollinger-Ellison syndrome, a rare GIT pathology caused by the presence of gastrin-secreting tumours in pancreas and/or intestine, involves both synthesis and secretion of HIS [29].

8. Conclusions and Future Prospects

Previous sections demonstrate that BAs play important physiological roles in the entire GIT. Consequently, aberrant functions of the metabolic pathways are involved in the most important gastrointestinal pathologies. HIS metabolism seems to be mainly important in gastric physiopathology, as well as in inflammatory intestinal diseases. 5-HT plays major pathophysiological roles in the intestine, as an immune modulator and regulator of the intestinal smooth muscle contraction/relaxation. Its involvement in inflammatory diseases needs further clarification. CAs are modulators of intestinal absorption, blood flux and motility and they have also been proposed as immune modulators in the GIT system. Some of these functions may be regulated by 5-HT [104]. This is a very interesting hypothesis that would also require more research efforts to be fully validated. PAs, being essential biomolecules for cell growth, are important for both epithelial reparation and proliferation. Thus, they are beneficial for a healthy intestinal epithelium and have been described as both immune and epithelial permeability modulators in GIT [199], but also proposed as a promising target for colon cancer chemoprevention. From a phenomenological point of view, all B/A-BAs have been described as modulators of both immunity and epithelial cell growth in the GIT, but many of the underlying molecular mechanisms are not fully characterized, yet. In any case, these results point to the GIT as an interesting scenario to study BA metabolic and functional interplay.

In spite of the specialization of the amine effects along the different segments of the GIT, in some cases, their metabolic routes are coincident in a given GIT segment, so they can share/compete for common elements, thus establishing a crosstalk among their metabolism and consequently their physiological missions. For instance, enzymes such as decarboxylases (i.e., DDC), amine oxidases (i.e., MAO and DAO), cofactors (i.e., PLP, BH_4), and metabolites (i.e., SAM), among others. Moreover, at least in neurons, heteroreceptor complexes have also been detected among HIS and DA receptors [200], and both Spd and HIS are ligands and modulate N-methylaspartate receptor activity [201]. In mouse mast cells, synthesis of PA, HIS and 5-HT seem to be antagonistic processes in the mast cell differentiation process. Thus, the pathophysiological consequences of this cross-talk among BA metabolic elements still present many gaps and open questions (mentioned throughout this review) in the GIT context, which deserve deeper molecular characterization, as this information could provide valuable insights useful for new diagnosis and intervention initiatives in the gastroenterology field as well as more personalized nutritional advices and preventive actions.

It is clear that the pathophysiological effects of B/A-BAs in GIT is a very complex issue that interacts with and is modified by a very extensive list of endogenous and exogenous factors, from metabolic interactions with other immune and/or neuroendocrine compounds/systems to the influx of diet and microbiota composition. The full characterization of the entire involved interactions still requires filling many gaps on specific biochemical and molecular details. This research objective should be helped by systemic approaches to provide integrative views (and even predictive models) of the multiple pathophysiological processes associated with BAs in the GIT. In fact, several independent groups have proposed to approach the problem, or subsets of the problem, by using integrative high-throughput and Systems Biology strategies currently successfully used with many other complex biological systems [1,202].

The translational benefits of the final objective are clear taking into account that the topic involves three of the most important but complex physiological systems for a human being, neurotransmission/neuroendocrine, immune and digestive systems. Consequently, life quality and/or span life of many human beings can depend on research advances in the topic. Thus, it should be considered among the research priorities not only for nutrition but for biomedicine, in general. In addition, different companies have developed a wide spectrum of drugs capable of modulating different elements of BA metabolism and signalling. The usefulness and/or efficiency of these compounds (or their analogues/derivatives) will probably increase when a deeper degree of integrative knowledge about the molecular basis and the roles of all B/A-BAs in the GIT system is achieved.

Author Contributions: Conceptualization, A.F.-R., J.L.U. and F.S.-J.; Investigation, A.F.-R., J.L.U. and F.S.-J.; Writing-Original Draft Preparation, A.F.-R. and F.S.-J.; Writing-Review & Editing, J.L.U. and F.S.-J.; Visualization, A.F.-R., J.L.U. and F.S.-J.; Supervision, J.L.U. and F.S.-J.; Funding Acquisition, F.S.-J.

Acknowledgments: Thanks are due to Consejería de Economía, Innovación, Ciencia y Empleo, Junta de Andalucía. This work also takes part of the activities of the group in "CIBER de enfermedades raras" (CIBERER), and IBIMA, both institutions being part of Instituto de Salud Carlos III (MINECO, Spain).

References

1. Sánchez-Jiménez, F.; Ruiz-Perez, M.V.; Urdiales, J.L.; Medina, M.A. Pharmacological potential of biogenic amine-polyamine interactions beyond neurotransmission. *Br. J. Pharmacol.* **2013**, *170*, 4–16. [CrossRef] [PubMed]
2. Okada, K.; Hidese, R.; Fukuda, W.; Niitsu, M.; Takao, K.; Horai, Y.; Umezawa, N.; Higuchi, T.; Oshima, T.; Yoshikawa, Y.; et al. Identification of a novel aminopropyltransferase involved in the synthesis of branched-chain polyamines in hyperthermophiles. *J. Bacteriol.* **2014**, *196*, 1866–1876. [CrossRef] [PubMed]
3. Suzzi, G.; Torriani, S. Editorial: Biogenic amines in foods. *Front. Microbiol.* **2015**, *6*, 472. [CrossRef] [PubMed]
4. Bodmer, S.; Imark, C.; Kneubühl, M. Biogenic amines in foods: Histamine and food processing. *Inflamm. Res.* **1999**, *48*, 296–300. [CrossRef] [PubMed]
5. Moya-Garcia, A.A.; Pino-Angeles, A.; Gil-Redondo, R.; Morreale, A.; Sanchez-Jimenez, F. Structural features of mammalian histidine decarboxylase reveal the basis for specific inhibition. *Br. J. Pharmacol.* **2009**, *157*, 4–13. [CrossRef] [PubMed]
6. Giardina, G.; Montioli, R.; Gianni, S.; Cellini, B.; Paiardini, A.; Voltattorni, C.B.; Cutruzzolà, F. Open conformation of human DOPA decarboxylase reveals the mechanism of PLP addition to Group II decarboxylases. *Proc. Natl. Acad. Sci. USA* **2011**, *108*, 20514–20519. [CrossRef] [PubMed]
7. Sandmeier, E.; Hale, T.I.; Christen, P. Multiple evolutionary origin of pyridoxal-5′-phosphate-dependent amino acid decarboxylases. *Eur. J. Biochem.* **1994**, *221*, 997–1002. [CrossRef] [PubMed]
8. Sánchez-Jiménez, F.; Moya-García, A.A.; Pino-Ángeles, A. New structural insights to help in the search for selective inhibitors of mammalian pyridoxal 5′-phosphate-dependent histidine decarboxylase. *Inflamm. Res.* **2006**, *55*, S55–S56. [CrossRef] [PubMed]
9. Fonda, M.L.; Eggers, D.K.; Mehta, R. Vitamin B-6 metabolism in the livers of young adult and senescent mice. *Exp. Gerontol.* **1980**, *15*, 457–463. [CrossRef]
10. Jalkanen, S.; Salmi, M. Cell surface monoamine oxidases: Enzymes in search of a function. *EMBO J.* **2001**, *20*, 3893–3901. [CrossRef] [PubMed]
11. Finney, J.; Moon, H.-J.; Ronnebaum, T.; Lantz, M.; Mure, M. Human copper-dependent amine oxidases. *Arch. Biochem. Biophys.* **2014**, *546*, 19–32. [CrossRef] [PubMed]
12. Edmondson, D.E.; Binda, C.; Mattevi, A. Structural insights into the mechanism of amine oxidation by monoamine oxidases A and B. *Arch. Biochem. Biophys.* **2007**, *464*, 269–276. [CrossRef] [PubMed]
13. Rodríguez-López, R.; Morales, M.; Sánchez-Jiménez, F. Histamine and its receptors as a module of the biogenic amine diseasome. In *Histamine Receptors*; Springer International Publishing: Cham, Switzerland, 2016; pp. 173–214.
14. Schubert, M.L. Gastric acid secretion. *Curr. Opin. Gastroenterol.* **2016**, *32*, 452–460. [CrossRef] [PubMed]
15. Barcik, W.; Wawrzyniak, M.; Akdis, C.A.; O'Mahony, L. Immune regulation by histamine and histamine-secreting bacteria. *Curr. Opin. Immunol.* **2017**, *48*, 108–113. [CrossRef] [PubMed]
16. Schneider, E.; Rolli-Derkinderen, M.; Arock, M.; Dy, M. Trends in histamine research: New functions during immune responses and hematopoiesis. *Trends Immunol.* **2002**, *23*, 255–263. [CrossRef]
17. Peters, L.J.; Kovacic, J.P. Histamine: Metabolism, physiology, and pathophysiology with applications in veterinary medicine. *J. Vet. Emerg. Crit. Care* **2009**, *19*, 311–328. [CrossRef] [PubMed]
18. Panula, P.; Chazot, P.L.; Cowart, M.; Gutzmer, R.; Leurs, R.; Liu, W.L.S.; Stark, H.; Thurmond, R.L.; Haas, H.L. International union of basic and clinical pharmacology. XCVIII. Histamine receptors. *Pharmacol. Rev.* **2015**, *67*, 601–655. [CrossRef] [PubMed]
19. Gardini, F.; Özogul, Y.; Suzzi, G.; Tabanelli, G.; Özogul, F. Technological factors affecting biogenic amine content in foods: A review. *Front. Microbiol.* **2016**, *7*, 1218. [CrossRef] [PubMed]
20. Piletz, J.E.; Aricioglu, F.; Cheng, J.-T.; Fairbanks, C.A.; Gilad, V.H.; Haenisch, B.; Halaris, A.; Hong, S.; Lee, J.E.; Li, J.; et al. Agmatine: Clinical applications after 100 years in translation. *Drug Discov. Today* **2013**, *18*, 880–893. [CrossRef] [PubMed]
21. López-Contreras, A.J.; López-Garcia, C.; Jiménez-Cervantes, C.; Cremades, A.; Peñafiel, R. Mouse ornithine decarboxylase-like gene encodes an antizyme inhibitor devoid of ornithine and arginine decarboxylating activity. *J. Biol. Chem.* **2006**, *281*, 30896–30906. [CrossRef] [PubMed]
22. Biji, K.B.; Ravishankar, C.N.; Venkateswarlu, R.; Mohan, C.O.; Gopal, T.K.S. Biogenic amines in seafood: A review. *J. Food Sci. Technol.* **2016**, *53*, 2210–2218. [CrossRef] [PubMed]

23. Morgan, D.M.L.; White, A.; Sánchez-Jiménez, F.; Bardócz, S. *COST 917—Biogenically Active Amines in Food. Volume IV, First General Workshop*; Office for Official Publicationsn of European Communities: Luxemburg, 2000.

24. Wallace, H.M.; Hughes, A. *COST Action 922. Health Implications of Dietary Amines*; Office for Official Publicationsn of European Communities: Luxemburg, 2004.

25. Naila, A.; Flint, S.; Fletcher, G.; Bremer, P.; Meerdink, G. Control of biogenic amines in food-existing and emerging approaches. *J. Food Sci.* **2010**, *75*, R139–R150. [CrossRef] [PubMed]

26. Sanchez-Jiménez, F.; Pino-Ángeles, A.; Rodríguez-López, R.; Morales, M.; Urdiales, J.L. Structural and functional analogies and differences between histidine decarboxylase and aromatic L-amino acid decarboxylase molecular networks: Biomedical implications. *Pharmacol. Res.* **2016**, *114*, 90–102. [CrossRef] [PubMed]

27. Rodríguez-López, R.; Reyes-Palomares, A.; Sánchez-Jiménez, F.; Medina, M.Á. PhenUMA: A tool for integrating the biomedical relationships among genes and diseases. *BMC Bioinform.* **2014**, *15*, 375. [CrossRef] [PubMed]

28. Reyes-Palomares, A.; Bueno, A.; Rodríguez-López, R.; Medina, M.Á.; Sánchez-Jiménez, F.; Corpas, M.; Ranea, J.A.G. Systematic identification of phenotypically enriched loci using a patient network of genomic disorders. *BMC Genom.* **2016**, *17*, 232. [CrossRef] [PubMed]

29. Pino-Ángeles, A.; Reyes-Palomares, A.; Melgarejo, E.; Sánchez-Jiménez, F. Histamine: An undercover agent in multiple rare diseases? *J. Cell. Mol. Med.* **2012**, *16*, 1947–1960. [CrossRef] [PubMed]

30. Ruiz-Pérez, M.V.; Medina, M.Á.; Urdiales, J.L.; Keinänen, T.A.; Tuomo, A.; Sánchez-Jiménez, F. Polyamine metabolism is sensitive to glycolysis inhibition in human neuroblastoma cells. *J. Biol. Chem.* **2015**, *290*, 6106–6119. [CrossRef] [PubMed]

31. Watanabe, T.; Ohtsu, H. L-histidine decarboxylase as a probe in studies on histamine. *Chem. Rec.* **2002**, *2*, 369–376. [CrossRef] [PubMed]

32. Stark, H. *Histamine H4 Receptor: A Novel Drug Target for Immunoregulation and Inflammation*; Versita: Berlin, Germany, 2013.

33. Nakanishi, T.; Kekuda, R.; Fei, Y.J.; Hatanaka, T.; Sugawara, M.; Martindale, R.G.; Leibach, F.H.; Prasad, P.D.; Ganapathy, V. Cloning and functional characterization of a new subtype of the amino acid transport system N. *Am. J. Physiol. Cell Physiol.* **2001**, *281*, C1757–C1768. [CrossRef] [PubMed]

34. Moya-Garcia, A.A.; Medina, M.A.; Sánchez-Jiménez, F. Mammalian histidine decarboxylase: From structure to function. *Bioessays* **2005**, *27*, 57–63. [CrossRef] [PubMed]

35. Schwelberger, H.G.; Ahrens, F.; Fogel, W.A.; Sánchez-Jiménez, F. Histamine metabolism. In *Histamine H4 Receptor: A Novel Drug Target in Immunoregulation and Inflammation*; Stark, H., Ed.; Versita: Berlin, Germany, 2013; pp. 63–102.

36. Metcalfe, D.D. Mast cells and mastocytosis. *Blood* **2008**, *112*, 946–956. [CrossRef] [PubMed]

37. Ennis, M.; Ciz, M.; Dib, K.; Friedman, S.; Gangwar, R.S.; Gibbs, B.F.; Levi-Schaffer, F.; Lojek, A.; Migalovich-Sheikhet, H.; O'Mahony, L.; et al. Histamine receptors and inflammatory cells. In *Histamine H4 Receptor: A Novel Drug Target in Immunoregulation and Inflammation*; Stark, H., Ed.; Versita: London, UK, 2013; pp. 103–144.

38. Dwyer, D.F.; Barrett, N.A.; Austen, K.F.; Immunological Genome Project Consortium. Expression profiling of constitutive mast cells reveals a unique identity within the immune system. *Nat. Immunol.* **2016**, *17*, 878–887. [CrossRef] [PubMed]

39. Fabbri, R.; Furini, C.R.G.; Passani, M.B.; Provensi, G.; Baldi, E.; Bucherelli, C.; Izquierdo, I.; de Carvalho Myskiw, J.; Blandina, P. Memory retrieval of inhibitory avoidance requires histamine H1 receptor activation in the hippocampus. *Proc. Natl. Acad. Sci. USA* **2016**, *113*, E2714–E2720. [CrossRef] [PubMed]

40. Bernsand, M.; Ericsson, P.; Bjorkqvist, M.; Zhao, C.-M.; Hakanson, R.; Norlen, P. Submucosal microinfusion of endothelin and adrenaline mobilizes ECL-cell histamine in rat stomach, and causes mucosal damage: A microdialysis study. *Br. J. Pharmacol.* **2003**, *140*, 707–717. [CrossRef] [PubMed]

41. Krauth, M.-T.T.; Agis, H.; Aichberger, K.J.; Simonitsch-Klupp, I.; Müllauer, L.; Mayerhofer, M.; Böhm, A.; Horny, H.-P.P.; Valent, P. Immunohistochemical detection of histidine decarboxylase in neoplastic mast cells in patients with systemic mastocytosis. *Hum. Pathol.* **2006**, *37*, 439–447. [CrossRef] [PubMed]

42. Osefo, N.; Ito, T.; Jensen, R.T. Gastric acid hypersecretory states: Recent insights and advances. *Curr. Gastroenterol. Rep.* **2009**, *11*, 433–441. [CrossRef] [PubMed]

43.   Kuramasu, A.; Saito, H.; Suzuki, S.; Watanabe, T.; Ohtsu, H. Mast cell-/basophil-specific transcriptional regulation of human L-histidine decarboxylase gene by CpG methylation in the promoter region. *J. Biol. Chem.* **1998**, *273*, 31607–31614. [CrossRef] [PubMed]

44.   Correa-Fiz, F.; Reyes-Palomares, A.; Fajardo, I.; Melgarejo, E.; Gutiérrez, A.; García-Ranea, J.A.; Medina, M.A.; Sánchez-Jiménez, F. Regulatory cross-talk of mouse liver polyamine and methionine metabolic pathways: A systemic approach to its physiopathological consequences. *Amino Acids* **2011**, *42*, 577–595. [CrossRef] [PubMed]

45.   Smolinska, S.; Jutel, M.; Crameri, R.; O'Mahony, L. Histamine and gut mucosal immune regulation. *Allergy* **2014**, *69*, 273–281. [CrossRef] [PubMed]

46.   Mamune-Sato, R.; Yamauchi, K.; Tanno, Y.; Ohkawara, Y.; Ohtsu, H.; Katayose, D.; Maeyama, K.; Watanabe, T.; Shibahara, S.; Takishima, T. Functional analysis of alternatively spliced transcripts of the human histidine decarboxylase gene and its expression in human tissues and basophilic leukemia cells. *Eur. J. Biochem.* **1992**, *209*, 533–539. [CrossRef] [PubMed]

47.   Abrighach, H.; Fajardo, I.; Sánchez-Jiménez, F.; Urdiales, J.L. Exploring polyamine regulation by nascent histamine in a human-transfected cell model. *Amino Acids* **2010**, *38*, 561–573. [CrossRef] [PubMed]

48.   Olmo, M.T.; Urdiales, J.L.; Pegg, A.E.; Medina, M.A.; Sánchez-Jiménez, F. In vitro study of proteolytic degradation of rat histidine decarboxylase. *Eur. J. Biochem.* **2000**, *267*, 1527–1531. [CrossRef] [PubMed]

49.   Fleming, J.V.; Fajardo, I.; Langlois, M.R.; Sanchez-Jimenez, F.; Wang, T.C. The C-terminus of rat L-histidine decarboxylase specifically inhibits enzymic activity and disrupts pyridoxal phosphate-dependent interactions with L-histidine substrate analogues. *Biochem. J.* **2004**, *381*, 769–778. [CrossRef] [PubMed]

50.   Furuta, K.; Nakayama, K.; Sugimoto, Y.; Ichikawa, A.; Tanaka, S. Activation of histidine decarboxylase through post-translational cleavage by caspase-9 in a mouse mastocytoma P-815. *J. Biol. Chem.* **2007**, *282*, 13438–13446. [CrossRef] [PubMed]

51.   Olmo, M.T.; Rodríguez-Agudo, D.; Medina, M.A.; Sánchez-Jiménez, F. The pest regions containing C-termini of mammalian ornithine decarboxylase and histidine decarboxylase play different roles in protein degradation. *Biochem. Biophys. Res. Commun.* **1999**, *257*, 269–272. [CrossRef] [PubMed]

52.   Rodriguez-Agudo, D.; Olmo, M.T.; Sanchez-Jimenez, F.; Medina, M.A. Rat histidine decarboxylase is a substrate for m-calpain in vitro. *Biochem. Biophys. Res. Commun.* **2000**, *271*, 777–781. [CrossRef] [PubMed]

53.   Pino-Angeles, A.; Morreale, A.; Negri, A.; Sánchez-Jiménez, F.; Moya-García, A.A. Substrate uptake and protein stability relationship in mammalian histidine decarboxylase. *Proteins* **2010**, *78*, 154–161. [CrossRef] [PubMed]

54.   Olmo, M.T.; Sanchez-Jimenez, F.; Medina, M.A.; Hayashi, H. Spectroscopic analysis of recombinant rat histidine decarboxylase. *J. Biochem.* **2002**, *132*, 433–439. [CrossRef] [PubMed]

55.   Moya-García, A.A.; Ruiz-Pernía, J.; Martí, S.; Sánchez-Jiménez, F.; Tuñón, I. Analysis of the decarboxylation step in mammalian histidine decarboxylase. A computational study. *J. Biol. Chem.* **2008**, *283*, 12393–12401. [CrossRef] [PubMed]

56.   Rodríguez-Caso, C.; Rodríguez-Agudo, D.; Moya-García, A.A.; Fajardo, I.; Medina, M.A.; Subramaniam, V.; Sánchez-Jiménez, F. Local changes in the catalytic site of mammalian histidine decarboxylase can affect its global conformation and stability. *Eur. J. Biochem.* **2003**, *270*, 4376–4387. [CrossRef] [PubMed]

57.   Fleming, J.V.; Sánchez-Jiménez, F.; Moya-García, A.A.; Langlois, M.R.; Wang, T.C. Mapping of catalytically important residues in the rat L-histidine decarboxylase enzyme using bioinformatic and site-directed mutagenesis approaches. *Biochem. J.* **2004**, *379*, 253–261. [CrossRef] [PubMed]

58.   Morgan, D.M.L.; Milovic, V.; Krizek, M.; White, A. *COST Action 917. Biogenically Active Amines in Food. Volume V. Polyamines and Tumor Growth, Biologically Active Amines in Food Processing and Amines Produced by Bacteria*; European Commission: Luxemburg, 2001.

59.   Wallace, H.M. Health implications of dietary amines: An overview of COST Action 922 (2001–2006). *Biochem. Soc. Trans.* **2007**, *35*, 293–294. [CrossRef] [PubMed]

60.   Schwelberger, H.G. Histamine N-methyltransferase (HNMT) enzyme and gene. In *Histamine: Biology and Medical Aspects*; Falus, A., Grosman, N., Darvas, Z., Eds.; SpringMed Publishing: Budapest, Hungary, 2004; pp. 53–59.

61.   Fagerberg, L.; Hallström, B.M.; Oksvold, P.; Kampf, C.; Djureinovic, D.; Odeberg, J.; Habuka, M.; Tahmasebpoor, S.; Danielsson, A.; Edlund, K.; et al. Analysis of the human tissue-specific expression

by genome-wide integration of transcriptomics and antibody-based proteomics. *Mol. Cell. Proteom.* **2014**, *13*, 397–406. [CrossRef] [PubMed]

62. Duff, M.O.; Olson, S.; Wei, X.; Garrett, S.C.; Osman, A.; Bolisetty, M.; Plocik, A.; Celniker, S.E.; Graveley, B.R. Genome-wide identification of zero nucleotide recursive splicing in Drosophila. *Nature* **2015**, *521*, 376–379. [CrossRef] [PubMed]

63. Prell, G.D.; Green, J.P. Measurement of histamine metabolites in brain and cerebrospinal fluid provides insights into histaminergic activity. *Agents Actions* **1994**, *41*, C5–C8. [CrossRef] [PubMed]

64. Song, W.-B.; Lv, Y.-H.; Zhang, Z.-S.; Li, Y.-N.; Xiao, L.-P.; Yu, X.-P.; Wang, Y.-Y.; Ji, H.-L.; Ma, L. Soluble intercellular adhesion molecule-1, D-lactate and diamine oxidase in patients with inflammatory bowel disease. *World J. Gastroenterol.* **2009**, *15*, 3916–3919. [CrossRef] [PubMed]

65. Koepsell, H.; Lips, K.; Volk, C. Polyspecific organic cation transporters: Structure, function, physiological roles, and biopharmaceutical implications. *Pharm. Res.* **2007**, *24*, 1227–1251. [CrossRef] [PubMed]

66. Eiden, L.E.; Weihe, E. VMAT2: A dynamic regulator of brain monoaminergic neuronal function interacting with drugs of abuse. *Ann. N. Y. Acad. Sci.* **2011**, *1216*, 86–98. [CrossRef] [PubMed]

67. Garcia-Faroldi, G.; Rodriguez, C.E.; Urdiales, J.L.; Perez-Pomares, J.M.; Davila, J.C.; Pejler, G.; Sanchez-Jimenez, F.; Fajardo, I. Polyamines are present in mast cell secretory granules and are important for granule homeostasis. *PLoS ONE* **2010**, *5*, e15071. [CrossRef] [PubMed]

68. Andersson, K.; Chen, D.; Håkanson, R.; Mattsson, H.; Sundler, F. Enterochromaffin-like cells in the rat stomach: Effect of alpha-fluoromethylhistidine-evoked histamine depletion. A chemical, histochemical and electron-microscopic study. *Cell Tissue Res.* **1992**, *270*, 7–13. [CrossRef] [PubMed]

69. Thurmond, R.L. The histamine H4 receptor: From orphan to the clinic. *Front. Pharmacol.* **2015**, *6*, 65. [CrossRef] [PubMed]

70. Kusters, J.G.; van Vliet, A.H.M.; Kuipers, E.J. Pathogenesis of Helicobacter pylori infection. *Clin. Microbiol. Rev.* **2006**, *19*, 449–490. [CrossRef] [PubMed]

71. Chen, D.; Aihara, T.; Zhao, C.-M.; Håkanson, R.; Okabe, S. Differentiation of the Gastric Mucosa I. Role of histamine in control of function and integrity of oxyntic mucosa: Understanding gastric physiology through disruption of targeted genes. *Am. J. Physiol. Gastrointest. Liver Physiol.* **2006**, *291*, G539–G544. [CrossRef] [PubMed]

72. Waldum, H.L.; Hauso, Ø.; Fossmark, R. The regulation of gastric acid secretion—Clinical perspectives. *Acta Physiol.* **2014**, *210*, 239–256. [CrossRef] [PubMed]

73. Coruzzi, G.; Adami, M.; Pozzoli, C.; de Esch, I.J.P.; Smits, R.; Leurs, R. Selective histamine H3 and H4 receptor agonists exert opposite effects against the gastric lesions induced by HCl in the rat stomach. *Eur. J. Pharmacol.* **2011**, *669*, 121–127. [CrossRef] [PubMed]

74. Rydning, A.; Lyng, O.; Falkmer, S.; Grønbech, J.E. Histamine is involved in gastric vasodilation during acid back diffusion via activation of sensory neurons. *Am. J. Physiol. Gastrointest. Liver Physiol.* **2002**, *283*, G603–G611. [CrossRef] [PubMed]

75. Ahmad, J.; Misra, M.; Rizvi, W.; Kumar, A. Histamine aspects in acid peptic diseases and cell proliferation. In *Biomedical Aspects of Histamine. Current Perspectives*; Shahid, M., Khardori, N., Khan, R.A., Tripathi, T., Eds.; Springer: Dordrecht, The Netherlands, 2010; pp. 175–198.

76. Sander, L.E.; Lorentz, A.; Sellge, G.; Coëffier, M.; Neipp, M.; Veres, T.; Frieling, T.; Meier, P.N.; Manns, M.P.; Bischoff, S.C. Selective expression of histamine receptors H1R, H2R, and H4R, but not H3R, in the human intestinal tract. *Gut* **2006**, *55*, 498–504. [CrossRef] [PubMed]

77. Schneider, E.; Leite-de-moraes, M.; Dy, M. Histamine, immune cells and autoimmunity. *Adv. Exp. Med. Biol.* **2010**, *709*, 81–94. [PubMed]

78. Xie, H.; He, S.-H. Roles of histamine and its receptors in allergic and inflammatory bowel diseases. *World J. Gastroenterol.* **2005**, *11*, 2851–2857. [CrossRef] [PubMed]

79. Gutzmer, R.; Diestel, C.; Mommert, S.; Köther, B.; Stark, H.; Wittmann, M.; Werfel, T. Histamine H4 receptor stimulation suppresses IL-12p70 production and mediates chemotaxis in human monocyte-derived dendritic cells. *J. Immunol.* **2005**, *174*, 5224–5232. [CrossRef] [PubMed]

80. Gutzmer, R.; Mommert, S.; Gschwandtner, M.; Zwingmann, K.; Stark, H.; Werfel, T. The histamine H4 receptor is functionally expressed on T(H)2 cells. *J. Allergy Clin. Immunol.* **2009**, *123*, 619–625. [CrossRef] [PubMed]

81. Coruzzi, G.; Adami, M.; Pozzoli, C. Role of histamine H4 receptors in the gastrointestinal tract. *Front. Biosci.* **2012**, *4*, 226–239. [CrossRef]

82. Mittal, R.; Debs, L.H.; Patel, A.P.; Nguyen, D.; Patel, K.; O'Connor, G.; Grati, M.; Mittal, J.; Yan, D.; Eshraghi, A.A.; et al. Neurotransmitters: The critical modulators regulating gut-brain axis. *J. Cell. Physiol.* **2017**, *232*, 2359–2372. [CrossRef] [PubMed]

83. Sikander, A.; Rana, S.V.; Prasad, K.K. Role of serotonin in gastrointestinal motility and irritable bowel syndrome. *Clin. Chim. Acta* **2009**, *403*, 47–55. [CrossRef] [PubMed]

84. Roberts, K.M.; Fitzpatrick, P.F. Mechanisms of tryptophan and tyrosine hydroxylase. *IUBMB Life* **2013**, *65*, 350–357. [CrossRef] [PubMed]

85. O'Mahony, S.M.; Clarke, G.; Borre, Y.E.; Dinan, T.G.; Cryan, J.F. Serotonin, tryptophan metabolism and the brain-gut-microbiome axis. *Behav. Brain Res.* **2015**, *277*, 32–48. [CrossRef] [PubMed]

86. Hayashi, H.; Tsukiyama, F.; Ishii, S.; Mizuguchi, H.; Kagamiyama, H. Acid-base chemistry of the reaction of aromatic L-amino acid decarboxylase and dopa analyzed by transient and steady-state kinetics: Preferential binding of the substrate with its amino group unprotonated. *Biochemistry* **1999**, *38*, 15615–15622. [CrossRef] [PubMed]

87. Bertoldi, M.; Voltattorni, C.B. Dopa decarboxylase exhibits low pH half-transaminase and high pH oxidative deaminase activities toward serotonin (5-hydroxytryptamine). *Protein Sci.* **2001**, *10*, 1178–1186. [CrossRef] [PubMed]

88. Bertoldi, M.; Gonsalvi, M.; Contestabile, R.; Voltattorni, C.B. Mutation of tyrosine 332 to phenylalanine converts dopa decarboxylase into a decarboxylation-dependent oxidative deaminase. *J. Biol. Chem.* **2002**, *277*, 36357–36362. [CrossRef] [PubMed]

89. Ruiz-Pérez, M.V.; Pino-Ángeles, A.; Medina, M.A.; Sánchez-Jiménez, F.; Moya-García, A.A. Structural perspective on the direct inhibition mechanism of EGCG on mammalian histidine decarboxylase and DOPA decarboxylase. *J. Chem. Inf. Model.* **2012**, *52*, 113–119. [CrossRef] [PubMed]

90. Montioli, R.; Cellini, B.; Dindo, M.; Oppici, E.; Voltattorni, C.B. Interaction of human Dopa decarboxylase with L-Dopa: Spectroscopic and kinetic studies as a function of pH. *BioMed Res. Int.* **2013**, *2013*, 161456. [CrossRef] [PubMed]

91. Bortolato, M.; Chen, K.; Shih, J.C. The degradation of serotonin: Role of MAO. *Handb. Behav. Neurosci.* **2010**, *21*, 203–218. [CrossRef]

92. Meiser, J.; Weindl, D.; Hiller, K. Complexity of dopamine metabolism. *Cell Commun. Signal.* **2013**, *11*, 34. [CrossRef] [PubMed]

93. Nagatsu, T. Progress in monoamine oxidase (MAO) research in relation to genetic engineering. *Neurotoxicology* **2004**, *25*, 11–20. [CrossRef]

94. Strege, P.R.; Knutson, K.; Eggers, S.J.; Li, J.H.; Wang, F.; Linden, D.; Szurszewski, J.H.; Milescu, L.; Leiter, A.B.; Farrugia, G.; et al. Sodium channel NaV1.3 is important for enterochromaffin cell excitability and serotonin release. *Sci. Rep.* **2017**, *7*, 15650. [CrossRef] [PubMed]

95. Mawe, G.M.; Hoffman, J.M. Serotonin signalling in the gut—Functions, dysfunctions and therapeutic targets. *Nat. Rev. Gastroenterol. Hepatol.* **2013**, *10*, 473–486. [CrossRef] [PubMed]

96. Gershon, M.D. 5-HT (serotonin) physiology and related drugs. *Curr. Opin. Gastroenterol.* **2000**, *16*, 113–120. [CrossRef] [PubMed]

97. Greenwood-van Meerveld, B. Importance of 5-hydroxytryptamine receptors on intestinal afferents in the regulation of visceral sensitivity. *Neurogastroenterol. Motil.* **2007**, *19*, 13–18. [CrossRef] [PubMed]

98. Versteeg, R.I.; Serlie, M.J.; Kalsbeek, A.; la Fleur, S.E. Serotonin, a possible intermediate between disturbed circadian rhythms and metabolic disease. *Neuroscience* **2015**, *301*, 155–167. [CrossRef] [PubMed]

99. Shortall, S.E.; Spicer, C.H.; Ebling, F.J.P.; Green, A.R.; Fone, K.C.F.; King, M.V. Contribution of serotonin and dopamine to changes in core body temperature and locomotor activity in rats following repeated administration of mephedrone. *Addict. Biol.* **2016**, *21*, 1127–1139. [CrossRef] [PubMed]

100. Arreola, R.; Becerril-Villanueva, E.; Cruz-Fuentes, C.; Velasco-Velázquez, M.A.; Garcés-Alvarez, M.E.; Hurtado-Alvarado, G.; Quintero-Fabian, S.; Pavón, L. Immunomodulatory Effects Mediated by Serotonin. *J. Immunol. Res.* **2015**, *2015*, 354957. [CrossRef] [PubMed]

101. Maggiorani, D.; Manzella, N.; Edmondson, D.E.; Mattevi, A.; Parini, A.; Binda, C.; Mialet-Perez, J. Monoamine oxidases, oxidative stress, and altered mitochondrial dynamics in cardiac ageing. *Oxid. Med. Cell. Longev.* **2017**, *2017*, 3017947. [CrossRef] [PubMed]

102. Margolis, K.G.; Gershon, M.D. Enteric neuronal regulation of intestinal inflammation. *Trends Neurosci.* **2016**, *39*, 614–624. [CrossRef] [PubMed]

103. Filip, M.; Bader, M. Overview on 5-HT receptors and their role in physiology and pathology of the central nervous system. *Pharmacol. Rep.* **2009**, *61*, 761–777. [CrossRef]

104. Li, Z.; Chalazonitis, A.; Huang, Y.-Y.; Mann, J.J.; Margolis, K.G.; Yang, Q.M.; Kim, D.O.; Côté, F.; Mallet, J.; Gershon, M.D. Essential roles of enteric neuronal serotonin in gastrointestinal motility and the development/survival of enteric dopaminergic neurons. *J. Neurosci.* **2011**, *31*, 8998–9009. [CrossRef] [PubMed]

105. Soyer, T.; Aktuna, Z.; Reşat Aydos, T.; Osmanoğlu, G.; Korkut, O.; Akman, H.; Cakmak, M. Esophageal and gastric smooth muscle activity after carbon dioxide pneumoperitoneum. *J. Surg. Res.* **2010**, *161*, 278–281. [CrossRef] [PubMed]

106. Flatmark, T. Catecholamine biosynthesis and physiological regulation in neuroendocrine cells. *Acta Physiol. Scand.* **2000**, *168*, 1–17. [CrossRef] [PubMed]

107. Wu, Q.; McLeish, M.J. Kinetic and pH studies on human phenylethanolamine *N*-methyltransferase. *Arch. Biochem. Biophys.* **2013**, *539*, 1–8. [CrossRef] [PubMed]

108. Eisenhofer, G.; Kopin, I.J.; Goldstein, D.S. Catecholamine metabolism: A contemporary view with implications for physiology and medicine. *Pharmacol. Rev.* **2004**, *56*, 331–349. [CrossRef] [PubMed]

109. Daubner, S.C.; Le, T.; Wang, S. Tyrosine hydroxylase and regulation of dopamine synthesis. *Arch. Biochem. Biophys.* **2011**, *508*, 1–12. [CrossRef] [PubMed]

110. Tank, A.W.; Lee Wong, D. Peripheral and Central Effects of Circulating Catecholamines. *Compr. Physiol.* **2014**, *5*, 1–15. [CrossRef]

111. Kobayashi, K. Role of catecholamine signaling in brain and nervous system functions: New insights from mouse molecular genetic study. *J. Investig. Dermatol. Symp. Proc.* **2001**, *6*, 115–121. [CrossRef] [PubMed]

112. Arreola, R.; Alvarez-Herrera, S.; Pérez-Sánchez, G.; Becerril-Villanueva, E.; Cruz-Fuentes, C.; Flores-Gutierrez, E.O.; Garcés-Alvarez, M.E.; de la Cruz-Aguilera, D.L.; Medina-Rivero, E.; Hurtado-Alvarado, G.; et al. Immunomodulatory effects mediated by dopamine. *J. Immunol. Res.* **2016**, *2016*, 3160486. [CrossRef] [PubMed]

113. Beaulieu, J.-M.; Espinoza, S.; Gainetdinov, R.R. Dopamine receptors—IUPHAR review 13. *Br. J. Pharmacol.* **2015**, *172*, 1–23. [CrossRef] [PubMed]

114. Elenkov, I.J.; Wilder, R.L.; Chrousos, G.P.; Vizi, E.S. The sympathetic nerve—An integrative interface between two supersystems: The brain and the immune system. *Pharmacol. Rev.* **2000**, *52*, 595–638. [PubMed]

115. Rizzetto, L.; Fava, F.; Tuohy, K.M.; Selmi, C. Connecting the immune system, systemic chronic inflammation and the gut microbiome: The role of sex. *J. Autoimmun.* **2018**, in press. [CrossRef] [PubMed]

116. Natale, G.; Ryskalin, L.; Busceti, C.L.; Biagioni, F.; Fornai, F. The nature of catecholamine-containing neurons in the enteric nervous system in relationship with organogenesis, normal human anatomy and neurodegeneration. *Arch. Ital. Biol.* **2017**, *155*, 118–130. [PubMed]

117. Rodriguez-Caso, C.; Montañez, R.; Cascante, M.; Sanchez-Jimenez, F.; Medina, M.A. Mathematical modeling of polyamine metabolism in mammals. *J. Biol. Chem.* **2006**, *281*, 21799–21812. [CrossRef] [PubMed]

118. Pegg, A.E. Mammalian polyamine metabolism and function. *IUBMB Life* **2009**, *61*, 880–894. [CrossRef] [PubMed]

119. Pegg, A.E. Functions of Polyamines in Mammals. *J. Biol. Chem.* **2016**, *291*, 14904–14912. [CrossRef] [PubMed]

120. Perez-Leal, O.; Merali, S. Regulation of polyamine metabolism by translational control. *Amino Acids* **2012**, *42*, 611–617. [CrossRef] [PubMed]

121. Kahana, C. Protein degradation, the main hub in the regulation of cellular polyamines. *Biochem. J.* **2016**, *473*, 4551–4558. [CrossRef] [PubMed]

122. Pegg, A.E. Regulation of ornithine decarboxylase. *J. Biol. Chem.* **2006**, *281*, 14529–14532. [CrossRef] [PubMed]

123. Pegg, A.E. *S*-Adenosylmethionine decarboxylase. *Essays Biochem.* **2009**, *46*, 25–45. [CrossRef] [PubMed]

124. Pegg, A.E.; Casero, R.A. Current status of the polyamine research field. In *Polyamines. Methods and Protocols*; Pegg, A.E., Casero, R.A., Eds.; Humana Press: Newyork, NK, USA, 2011; pp. 3–35.

125. Casero, R.A.; Marton, L.J. Targeting polyamine metabolism and function in cancer and other hyperproliferative diseases. *Nat. Rev. Drug Discov.* **2007**, *6*, 373–390. [CrossRef] [PubMed]

126. Casero, R.A.; Pegg, A.E. Polyamine catabolism and disease. *Biochem. J.* **2009**, *421*, 323–338. [CrossRef] [PubMed]

127. Cecco, L.; Antoniello, S.; Auletta, M.; Cerra, M.; Bonelli, P. Pattern and concentration of free and acetylated polyamines in urine of cirrhotic patients. *Int. J. Biol. Mark.* **1992**, *7*, 52–58.

128. Qian, H.; Beard, D.A. Metabolic futile cycles and their functions: A systems analysis of energy and control. *Syst. Biol.* **2006**, *153*, 192–200. [CrossRef]

129. Reyes-Palomares, A.; Montañez, R.; Sánchez-Jiménez, F.; Medina, M.A.; Sanchez-Jimenez, F.; Medina, M.A. A combined model of hepatic polyamine and sulfur amino acid metabolism to analyze *S*-adenosyl methionine availability. *Amino Acids* **2012**, *42*, 597–610. [CrossRef] [PubMed]

130. Muth, A.; Madan, M.; Archer, J.J.; Ocampo, N.; Rodriguez, L.; Phanstiel, O. Polyamine transport inhibitors: Design, synthesis, and combination therapies with difluoromethylornithine. *J. Med. Chem.* **2014**, *57*, 348–363. [CrossRef] [PubMed]

131. Uimari, A.; Keinänen, T.A.; Karppinen, A.; Woster, P.; Uimari, P.; Jänne, J.; Alhonen, L. Spermine analogue-regulated expression of spermidine/spermine N1-acetyltransferase and its effects on depletion of intracellular polyamine pools in mouse fetal fibroblasts. *Biochem. J.* **2009**, *422*, 101–109. [CrossRef] [PubMed]

132. Soulet, D.; Gagnon, B.; Rivest, S.; Audette, M.; Poulin, R. A fluorescent probe of polyamine transport accumulates into intracellular acidic vesicles via a two-step mechanism. *J. Biol. Chem.* **2004**, *279*, 49355–49366. [CrossRef] [PubMed]

133. Poulin, R.; Casero, R.A.; Soulet, D. Recent advances in the molecular biology of metazoan polyamine transport. *Amino Acids* **2011**, *42*, 711–723. [CrossRef] [PubMed]

134. Belting, M.; Mani, K.; Jönsson, M.; Cheng, F.; Sandgren, S.; Jonsson, S.; Ding, K.; Delcros, J.-G.; Fransson, L.-A. Glypican-1 is a vehicle for polyamine uptake in mammalian cells: A pivital role for nitrosothiol-derived nitric oxide. *J. Biol. Chem.* **2003**, *278*, 47181–47189. [CrossRef] [PubMed]

135. Abdulhussein, A.A.; Wallace, H.M. Polyamines and membrane transporters. *Amino Acids* **2014**, *46*, 655–660. [CrossRef] [PubMed]

136. Uemura, T.; Yerushalmi, H.F.; Tsaprailis, G.; Stringer, D.E.; Pastorian, K.E.; Hawel, L.; Byus, C.V.; Gerner, E.W. Identification and characterization of a diamine exporter in colon epithelial cells. *J. Biol. Chem.* **2008**, *283*, 26428–26435. [CrossRef] [PubMed]

137. Uemura, T.; Gerner, E.W. Polyamine transport systems in mammalian cells and tissues. *Methods Mol. Biol.* **2011**, *720*, 339–348. [CrossRef] [PubMed]

138. Ruiz-Chica, J.; Medina, M.A.; Sánchez-Jiménez, F.; Ramírez, F.J. Fourier transform Raman study of the structural specificities on the interaction between DNA and biogenic polyamines. *Biophys. J.* **2001**, *80*, 443–454. [CrossRef]

139. Finger, S.; Schwieger, C.; Arouri, A.; Kerth, A.; Blume, A. Interaction of linear polyamines with negatively charged phospholipids: The effect of polyamine charge distance. *Biol. Chem.* **2014**, *395*, 769–778. [CrossRef] [PubMed]

140. Lightfoot, H.L.; Hall, J. Endogenous polyamine function—The RNA perspective. *Nucleic Acids Res.* **2014**, *42*, 11275–11290. [CrossRef] [PubMed]

141. Thomas, T.J.; Tajmir-Riahi, H.A.; Thomas, T. Polyamine-DNA interactions and development of gene delivery vehicles. *Amino Acids* **2016**, *48*, 2423–2431. [CrossRef] [PubMed]

142. Acosta-Andrade, C.; Artetxe, I.; Lete, M.G.; Monasterio, B.G.; Ruiz-Mirazo, K.; Goñi, F.M.; Sánchez-Jiménez, F. Polyamine-RNA-membrane interactions: From the past to the future in biology. *Colloids Surf. B Biointerfaces* **2017**, *155*, 173–181. [CrossRef] [PubMed]

143. Igarashi, K.; Kashiwagi, K. Modulation of protein synthesis by polyamines. *IUBMB Life* **2015**, *67*, 160–169. [CrossRef] [PubMed]

144. Ramani, D.; De Bandt, J.P.; Cynober, L. Aliphatic polyamines in physiology and diseases. *Clin. Nutr.* **2014**, *33*, 14–22. [CrossRef] [PubMed]

145. Miller-Fleming, L.; Olin-Sandoval, V.; Campbell, K.; Ralser, M. Remaining mysteries of molecular biology: The role of polyamines in the cell. *J. Mol. Biol.* **2015**, *427*, 3389–3406. [CrossRef] [PubMed]

146. Murphy, G.M. Polyamines in the human gut. *Eur. J. Gastroenterol. Hepatol.* **2001**, *13*, 1011–1014. [CrossRef] [PubMed]

147. Yuan, Q.; Ray, R.M.; Viar, M.J.; Johnson, L.R. Polyamine regulation of ornithine decarboxylase and its antizyme in intestinal epithelial cells. *Am. J. Physiol. Gastrointest. Liver Physiol.* **2001**, *280*, G130–G138. [CrossRef] [PubMed]

148. Seiler, N.; Raul, F. Polyamines and the intestinal tract. *Crit. Rev. Clin. Lab. Sci.* **2007**, *44*, 365–411. [CrossRef] [PubMed]

149. Deloyer, P.; Peulen, O.; Dandrifosse, G. Dietary polyamines and non-neoplastic growth and disease. *Eur. J. Gastroenterol. Hepatol.* **2001**, *13*, 1027–1032. [CrossRef] [PubMed]

150. Arndt, M.A.; Battaglia, V.; Parisi, E.; Lortie, M.J.; Isome, M.; Baskerville, C.; Pizzo, D.P.; Ientile, R.; Colombatto, S.; Toninello, A.; et al. The arginine metabolite agmatine protects mitochondrial function and confers resistance to cellular apoptosis. *Am. J. Physiol. Cell Physiol.* **2009**, *296*, C1411–C1419. [CrossRef] [PubMed]

151. Satriano, J.; Isome, M.; Casero, R.A.; Thomson, S.C.; Blantz, R.C. Polyamine transport system mediates agmatine transport in mammalian cells. *Am. J. Physiol. Cell Physiol.* **2001**, *281*, C329–C334. [CrossRef] [PubMed]

152. Yang, X.C.; Reis, D.J. Agmatine selectively blocks the N-methyl-D-aspartate subclass of glutamate receptor channels in rat hippocampal neurons. *J. Pharmacol. Exp. Ther.* **1999**, *288*, 544–549. [PubMed]

153. Molderings, G.J.; Heinen, A.; Menzel, S.; Lübbecke, F.; Homann, J.; Göthert, M. Gastrointestinal uptake of agmatine: Distribution in tissues and organs and pathophysiologic relevance. *Ann. N. Y. Acad. Sci.* **2003**, *1009*, 44–51. [CrossRef] [PubMed]

154. Halaris, A.; Plietz, J. Agmatine: Metabolic pathway and spectrum of activity in brain. *CNS Drugs* **2007**, *21*, 885–900. [CrossRef] [PubMed]

155. Winter, T.N.; Elmquist, W.F.; Fairbanks, C.A. OCT2 and MATE1 provide bidirectional agmatine transport. *Mol. Pharm.* **2011**, *8*, 133–142. [CrossRef] [PubMed]

156. Sugiyama, Y.; Nara, M.; Sakanaka, M.; Gotoh, A.; Kitakata, A.; Okuda, S.; Kurihara, S. Comprehensive analysis of polyamine transport and biosynthesis in the dominant human gut bacteria: Potential presence of novel polyamine metabolism and transport genes. *Int. J. Biochem. Cell Biol.* **2017**, *93*, 52–61. [CrossRef] [PubMed]

157. Landete, J.M.; De las Rivas, B.; Marcobal, A.; Muñoz, R. Updated molecular knowledge about histamine biosynthesis by bacteria. *Crit. Rev. Food Sci. Nutr.* **2008**, *48*, 697–714. [CrossRef] [PubMed]

158. Ganesh, B.P.; Hall, A.; Ayyaswamy, S.; Nelson, J.W.; Fultz, R.; Major, A.; Haag, A.; Esparza, M.; Lugo, M.; Venable, S.; et al. Diacylglycerol kinase synthesized by commensal *Lactobacillus reuteri* diminishes protein kinase C phosphorylation and histamine-mediated signaling in the mammalian intestinal epithelium. *Mucosal. Immunol.* **2018**, *11*, 380–393. [CrossRef] [PubMed]

159. Williams, B.B.; Van Benschoten, A.H.; Cimermancic, P.; Donia, M.S.; Zimmermann, M.; Taketani, M.; Ishihara, A.; Kashyap, P.C.; Fraser, J.S.; Fischbach, M.A. Discovery and characterization of gut microbiota decarboxylases that can produce the neurotransmitter tryptamine. *Cell Host Microbe* **2014**, *16*, 495–503. [CrossRef] [PubMed]

160. Westfall, S.; Lomis, N.; Kahouli, I.; Dia, S.Y.; Singh, S.P.; Prakash, S. Microbiome, probiotics and neurodegenerative diseases: Deciphering the gut brain axis. *Cell. Mol. Life Sci.* **2017**, *74*, 3769–3787. [CrossRef] [PubMed]

161. Ge, X.; Pan, J.; Liu, Y.; Wang, H.; Zhou, W.; Wang, X. Intestinal crosstalk between microbiota and serotonin and its impact on gut motility. *Curr. Pharm. Biotechnol.* **2018**, in press. [CrossRef] [PubMed]

162. Gao, C.; Ganesh, B.P.; Shi, Z.; Shah, R.R.; Fultz, R.; Major, A.; Venable, S.; Lugo, M.; Hoch, K.; Chen, X.; et al. Gut microbe-mediated suppression of inflammation-associated colon carcinogenesis by luminal histamine production. *Am. J. Pathol.* **2017**, *187*, 2323–2336. [CrossRef] [PubMed]

163. Russo, F.; Linsalata, M.; Orlando, A. Probiotics against neoplastic transformation of gastric mucosa: Effects on cell proliferation and polyamine metabolism. *World J. Gastroenterol.* **2014**, *20*, 13258–13272. [CrossRef] [PubMed]

164. Ramakrishnan, K.; Salinas, R.C. Peptic ulcer disease. *Am. Fam. Physician* **2007**, *76*, 1005–1012. [PubMed]

165. Calam, J.; Baron, J.H. ABC of the upper gastrointestinal tract: Pathophysiology of duodenal and gastric ulcer and gastric cancer. *BMJ* **2001**, *323*, 980–982. [CrossRef] [PubMed]

166. Lai, L.H.; Sung, J.J.Y. Helicobacter pylori and benign upper digestive disease. *Best Pract. Res. Clin. Gastroenterol.* **2007**, *21*, 261–279. [CrossRef] [PubMed]

167. Konturek, S.J.; Konturek, P.C.; Brzozowski, T.; Konturek, J.W.; Pawlik, W.W. From nerves and hormones to

bacteria in the stomach; Nobel prize for achievements in gastrology during last century. *J. Physiol. Pharmacol.* **2005**, *56*, 507–530. [PubMed]

168. Singh, V.; Gohil, N.; Ramírez-García, R. New insight into the control of peptic ulcer by targeting the histamine H2 receptor. *J. Cell. Biochem.* **2018**, *119*, 2003–2011. [CrossRef] [PubMed]

169. Safavi, M.; Sabourian, R.; Foroumadi, A. Treatment of Helicobacter pylori infection: Current and future insights. *World J. Clin. Cases* **2016**, *4*, 5–19. [CrossRef] [PubMed]

170. Figueiredo, C.A.; Marques, C.R.; dos Santos Costa, R.; da Silva, H.B.F.; Alcantara-Neves, N.M. Cytokines, cytokine gene polymorphisms and *Helicobacter pylori* infection: Friend or foe? *World J. Gastroenterol.* **2014**, *20*, 5235–5243. [CrossRef] [PubMed]

171. Wessler, S.; Höcker, M.; Fischer, W.; Wang, T.C.; Rosewicz, S.; Haas, R.; Wiedenmann, B.; Meyer, T.F.; Naumann, M. Helicobacter pylori activates the histidine decarboxylase promoter through a mitogen-activated protein kinase pathway independent of pathogenicity island-encoded virulence factors. *J. Biol. Chem.* **2000**, *275*, 3629–3636. [CrossRef] [PubMed]

172. Wadhwa, R.; Song, S.; Lee, J.-S.; Yao, Y.; Wei, Q.; Ajani, J.A. Gastric cancer-molecular and clinical dimensions. *Nat. Rev. Clin. Oncol.* **2013**, *10*, 643–655. [CrossRef] [PubMed]

173. Takaishi, S.; Cui, G.; Frederick, D.M.; Carlson, J.E.; Houghton, J.; Varro, A.; Dockray, G.J.; Ge, Z.; Whary, M.T.; Rogers, A.B.; et al. Synergistic inhibitory effects of gastrin and histamine receptor antagonists on Helicobacter-induced gastric cancer. *Gastroenterology* **2005**, *128*, 1965–1983. [CrossRef] [PubMed]

174. Langman, M.J.; Dunn, J.A.; Whiting, J.L.; Burton, A.; Hallissey, M.T.; Fielding, J.W.; Kerr, D.J. Prospective, double-blind, placebo-controlled randomized trial of cimetidine in gastric cancer. *Br. J. Cancer* **1999**, *81*, 1356–1362. [CrossRef] [PubMed]

175. Garvin, B.; Wiley, J.W. The role of serotonin in irritable bowel syndrome: Implications for management. *Curr. Gastroenterol. Rep.* **2008**, *10*, 363–368. [CrossRef] [PubMed]

176. Crowell, M.D. Role of serotonin in the pathophysiology of the irritable bowel syndrome. *Br. J. Pharmacol.* **2004**, *141*, 1285–1293. [CrossRef] [PubMed]

177. Foley, S.; Garsed, K.; Singh, G.; Duroudier, N.P.; Swan, C.; Hall, I.P.; Zaitoun, A.; Bennett, A.; Marsden, C.; Holmes, G.; et al. Impaired uptake of serotonin by platelets from patients with irritable bowel syndrome correlates with duodenal immune activation. *Gastroenterology* **2011**, *140*, 1434–1443. [CrossRef] [PubMed]

178. Mawe, G.M.; Coates, M.D.; Moses, P.L. Intestinal serotonin signalling in irritable bowel syndrome. *Aliment. Pharmacol. Ther.* **2006**, *23*, 1067–1076. [CrossRef] [PubMed]

179. Konturek, P.C.; Brzozowski, T.; Konturek, S.J. Stress and the gut: Pathophysiology, clinical consequences, diagnostic approach and treatment options. *J. Physiol. Pharmacol.* **2011**, *62*, 591–599. [PubMed]

180. Liu, Q.; Yang, Q.; Sun, W.; Vogel, P.; Heydorn, W.; Yu, X.-Q.; Hu, Z.; Yu, W.; Jonas, B.; Pineda, R.; et al. Discovery and characterization of novel tryptophan hydroxylase inhibitors that selectively inhibit serotonin synthesis in the gastrointestinal tract. *J. Pharmacol. Exp. Ther.* **2008**, *325*, 47–55. [CrossRef] [PubMed]

181. Matthes, S.; Bader, M. Peripheral Serotonin Synthesis as a New Drug Target. *Trends Pharmacol. Sci.* **2018**, *39*, 560–572. [CrossRef] [PubMed]

182. Fogel, W.A.; Lewiński, A.; Jochem, J. Histamine in idiopathic inflammatory bowel diseases—Not a standby player. *Folia Med. Cracov.* **2005**, *46*, 107–118. [PubMed]

183. Neumann, D.; Seifert, R. The therapeutic potential of histamine receptor ligands in inflammatory bowel disease. *Biochem. Pharmacol.* **2014**, *91*, 12–17. [CrossRef] [PubMed]

184. Deiteren, A.; De Man, J.G.; Pelckmans, P.A.; De Winter, B.Y. Histamine H_4 receptors in the gastrointestinal tract. *Br. J. Pharmacol.* **2015**, *172*, 1165–1178. [CrossRef] [PubMed]

185. Xie, Q.; Gan, H.-T. Controversies about the use of serological markers in diagnosis of inflammatory bowel disease. *World J. Gastroenterol.* **2010**, *16*, 279–280. [CrossRef] [PubMed]

186. Thomas, H. IBD: Probiotics for IBD: A need for histamine? *Nat. Rev. Gastroenterol. Hepatol.* **2016**, *13*, 62–63. [CrossRef] [PubMed]

187. Milano, A.F.; Singer, R.B. The cancer mortality risk project—Cancer mortality risks by anatomic site: Part 1—Introductory overview; part II—Carcinoma of the Colon: 20-Year mortality follow-up derived from 1973-2013 (NCI) SEER*Stat Survival Database. *J. Insur. Med.* **2017**, *47*, 65–94. [CrossRef] [PubMed]

188. Babbar, N.; Gerner, E.W. Targeting polyamines and inflammation for cancer prevention. *Recent Results Cancer Res.* **2010**, *188*, 49–64. [CrossRef]

189. Shen, P.; Pichler, M.; Chen, M.; Calin, G.A.; Ling, H. To WNT or Lose: The Missing Non-Coding Linc in Colorectal Cancer. *Int. J. Mol. Sci.* **2017**, *18*, 2003. [CrossRef] [PubMed]

190. Gerner, E.W.; Meyskens, F.L. Combination chemoprevention for colon cancer targeting polyamine synthesis and inflammation. *Clin. Cancer Res.* **2009**, *15*, 758–761. [CrossRef] [PubMed]

191. Ogino, S.; Shima, K.; Baba, Y.; Nosho, K.; Irahara, N.; Kure, S.; Chen, L.; Toyoda, S.; Kirkner, G.J.; Wang, Y.L.; et al. Colorectal cancer expression of peroxisome proliferator-activated receptor gamma (PPARG, PPARgamma) is associated with good prognosis. *Gastroenterology* **2009**, *136*, 1242–1250. [CrossRef] [PubMed]

192. Babbar, N.; Ignatenko, N.A.; Casero, R.A.; Gerner, E.W. Cyclooxygenase-independent induction of apoptosis by sulindac sulfone is mediated by polyamines in colon cancer. *J. Biol. Chem.* **2003**, *278*, 47762–47775. [CrossRef] [PubMed]

193. Gerner, E.W.; Meyskens, F.L. Polyamines and cancer: Old molecules, new understanding. *Nat. Rev. Cancer* **2004**, *4*, 781–792. [CrossRef] [PubMed]

194. Vargas, A.J.; Ashbeck, E.L.; Thomson, C.A.; Gerner, E.W.; Thompson, P.A. Dietary polyamine intake and polyamines measured in urine. *Nutr. Cancer* **2014**, *66*, 1144–1153. [CrossRef] [PubMed]

195. Simon, T.; László, V.; Falus, A. Impact of histamine on dendritic cell functions. *Cell Biol. Int.* **2011**, *35*, 997–1000. [CrossRef] [PubMed]

196. Martinel Lamas, D.J.; Rivera, E.S.; Medina, V.A. Histamine H_4 receptor: Insights into a potential therapeutic target in breast cancer. *Front. Biosci.* **2015**, *7*, 1–9.

197. Palucka, A.K.; Coussens, L.M. The Basis of Oncoimmunology. *Cell* **2016**, *164*, 1233–1247. [CrossRef] [PubMed]

198. Medina, V.A.; Coruzzi, G.; Martinel Lamas, D.J.; Massari, N.; Adami, M.; Levi-Schaffer, F.; Ben-Zimra, M.; Schwelberger, H.G.; Rivera, E.S. Histamine in cancer. In *Histamine H4 Receptor: A Novel Drug Target in Immunoregulation and Inflammation*; Stark, H., Ed.; Versita: London, UK, 2013; pp. 259–308.

199. Barilli, A.; Rotoli, B.M.; Visigalli, R.; Ingoglia, F.; Cirlini, M.; Prandi, B.; Dall'Asta, V. Gliadin-mediated production of polyamines by RAW264.7 macrophages modulates intestinal epithelial permeability in vitro. *Biochim. Biophys. Acta* **2015**, *1852*, 1779–1786. [CrossRef] [PubMed]

200. Rodríguez-Ruiz, M.; Moreno, E.; Moreno-Delgado, D.; Navarro, G.; Mallol, J.; Cortés, A.; Lluís, C.; Canela, E.I.; Casadó, V.; McCormick, P.J.; et al. Heteroreceptor complexes formed by dopamine d1, histamine h3, and n-methyl-d-aspartate glutamate receptors as targets to prevent neuronal death in Alzheimer's disease. *Mol. Neurobiol.* **2017**, *54*, 4537–4550. [CrossRef] [PubMed]

201. Burban, A.; Faucard, R.; Armand, V.; Bayard, C.; Vorobjev, V.; Arrang, J.-M. Histamine potentiates N-methyl-D-aspartate receptors by interacting with an allosteric site distinct from the polyamine binding site. *J. Pharmacol. Exp. Ther.* **2010**, *332*, 912–921. [CrossRef] [PubMed]

202. Hornung, B.; Martins Dos Santos, V.A.P.; Smidt, H.; Schaap, P.J. Studying microbial functionality within the gut ecosystem by systems biology. *Genes Nutr.* **2018**, *13*, 5. [CrossRef] [PubMed]

9

The Role of *Enterococcus faecium* as a Key Producer and Fermentation Condition as an Influencing Factor in Tyramine Accumulation in *Cheonggukjang*

Young Kyoung Park [1], Young Hun Jin [1], Jun-Hee Lee [1], Bo Young Byun [1], Junsu Lee [1], Kwangcheol Casey Jeong [2,3] and Jae-Hyung Mah [1,*]

[1]   Department of Food and Biotechnology, Korea University, 2511 Sejong-ro, Sejong 30019, Korea;
      eskimo@korea.ac.kr (Y.K.P.); younghoonjin3090@korea.ac.kr (Y.H.J.); bory92@korea.ac.kr (J.-H.L.);
      by-love23@hanmail.net (B.Y.B.); jpang@korea.ac.kr (J.L.)
[2]   Department of Animal Sciences, University of Florida, Gainesville, FL 32611, USA; kcjeong@ufl.edu
[3]   Emerging Pathogens Institute, University of Florida, Gainesville, FL 32611, USA
*   Correspondence: nextbio@korea.ac.kr

Abstract: The study evaluated the role of *Enterococcus faecium* in tyramine production and its response to fermentation temperature in a traditional Korean fermented soybean paste, *Cheonggukjang*. Tyramine content was detected in retail *Cheonggukjang* products at high concentrations exceeding the recommended limit up to a factor of 14. All retail *Cheonggukjang* products contained *Enterococcus* spp. at concentrations of at least 6 Log CFU/g. Upon isolation of *Enterococcus* strains, approximately 93% (157 strains) produced tyramine at over 100 µg/mL. The strains that produced the highest concentrations of tyramine (301.14–315.29 µg/mL) were identified as *E. faecium* through 16S rRNA sequencing. The results indicate that *E. faecium* is one of the major contributing factors to high tyramine content in *Cheonggukjang*. During fermentation, tyramine content in *Cheonggukjang* groups co-inoculated with *E. faecium* strains was highest at 45 °C, followed by 37 °C and 25 °C. The tyramine content of most *Cheonggukjang* groups continually increased as fermentation progressed, except groups fermented at 25 °C. At 45 °C, the tyramine content occasionally exceeded the recommended limit within 3 days of fermentation. The results suggest that lowering fermentation temperature and shortening duration may reduce the tyramine content of *Cheonggukjang*, thereby reducing the safety risks that may arise when consuming food with high tyramine concentrations.

Keywords: *Cheonggukjang*; *Enterococcus faecium*; tyramine; biogenic amines; fermentation temperature; fermentation duration; tyrosine decarboxylase gene (*tdc*)

---

1. Introduction

   Cheonggukjang is a traditional Korean soybean paste produced by fermenting soybeans with *Bacillus subtilis*. Traditional methods of *Cheonggukjang* production utilize rice straw added to steamed soybeans for a short fermentation period of approximately 2–3 days, while starter cultures are used instead of rice straw for modern methods of production [1,2]. Fermentation of *Cheonggukjang* is a process involving microbial enzymatic proteolysis resulting in uniquely characteristic savory aromatic and flavor properties [3]. Consumption of *Cheonggukjang* has been reported to be associated with numerous benefits such as antioxidative, antihypertensive, thrombolytic, and antimicrobial properties [4,5]. However, despite the beneficial properties of *Cheonggukjang*, potentially hazardous biogenic amines (BAs) may be produced during fermentation of the proteinous food rich in precursor amino acids.

   The majority of BAs are formed through the reductive amination of ketones and aldehydes, as well as the decarboxylation of amino acids by microbially produced enzymes [6]. Though BAs are essential for the regulation of protein synthesis, nucleic acid functions, and membrane stabilization in living cells, consumption of food products with high concentrations of BAs may result in toxicological effects [7–10]. The excessive intake of food products such as mackerel, pacific saury, sardines, and tuna may result in "scombroid poisoning" owing to potentially high concentrations of toxic histamine that may cause symptoms similar to an allergic reaction including diarrhea, dyspnea, headache, hives, and hypotension [10–13]. Overconsumption of foods with high concentrations of tyramine may potentially result in a "cheese crisis" with various symptoms including heart failure, hemorrhages, hypertensive crisis, high blood pressure, and severe headaches [9,10,14,15]. Such a high content of tyramine produced by microbial tyrosine decarboxylase activity has occasionally been found in tyrosine-rich foods such as cheese [16,17] and soybean-based fermented products [18–20]. Therefore, Ten Brink, et al. [21] suggested BA toxicity limits of 30 mg/kg for β-phenylethylamine, 100 mg/kg for histamine, and 100–800 mg/kg for tyramine in foods.

Previous studies by Ko, et al. [18], Jeon, et al. [19], and Seo, et al. [20] on the BA content of *Cheonggukjang* have shown that tyramine in particular has been detected in high concentrations up to 1913.51, 251.66, and 905.0 mg/kg, respectively. Ibe, et al. [22] suggested that *Enterococcus faecium* may be largely responsible for the BA content of *Miso* (a Japanese fermented soybean paste). Notably, numerous studies have reported that *Enterococcus* spp. possess the tyrosine decarboxylase gene (*tdc*) [23,24]. Moreover, in particular Kang and Park [25] and Kang, et al. [26] confirmed the presence of *E. faecium* in *Cheonggukjang*, while a previous study by Jeon, et al. [19] showed that *Enterococcus* spp. isolated from *Cheonggukjang* exhibited tyramine production at concentrations of at least 351.59 µg/mL. Taken together, the previous reports imply that *E. faecium* may also be responsible for the BA content of *Cheonggukjang*. Meanwhile, the growth of *Enterococcus* spp. has been reported to occur at temperatures ranging from 10 °C up to 45 °C that overlap with *Cheonggukjang* fermentation temperatures ranging from 25 to 50 °C [27–30]. The corresponding range in temperature may be beneficial for *E. faecium* growth and tyramine production during the fermentation of contaminated *Cheonggukjang* products. Furthermore, a previous study reported that tyramine content increases in fermented soybeans as fermentation progresses [19]. According to Bhardwaj, et al. [31], the production of tyramine by *E. faecium* strains may be affected by incubation conditions such as temperature and time. Therefore, the current study assessed the safety risk of BAs (particularly tyramine) in *Cheonggukjang*, clarified the microorganism responsible for tyramine accumulation, and evaluated the effect of fermentation temperature/duration on *E. faecium* growth and subsequent tyramine production in the food.

2. Materials and Methods

2.1. Cheonggukjang Products

Six representative, but different *Cheonggukjang* products were purchased from various retail markets in the Republic of Korea and stored at 4 °C until further experimentation. Within a day of storage, the BA content of *Cheonggukjang* products was measured, followed by physicochemical and microbial analyses.

2.2. Physicochemical Analyses

To investigate the influencing factors such as pH, salinity, and water activity on BA content in *Cheonggukjang*, the physicochemical properties of *Cheonggukjang* samples (retail *Cheonggukjang* products purchased and *Cheonggukjang* groups fermented in this study) were measured as described below. Samples weighing 10 g using an analytical balance (Ohaus Adventurer™, Ohaus Corporation, Parsippany, NJ, USA) were homogenized with 90 mL of distilled water using a stomacher (Laboratory Blender Stomacher 400, Seward, Ltd., Worthing, UK). The pH of the homogenates was measured using a pH meter (Orion 3-star pH Benchtop Thermo Scientific, Waltham, MA, USA), while salinity was measured using the procedure described by the Association of Official Analytical Chemists

(AOAC; Official Method 960.29) [32]. The water activity of the samples was measured using an electric hygrometer (AquaLab Pre, Meter Group, Inc., Pullman, WA, USA).

2.3. Microbial Analyses

The analysis of the microbial community in *Cheonggukjang* samples was conducted using Plate Count Agar (PCA; Difco, Becton Dickinson, Sparks, MD, USA); de Man, Rogosa, and Sharpe (MRS; Conda, Madrid, Spain) agar; and m-Enterococcus Agar (m-EA; MB Cell, Seoul, Korea) for the enumeration of total mesophilic viable bacteria, lactic acid bacteria, and *Enterococcus* spp., respectively. Samples weighing 10 g were homogenized with 90 mL of sterile 0.1% peptone saline using a stomacher. The homogenates were 10-fold serially diluted with 0.1% peptone saline up to 10^{-5}, and 100 µL of each dilution was spread on PCA, MRS agar, and m-EA in duplicates. Incubation conditions were set according to the manufacturer's instructions: PCA at 37 °C for 24 h and m-EA at 37 °C for 48 h under aerobic condition; MRS agar at 37 °C for 48 h under anaerobic condition. Anaerobic condition was achieved using an anaerobic chamber (Coy Lab. Products, Inc., Grass Lake, MI, USA) containing an atmosphere of 95% nitrogen and 5% hydrogen. After incubation, the bacterial concentrations of the *Cheonggukjang* samples were calculated by enumerating the colony-forming units (CFU) on the plates of respective media with approximately 10 to 300 colonies [33] and adjusting for the dilution.

2.4. Isolation and Identification of Enterococcus Strains from Retail Cheonggukjang Products

A total of 169 *Enterococcus* strains were isolated from retail *Cheonggukjang* products according to the method described by Mareková, et al. [34], with minor modifications. Upon enumeration of colonies on m-EA, individual colonies were streaked on MRS agar and incubated at 37 °C for 48 h under anaerobic condition. Single colonies were streaked again on MRS agar and incubated under the same conditions. The pure single colonies were inoculated in MRS broth, incubated at 37 °C for 48 h, and stored at −70 °C using glycerol (20%, *v/v*).

The identities (at species level) of the individual *Enterococcus* strains that displayed the highest tyramine production were further investigated through sequence analysis of 16S rRNA gene amplified with the universal bacterial primer pair 518F (5'-CCAGCAGCCGCGGTAATACG-3') and 805R (5'-GACTACCAGGGTATCTAAT-3') (Solgent Co., Daejeon, Korea). The identities of sequences were determined using the basic local alignment search tool (BLAST) of the National Center for Biotechnology Information (NCBI; http://www.ncbi.nlm.nih.gov/BLAST/).

2.5. Preparation of Cheonggukjang

To investigate the effect of fermentation temperature on tyramine production by *E. faecium*, several temperatures were set for in situ *Cheonggukjang* fermentation experiments. The temperature for *Cheonggukjang* fermentation (intermediate-temperature group) was determined based upon previous studies in which 37 °C was reported as the temperature commonly used for *Cheonggukjang* production [19,35,36]. In addition, the temperatures of 25 °C and 45 °C used by other studies for *Cheonggukjang* fermentation were utilized for the low and high temperature groups, respectively [29,30].

White soybeans (*Glycine* max Merrill) were purchased from a retail market in the Republic of Korea. The soybeans were soaked in distilled water at 4 °C for 12 h, and subsequently drained for 1 h. Approximately 200 g of soybeans were adjusted to a final salinity of 2.40% according to the salinity of *Cheonggukjang* outlined in the 9th revision of the Korean food composition table [37] and subsequently steamed at 125 °C for 30 min using an autoclave. The steamed soybeans were cooled to 50 °C and inoculated with bacterial inocula in M/15 Sörensen's phosphate buffer (pH 7) to final concentrations of approximately 6 Log CFU/g of *B. subtilis* KCTC 3135 (also designated as ATCC 6051; type strain) and 4 Log CFU/g of *E. faecium* KCCM 12118 (ATCC 19434; type strain) or *E. faecium* CJE 216 (strain isolated from *Cheonggukjang* and selected owing to both strong tyramine production and *tdc* gene expression). The control group (without any *E. faecium* strains) was inoculated with only *B. subtilis* KCTC 3135 to a final concentration of 6 Log CFU/g. The sizes of inocula were selected with consideration of the cell count of each microorganism in *Cheonggukjang* products determined in our previous study [19]. The inoculated steamed soybeans were then fermented at 25 °C, 37 °C, or 45 °C for 3 days. Approximately 20 g of the fermented soybeans were collected daily during fermentation to measure the BA content as well as physicochemical and microbial properties. Fermented soybeans sampled during fermentation were stored at −70 °C for further testing, as required.

2.6. BA Analyses in Cheonggukjang Samples and Bacterial Cultures

2.6.1. BA Extraction from Cheonggukjang Samples and Bacterial Cultures

Quantification of the BA content of *Cheonggukjang* was conducted as previously described by Ben-Gigirey, et al. [38]. Five grams of *Cheonggukjang* with 20 mL of 0.4 M perchloric acid (Sigma-Aldrich, St. Louis, MO, USA) were homogenized by vortex (Vortex-Genie, Scientific industries, Bohemia, NY, USA) and stored at 4 °C for 2 h. The mixture was then centrifuged at 3000× g for 10 min at 4 °C (1736R, Labogene, Seoul, Korea), and the supernatant was collected. Upon resuspension of the pellet with 20 mL of 0.4 M perchloric acid, the mixture was stored at 4 °C for 2 h and centrifuged again at 3000× g at 4 °C for 10 min. The supernatant was combined with the previously collected supernatant and adjusted to a final volume of 50 mL with 0.4 M perchloric acid. Then, the extract was filtered through Whatman paper No. 1 (Whatman International Ltd., Maidstone, UK).

The bacterial production of BAs was measured using the procedures described by Eerola, et al. [39], modified by Ben-Gigirey, et al. [38,40], and further modified in the present study to culture *Enterococcus* spp. based on Marcobal, et al. [41]. A loopful (10 µL) of glycerol stock of each enterococcal strain was inoculated in 5 mL of MRS broth supplemented with 0.5% of each amino acid, including L-histidine monohydrochloride monohydrate, L-tyrosine disodium salt hydrate, L-ornithine monohydrochloride, L-lysine monohydrochloride (pH 5.8), and 0.0005% pyridoxal-HCl (all from Sigma-Aldrich) and incubated at 37 °C for 48 h. Approximately 100 µL of the broth culture was then transferred to another tube containing 5 mL of the same medium. Upon incubation at 37 °C for 48 h, the broth culture was filtered using a sterile syringe with a 0.2 µm membrane (Millipore Co., Bedford, MA, USA). Then, 9 mL of 0.4 M perchloric acid were added to 1 mL of the filtered broth culture and mixed by a vortex mixer. The mixture was reacted in a cold chamber at 4 °C for 2 h and centrifuged at 3000× g at 4 °C for 10 min. The extract was filtered through Whatman paper No. 1.

2.6.2. Preparation of Standard Solutions for High Performance Liquid Chromatography (HPLC) Analysis

Standard solutions with concentrations of 0, 10, 50, 100, and 1000 ppm were prepared for tryptamine, β-phenylethylamine hydrochloride, putrescine dihydrochloride, cadaverine dihydrochloride, histamine dihydrochloride, tyramine hydrochloride, spermidine trihydrochloride, and spermine tetrahydrochloride (all from Sigma-Aldrich). Internal standard solution with the same concentrations was prepared using 1,7-diaminoheptane (Sigma-Aldrich).

2.6.3. Derivatization of Extracts and Standards

Derivatization of BAs was conducted according to the method described by Eerola, et al. [39]. One milliliter of extract or standard solution prepared as aforementioned was mixed with 200 µL of 2 M sodium hydroxide and 300 µL of saturated sodium bicarbonate (all from Sigma-Aldrich). Two milliliters of dansyl chloride (Sigma-Aldrich) solution (10 mg/mL) in acetone were added to the mixture and incubated at 40 °C for 45 min. The residual dansyl chloride was removed by adding 100 µL of 25% ammonium hydroxide and incubating for 30 min at 25 °C. Using acetonitrile, the mixture was adjusted to a final volume of 5 mL and centrifuged at 3000× g for 5 min. After filtration using 0.2 µm pore-size filters (Millipore), the filtered supernatant was kept at 4 °C until further analysis using HPLC.

2.6.4. Chromatographic Separations

Chromatographic separation of BAs was conducted according to the method previously developed by Eerola, et al. [39] and modified by Ben-Gigirey, et al. [40]. An HPLC unit (YL9100, YL Instruments Co., Ltd., Anyang, Korea) equipped with a UV/vis detector (YL Instruments) and Autochro-3000 data system (YL Instruments) was used. For chromatographic separation, a Nova-Pak C_{18} 4 μm column (150 mm × 4.6 mm, Waters, Milford, MA, USA) held at 40 °C was utilized. The mobile phases were 0.1 M ammonium acetate dissolved in deionized water (solvent A; Sigma-Aldrich) and acetonitrile (solvent B; SK chemicals, Ulsan, Korea) adjusted to a flow rate of 1 mL/min with a linear gradient starting from 50% of solvent B reaching 90% by 19 min. A 10 μL sample was injected and monitored at 254 nm. The limits of detection were approximately 0.1 μg/mL for all BAs in standard solutions and bacterial cultures, and about 0.1 mg/kg for all BAs in food matrices [42].

2.7. Gene Expression Analyses in Bacterial Cultures and Cheonggukjang

2.7.1. RNA Extraction and Reverse Transcription

Expression analysis of tyrosine decarboxylase gene (*tdc*) involved RNA extraction from bacterial cultures (for in vitro experiments) and *Cheonggukjang* samples (viz., *Cheonggukjang* groups prepared through fermentation; for in situ fermentation experiments) with a Ribo-Ex Total RNA isolation solution (Geneall, Seoul, Korea). The extraction was conducted according to the manufacturer's instructions with minor modifications as follows. To prepare bacterial culture for in vitro gene expression analysis, a loopful (10 μL) of glycerol stock of each enterococcal strain was inoculated in 5 mL of MRS broth supplemented with 0.5% L-histidine monohydrochloride monohydrate, L-tyrosine disodium salt hydrate, L-ornithine monohydrochloride, L-lysine monohydrochloride (pH 5.8), and 0.0005% pyridoxal-HCl (all from Sigma-Aldrich) and incubated at 37 °C for 48 h. Approximately 100 μL of the broth culture was then transferred to another tube containing 5 mL of the same medium and incubated under the same conditions. As for *Cheonggukjang* samples, 10 g of *Cheonggukjang* were gently mixed with 40 mL of phosphate buffer in a sterile bag, and the liquid part was collected. Subsequently, 3 mL of the bacterial culture or all liquid part of the mixture were immediately transferred into a 50 mL conical tube and centrifuged at 10,000× *g* at 4 °C for 5 min. After removing the supernatant, the pellet was suspended with 7 mL of phosphate buffer and centrifuged under the same conditions. Then, the pellet was homogenized with 800 μL of Ribo-Ex reagent in a bacterial lysing tube (Lysing Matrix B; MP Biomedicals, Santa Ana, CA, USA) using a Precellys 24 homogenizer (Bertin Technologies, Montigny, France) with two cycles for 30 s at 6800 rpm, pausing for 90 s between cycles. Approximately 200 μL of chloroform were added to the lysate, vortexed, and centrifuged at 10,000× *g* for 1 min. Approximately 400 μL of the supernatant were mixed with 600 μL of chilled absolute ethanol. The mixture was reacted at −70 °C for 15 min and purified with a Nucleospin RNA kit (Macherey-Nagel, Düren, Germany) according to the manufacturer's instructions. The quality of the extracted RNA was evaluated using a NanoDrop 1000 spectrophotometer (Thermo Fisher, Waltham, MA, USA).

ReverTra Ace qPCR RT Master Mix with gDNA Remover kit (Toyobo, Osaka, Japan) containing reverse transcriptase, RNase inhibitor, oligo (dT) primers, random primers, and deoxynucleoside triphosphates (dNTPs) was used to synthesize cDNA from 1 μL of extracted RNA according to the manufacturer's instructions. Reverse transcription was conducted under the following conditions: 37 °C for 15 min, 50 °C for 5 min, and 98 °C for 5 min. After the reaction, the resulting cDNA was stored at −70 °C until quantitative PCR analysis.

2.7.2. Quantitative PCR Analysis

As designed by Kang, et al. [43], *q-tdc* F (5′-AGACCAAGTAATTCCAGTGCC-3′) and *q-tdc* R (5′-CACCGACTACACCTAAGATTGG-3′) primers were used for the quantitation of *tdc* gene expression by *E. faecium*. The primers for reference genes including *q-gap* F (5′-ATACGACACAACTCAAGGACG-3′) and *q-gap* R (5′-GATATCTACGCCTAGTTCGCC-3′) [34], along with *tufA*-RT F (5′-TACACGCCACTAC

GCTCAC-3') and *tufA*-RT R (5'-AGCTCCGTCCATTTGAGCAG-3') [44] were used for the normalization of *tdc* gene expression. The efficiency of each set of primers for reverse transcription quantitative polymerase chain reaction (RT-qPCR) was determined by the following equation: $E = 10^{(-1/S)} - 1$, where E is the amplification efficiency and S is the slope of standard curves generated through threshold cycle (Ct) values of serial dilutions of cDNA obtained from reverse-transcription of RNA from *E. faecium* KCCM 12118.

For the RT-qPCR analysis of *tdc* gene expression in bacterial cultures and *Cheonggukjang* samples, 5 µL of a 10-fold diluted cDNA were added to 15 µL of a master mix containing 10 µL of Power SYBR Green PCR Master Mix (Applied Biosystems, Foster City, CA, USA), 3 µL of RNase free water, and 1 µL of each primer (forward and reverse; 500 nM). Subsequently, thermal cycling was conducted using an Applied Biosystems 7500 Real-Time PCR system (Applied Biosystems) with the thermal cycling conditions programmed as follows: initial denaturation at 95 °C for 10 min; 40 cycles at 95 °C for 15 s (denaturation step), and 60 °C for 60 s (annealing and elongation steps, unless otherwise mentioned). Annealing and elongation conditions for primer *tufA*-RT were set at 55 °C for 60 s. Melting curve analysis was conducted using the RT-PCR system to confirm the specificity and to analyze the amplified products. Ct values were detected when the emissions from fluorescence exceeded the fixed threshold automatically determined by thermocycler software. Relative expression of *tdc* genes was further calculated by the $2^{-(\Delta\Delta ct)}$ method, normalized to the expression levels detected in *E. faecium* KCCM 12118 (refer to Figure 2) or *Cheonggukjang* groups fermented at 37 °C (refer to Figure 6), and expressed as n-fold differences to compare gene expression levels in different bacterial cultures and *Cheonggukjang* samples.

2.8. Statistical Analyses

Data were presented as means and standard deviations of duplicates or triplicates. All measurements on retail products were performed in triplicates, while the other experiments were conducted in duplicate. The significance of differences was determined by one-way analysis of variance (ANOVA) with Fisher's pairwise comparison module of the Minitab statistical software, version 17 (Minitab Inc., State College, PA, USA), and differences with probability (p) value of <0.05 were considered statistically significant.

3. Results and Discussion

3.1. Physicochemical Properties of Retail Cheonggukjang Products

Physicochemical and microbial properties as well as BA content in retail *Cheonggukjang* products were analyzed to estimate the contributing factors to BA content (particularly tyramine) in *Cheonggukjang* (Sections 3.1–3.3). Table 1 displays the physicochemical properties of *Cheonggukjang* products purchased from retail markets in the Republic of Korea. The pH ranged from 6.39 to 7.05, with an average pH of 6.84 ± 0.23 (mean ± standard deviation). The results were similar to the study conducted by Lee, et al. [45], which reported the average pH of *Cheonggukjang* to be 7.0 ± 0.8. Jeon, et al. [19] and Yoo, et al. [46] also reported the average pH of *Cheonggukjang* to be pH 6.07 ± 0.72 (range of pH 4.62–8.14) and pH 7.21 ± 0.59 (range of pH 5.89–7.95), respectively. Such differences in the pH of the *Cheonggukjang* products may be owing to different fermentation conditions [47] and/or fermentation metabolites [48]. The salinity of retail *Cheonggukjang* products ranged from 1.95 to 9.36% with an average salinity of 5.16 ± 2.78%. In comparison, Ko, et al. [18], Jeon, et al. [19], and Kang, et al. [49] reported the average salinity of *Cheonggukjang* to be 2.12 ± 1.66% (0.12–11.51%), 1.56 ± 1.19% (0.10–5.33%), and 3.51 ± 2.45 (1.64–8.39%), respectively. Though the salinity of the *Cheonggukjang* products was found to vary substantially, Ko, et al. [18] suggested that the large differences in *Cheonggukjang* salinities may be traced to the production process, as some methods utilize the addition of different amounts of salt to preserve the fermented soybean product. The water activity of retail *Cheonggukjang* products ranged from 0.919 to 0.973 with an average of 0.951 ± 0.019. In a previous study by Kim, et al. [47], the average water activity was found to be 0.962 ± 0.028 (0.857–0.991). Overall, the physicochemical properties of retail *Cheonggukjang* products analyzed in the current study were mostly similar to the values reported in previous studies. Although the results of the current study did not show any correlation between physicochemical properties and BA content (especially tyramine) based on linear regression analyses (data not shown), it is noteworthy that the ranges of the physicochemical parameters were within the specific conditions for the growth of *E. faecium*, which are as follows: pH, from 4 to 10 [50]; salinity, up to 7% [50]; water activity, above 0.940 [51].

Table 1. Physicochemical properties of retail *Cheonggukjang* products.

Products [1]	pH	Salinity (%)	Water Activity
CJ1	6.91 ± 0.03 [2]	5.54 ± 0.09	0.948 ± 0.002
CJ2	6.84 ± 0.02	7.25 ± 0.06	0.919 ± 0.001
CJ3	6.87 ± 0.02	3.16 ± 0.06	0.968 ± 0.002
CJ4	6.99 ± 0.03	1.95 ± 0.09	0.973 ± 0.002
CJ5	7.05 ± 0.05	9.36 ± 0.59	0.944 ± 0.003
CJ6	6.39 ± 0.03	3.71 ± 0.34	0.954 ± 0.003
Average	6.84 ± 0.23	5.16 ± 2.78	0.951 ± 0.019

[1] CJ: *Cheonggukjang*; [2] Mean ± standard deviation were calculated from triplicate experiments.

3.2. Microbial Properties of Retail Cheonggukjang Products

Table 2 shows the microbial properties of retail *Cheonggukjang* products. The number of total mesophilic viable bacteria ranged from 8.54 to 9.81 Log CFU/g, with an average of 9.27 ± 0.45 Log CFU/g. Comparatively, Ko, et al. [18] and Jeon, et al. [19] reported the total counts of viable mesophilic bacteria of *Cheonggukjang* products to be 7.50 ± 1.01 Log CFU/g (5.30–9.98 Log CFU/g) and 9.65 ± 0.77 Log CFU/g (8.23–11.66 Log CFU/g), respectively. The wide range of total mesophilic viable bacteria may result from an insufficient standardization of *Cheonggukjang* manufacturing processes such as different fermentation materials and conditions [18,47]. *Enterococcus* spp. were detected at concentrations of 6.64–7.99 Log CFU/g, with an average of 7.17 ± 0.49 Log CFU/g. The number of lactic acid bacteria was found to be approximately 6.66–8.12 Log CFU/g, with an average of 7.09 ± 0.58 Log CFU/g (Table 2). For comparison, a study by Kang and Park [25] showed that *Enterococcus* spp. were detected in all 31 *Cheonggukjang* products at concentrations of 3.51–8.46 Log CFU/g, with an average of 5.95 ± 1.60 Log CFU/g. In the report, approximately 58% and 16.8% of the isolated *Enterococcus* strains were identified as *E. faecium* and *E. faecalis*, respectively. The presence of *E. faecium* in *Cheonggukjang* was also reported by Kang, et al. [26]. The reported results on the presence of *Enterococcus* spp. at high concentrations in *Cheonggukjang* concurred with the findings of the current study. As *E. faecium* has been reported as a pathogenic and/or tyramine-producing bacterium detected in some foods including Chinese and Japanese fermented soybean products, previous studies have mentioned that preventative measures are necessary to avoid contamination during the manufacturing of fermented foods [52–55]. The traditional *Cheonggukjang* production process may also be susceptible to contamination by harmful microbes owing to the reliance on rice straw containing *B. subtilis* for fermentation [56]. In fact, according to Heu, et al. [57], rice straw contains a variety of bacteria, including mesophiles, thermophiles, coliforms, and actinomycetes, as well as fungi. Moreover, as sterilization processes are not utilized in the manufacturing of *Cheonggukjang*, occasional contamination by tyramine-producing bacteria such as *E. faecium* may be present in the final product. The results of the current and previous studies suggest that further research appears to be necessary for the development of methods to inhibit *E. faecium* growth during the manufacturing of *Cheonggukjang* as well as other fermented soybean products described above.

Table 2. Microbial properties of retail *Cheonggukjang* products.

Products [1]	Total Mesophilic Viable Bacteria (Log CFU/g)	*Enterococcus* spp. (Log CFU/g)	Lactic Acid Bacteria (Log CFU/g)
CJ1	9.45 ± 0.06 [2]	7.32 ± 0.03	7.45 ± 0.10
CJ2	9.04 ± 0.09	6.78 ± 0.14	6.66 ± 0.10
CJ3	9.81 ± 0.25	6.64 ± 0.01	6.70 ± 0.15
CJ4	9.57 ± 0.15	7.27 ± 0.03	6.96 ± 0.09
CJ5	8.54 ± 0.48	7.00 ± 0.08	6.67 ± 0.18
CJ6	9.20 ± 0.44	7.99 ± 0.04	8.12 ± 0.01
Average	9.27 ± 0.45	7.17 ± 0.49	7.09 ± 0.58

[1] CJ: *Cheonggukjang;* [2] Mean ± standard deviation were calculated from triplicate experiments; CFU: colony-forming units.

3.3. BA Content of Retail Cheonggukjang Products

Cheonggukjang contains abundant BA precursor amino acids such as lysine, histidine, tyrosine, and phenylalanine [58]. The high amino acid content may pose a risk for conversion into BAs during *Cheonggukjang* fermentation. In the present study, tryptamine, β-phenylethylamine, putrescine, cadaverine, histamine, tyramine, spermidine, and spermine contents in retail *Cheonggukjang* products were detected at concentrations of 70.63 ± 44.74 mg/kg, 36.22 ± 29.55 mg/kg, 10.80 ± 5.07 mg/kg, 18.57 ± 9.08 mg/kg, 8.37 ± 2.40 mg/kg, 457.42 ± 573.15 mg/kg, 121.92 ± 19.69 mg/ kg, and 187.20 ± 110.27 mg/kg, respectively (Table 3). A previous study suggested toxicity limits of 30 mg/kg for β-phenylethylamine, 100 mg/kg for histamine, and 100–800 mg/kg for tyramine in foods [21].

Therefore, the evaluation of the BA content of the *Cheonggukjang* products was continued with regard to the aforementioned BA intake limits with the exception of tyramine (at 100 mg/kg). Evaluation of β-phenylethylamine content in two *Cheonggukjang* products revealed that concentrations exceeded the recommended limit of 30 mg/kg, with one product exceeding the limit by a factor of approximately 3. Though the histamine content of all *Cheonggukjang* products was found to be below the recommended limit of 100 mg/kg, the tyramine content of three products exceeded the 100 mg/kg limit by factors of 2, 9, and 14, respectively. Other studies have reported similarly high concentrations of tyramine in *Cheonggukjang*. Ko, et al. [18] and Seo, et al. [20] reported the highest concentrations of tyramine in *Cheonggukjang* products, exceeding the recommended limit by factors of 19 and 9, respectively. Furthermore, Cho, et al. [59], Han, et al. [60], and Jeon, et al. [19] reported that *Cheonggukjang* products contained high concentrations of tyramine, which exceeded the recommended limit by up to factors of 5, 5, and 3, respectively. Altogether, as β-phenylethylamine and tyramine content of several *Cheonggukjang* products exceeded the recommended limits, overconsumption of such fermented soybean products may occasionally result in adverse effects on the body. Moreover, *Cheonggukjang* was found to contain other BAs enhancing the toxicity of β-phenylethylamine and tyramine. Therefore, further research remains necessary for precautionary measures and remedial methods to reduce the BA content of *Cheonggukjang* to ensure the safety of the fermented soybean food.

Table 3. Biogenic amine (BA) content of retail *Cheonggukjang* products.

Products [1]	BA Content (mg/kg) [2]							
	TRP	PHE	PUT	CAD	HIS	TYR	SPD	SPM
CJ1	115.06 ± 19.72 [A,3]	31.66 ± 3.82 [B]	8.54 ± 3.45 [BC]	22.66 ± 0.82 [C]	8.60 ± 0.88 [B]	222.25 ± 15.1 [C]	137.88 ± 1.76 [A]	292.99 ± 27.86 [A]
CJ2	86.02 ± 3.12 [B]	16.06 ± 2.24 [B]	12.18 ± 0.73 [B]	29.95 ± 0.82 [A]	12.69 ± 0.28 [A]	57.14 ± 8.06 [D]	99.99 ± 4.25 [C]	206.32 ± 23.82 [B]
CJ3	118.27 ± 5.97 [A]	27.22 ± 2.00 [B]	18.33 ± 4.93 [A]	26.58 ± 1.65 [B]	6.50 ± 1.00 [C]	80.83 ± 3.91 [D]	96.82 ± 4.09 [C]	201.63 ± 5.77 [B]
CJ4	54.87 ± 3.18 [C]	22.20 ± 7.06 [B]	11.36 ± 2.28 [B]	8.28 ± 0.54 [F]	5.86 ± 0.37 [C]	898.41 ± 79.43 [B]	125.61 ± 4.64 [B]	91.09 ± 24.03 [C]
CJ5	47.85 ± 4.04 [C]	24.48 ± 1.97 [B]	11.61 ± 4.21 [B]	13.57 ± 0.21 [D]	8.65 ± 0.51 [B]	61.98 ± 6.30 [D]	125.98 ± 7.60 [B]	305.05 ± 17.35 [A]
CJ6	1.70 ± 2.94 [D]	95.58 ± 46.97 [A]	2.81 ± 1.77 [C]	10.19 ± 0.50 [E]	7.84 ± 0.33 [B]	1424.04 ± 62.43 [A]	145.18 ± 7.64 [A]	26.20 ± 8.51 [D]
Average	70.63 ± 44.74	36.22 ± 29.55	10.80 ± 5.07	18.57 ± 9.08	8.37 ± 2.40	457.42 ± 573.15	121.92 ± 19.69	187.20 ± 110.27

[1] CJ: *Cheonggukjang*; [2] TRP: tryptamine, PHE: β-phenylethylamine, PUT: putrescine, CAD: cadaverine, HIS: histamine, TYR: tyramine, SPD: spermidine, SPM: spermine; [3] Mean ± standard deviation were calculated from triplicate experiments. Mean values in the same column followed by different letters (A–F) are significantly different ($p < 0.05$).

3.4. In Vitro BA Production by Enterococcus Strains Isolated from Retail Cheonggukjang Products

Microbial decarboxylation of free amino acids is one of the main causing factors in the production of BAs, and various microorganisms, including *Bacillus*, *Clostridium*, Enterobacteriaceae, enterococci, *Lactobacillus*, and *Pseudomonas*, are capable of producing the decarboxylases responsible for the conversion of amino acids into BAs [21,61,62]. Considering previous studies in which *Enterococcus* spp. have been suggested to be responsible for tyramine accumulation in Chinese and Japanese fermented soybean products [52,53], the current study analyzed the BA production capabilities of 169 enterococcal strains isolated from retail *Cheonggukjang* products using an MRS broth-based assay medium. As shown in Figure 1, the production of tryptamine, β-phenylethylamine, putrescine, cadaverine, spermidine, and spermine was observed at concentrations lower than 10 μg/mL. Histamine production by 168 of the 169 strains was detected at quantities lower than 2 μg/mL; however, only one strain was capable of producing histamine at 96.06 μg/mL. Though tyramine production ranged from ND (not detected) to 315.29 μg/mL, 157 strains (about 93%) produced over 100 μg/mL. Through 16S rRNA sequencing, the seven strains (CJE 101, CJE 115, CJE 119, CJE 128, CJE 130, CJE 210, and CJE 216) that produced the highest levels of tyramine (301.14–315.29 μg/mL; refer to Figure 2) among the enterococcal strains were all identified as *E. faecium*. Novella-Rodríguez, et al. [63] suggested that the presence of Enterobacteriaceae or enterococci may result in the production of BAs in contaminated food products. Marcobal, et al. [64] demonstrated that *E. faecium* possesses a gene that codes an enzyme capable of L-tyrosine decarboxylation. According to Ibe, et al. [22], high levels of tyramine in *Miso* (a Japanese fermented soybean paste) products may partially result from tyramine production by *E. faecium*. Jeon, et al. [19] reported that *Enterococcus* spp. exhibited strong production of tyramine ranging from 0.41 μg/mL to 351.59 μg/mL in assay media. The author also found tyramine-producing *Bacillus* spp. (up to 123.08 μg/mL) and suggested that the species is one of the major tyramine producers in *Cheonggukjang* along with *Enterococcus* species based on the in situ fermentation experiment. Consequently, the present results suggest that *Enterococcus* spp. (particularly *E. faecium*) may be largely responsible for high tyramine concentrations in *Cheonggukjang*.

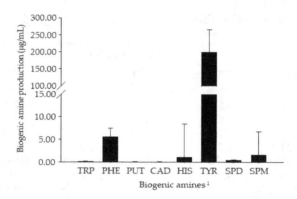

Figure 1. Biogenic amine (BA) production by *Enterococcus* strains (n = 169) isolated from retail *Cheonggukjang* products. Error bars indicate standard deviations calculated from duplicate experiments. [1] TRP: tryptamine, PHE: β-phenylethylamine, PUT: putrescine, CAD: cadaverine, HIS: histamine, TYR: tyramine, SPD: spermidine, SPM: spermine.

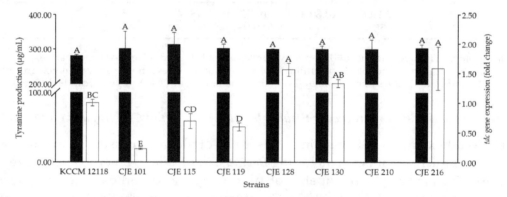

Figure 2. Comparison of tyramine production and *tdc* expression by *E. faecium* strains. ■: tyramine production, □: *tdc* gene expression. The expression levels observed in *E. faecium* strains isolated from retail *Cheonggukjang* products were normalized to that detected in *E. faecium* KCCM 12118 (type strain). The *tdc* gene expression was not detected in *E. faecium* strain CJE 210. Mean values followed by different letters are significantly different ($p < 0.05$). Error bars indicate standard deviations calculated from duplicate experiments.

3.5. Selection of Tyramine-Producing E. faecium Strain for Cheonggukjang Fermentation Based on Tyrosine Decarboxylase Gene Expression In Vitro

In the current study, the efficiency of primer sets *q-tdc* (for the quantitation of *tdc* gene expression) along with *q-gap* and *tufA*-RT (for the normalization of *tdc* gene expression) was calculated to be 100.71%, 94.84%, and 95.03%, respectively. An amplification efficiency between 90 and 110% indicates that the results of gene expression obtained using RT-qPCR are reproducible [65].

The aforementioned primer sets were used to detect *tdc* gene expression by the seven *E. faecium* strains (CJE 101, CJE 115, CJE 119, CJE 128, CJE 130, CJE 210, and CJE 216) with the highest tyramine production in vitro as described in the previous section (note that the primer sets were also used for in situ gene expression analysis). Among the strains, *E. faecium* strain CJE 216 showed the highest expression level of *tdc* gene (Figure 2). Considering the highest *tdc* gene expression as well as tyramine production in vitro, the CJE 216 strain was selected as an inoculant for fermentation experiments in the next section.

3.6. Tyramine Production by E. faecium during Cheonggukjang Fermentation at Various Temperatures

3.6.1. Changes in Physicochemical and Microbial Properties during Cheonggukjang Fermentation at Various Temperatures

As the results of the previous sections indicated that *E. faecium* was most likely one of the major contributing factors to high levels of tyramine in *Cheonggukjang*, in situ fermentation experiments were performed to empirically investigate the influence of *E. faecium* on tyramine content in *Cheonggukjang*. For the in situ fermentation experiments, three experimental groups of *Cheonggukjang* were used: control group inoculated with only *B. subtilis* KCTC 3135, and other two groups co-inoculated with the *B. subtilis* strain and each *E. faecium* strain (*E. faecium* KCCM 12118 or *E. faecium* CJE 216). Each group was further divided into three groups based on fermentation temperatures of 25 °C, 37 °C, and 45 °C (low-, intermediate-, and high-temperature groups, respectively). As shown in Figure 3, the changes in the physicochemical and microbial properties of *Cheonggukjang* were measured at 24-hour intervals for 3 days of fermentation. The pH of all *Cheonggukjang* groups was lowest on day 2, with progressively lower pH depending on the fermentation temperature, independent of which inoculum was used. The pH on day 2 of *Cheonggukjang* fermentation at 25 °C, 37 °C, and 45 °C ranged from pH 6.34 to 6.36, pH 5.90 to 6.09, and pH 5.49 to 5.88, respectively (Figure 3a). Loizzo, et al. [66] suggested that decarboxylases are produced by bacteria owing to a mechanism to neutralize acidic environments that restrict the growth of the bacteria. A previous study reported that low pH between 4.0 and 5.5 may result in the production of BAs [67]. Therefore, in this study, as the *Cheonggukjang* groups fermented at 45 °C (high-temperature group) resulted in a lower pH than other groups fermented at 37 °C and 25 °C (intermediate- and low-temperature groups, respectively), regardless of inoculum, the BA content was expected to be detected at the highest concentration among all *Cheonggukjang* groups. As for water activity, all *Cheonggukjang* groups remained within 0.95–0.97 during fermentation (Figure 3b).

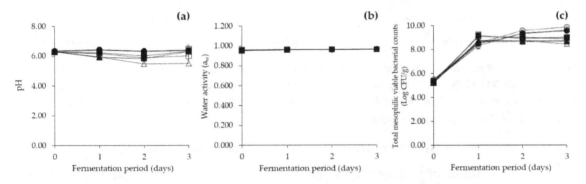

Figure 3. Physicochemical and microbial properties during *Cheonggukjang* fermentation at various temperatures. (a) pH, (b) water activity, (c) total mesophilic viable bacterial counts. ●: 25 °C, ■: 37 °C, ▲: 45 °C (inoculated with only *B. subtilis* KCTC 3135); ●: 25 °C, ■: 37 °C, ▲: 45 °C (inoculated with *B. subtilis* KCTC 3135 and *E. faecium* KCCM 12118); ○: 25 °C, □: 37 °C, △: 45 °C (inoculated with *B. subtilis* KCTC 3135 and *E. faecium* CJE 216). Error bars indicate standard deviations calculated from duplicate experiments.

The counts of total mesophilic viable bacteria, most probably attributed to *B. subtilis* inoculated, showed that microbial concentrations started from approximately 6 Log CFU/g on day 0 and remained at approximately 8–9 Log CFU/g throughout *Cheonggukjang* fermentation at all three temperatures, regardless of the presence or absence of *E. faecium* inoculum (Figure 3c). The total mesophilic viable bacteria in *Cheonggukjang* increased as fermentation temperature decreased; however, on day 1, those in the groups fermented at 25 °C and 45 °C showed growth up to 8 Log CFU/g, while those in the groups fermented at 37 °C exhibited the highest counts at 9 Log CFU/g. The results concurred with a previous finding that 37 °C is the optimal in situ growth temperature for *B. subtilis* during *Cheonggukjang* fermentation [29]. Similarly, Mann, et al. [68] reported the optimal in vitro growth temperature for *B. subtilis* strains isolated from *Cheonggukjang* to be 37 °C.

The enterococcal count in *Cheonggukjang* co-inoculated with *E. faecium* KCCM 12118 at approximately 4 Log CFU/g (and *B. subtilis* KCTC 3135 at 6 Log CFU/g as well) increased by 1.63 Log CFU/g, 3.52 Log CFU/g, and 4.06 Log CFU/g after 3 days of fermentation at 25 °C, 37 °C, and 45 °C, respectively (Figure 4). In *Cheonggukjang* co-inoculated with *E. faecium* CJE 216 at 4 Log CFU/g (and *B. subtilis* KCTC 3135), enterococcal count increased at all fermentation temperatures of 25 °C,

37 °C, and 45 °C by 3.48 Log CFU/g, 4.78 Log CFU/g, and 4.80 Log CFU/g, respectively, by day 3. *Enterococcus* spp. were not detected in the control group for the duration of the fermentation period. The results displayed progressively higher enterococcal counts that increased alongside rising fermentation temperatures with the highest enterococcal counts in *Cheonggukjang* fermented at 45 °C (high-temperature group). The findings were comparable to a previous study by Morandi, et al. [69], which showed that lower fermentation temperatures weakened *E. faecium* growth as the reported generation time at 25 °C was nearly two times longer than at 37 °C. *E. faecium* has been reported to display active growth within the temperature range of 37–53 °C, with an optimal growth temperature of 42.7 °C [50,70]. The current and previous studies, therefore, indicate that the use of high fermentation temperatures such as 45 °C may enhance *E. faecium* growth, thereby increasing the potential for high tyramine production during *Cheonggukjang* fermentation.

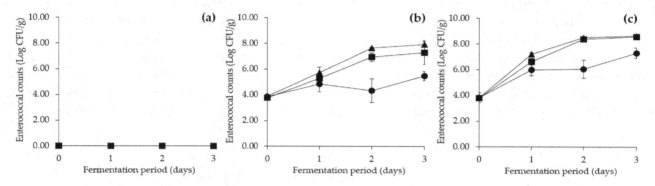

Figure 4. Effect of fermentation temperature on enterococcal counts in *Cheonggukjang* inoculated with (**a**) *B. subtilis* KCTC 3135, (**b**) *B. subtilis* KCTC 3135 and *E. faecium* KCCM 12118, and (**c**) *B. subtilis* KCTC 3135 and *E. faecium* CJE 216. ●: 25 °C, ■: 37 °C, ▲: 45 °C. Error bars indicate standard deviations calculated from duplicate experiments.

3.6.2. Effect of Fermentation Temperature on Tyramine Content in *Cheonggukjang*

The tyramine content of *Cheonggukjang* co-inoculated with either *E. faecium* KCCM 12118 or *E. faecium* CJE 216, along with *B. subtilis* KCTC 3135, was measured during fermentation, as seen in Figure 5. The tyramine content of the control group without *E. faecium* was detected at concentrations that did not exceed 10 mg/kg in all fermentation conditions (Figure 5a). In contrast, other groups with *E. faecium* contained higher levels of tyramine, which indicated that *E. faecium* was capable of and responsible for producing tyramine in *Cheonggukjang*. In *Cheonggukjang* groups co-inoculated with *E. faecium* KCCM 12118, initial tyramine content increased by 0.78 mg/kg, 33.36 mg/kg, and 101.17 mg/kg at 3 days of fermentation at 25 °C, 37 °C, and 45 °C, respectively (Figure 5b). As for *Cheonggukjang* groups co-inoculated with *E. faecium* CJE 216, initial tyramine content increased by 1.59 mg/kg, 74.11 mg/kg, and 85.14 mg/kg at 25 °C, 37 °C, and 45 °C, respectively, by day 3 of fermentation (Figure 5c). All *Cheonggukjang* groups fermented at 25 °C contained the lowest tyramine concentrations at less than 10 mg/kg during the entire fermentation duration. However, at 45 °C, the *Cheonggukjang* group co-inoculated with *E. faecium* KCCM 12118 displayed an exceptionally high tyramine content detected at 105.13 ± 5.68 mg/kg, exceeding the recommended limit, as expected owing to the acidic pH described in Section 3.6.1. Both *E. faecium* strains appeared to continuously produce tyramine during *Cheonggukjang* fermentation at 37 °C and 45 °C (Figure 5b,c). The results of the current study are in agreement with findings reported by Kalhotka, et al. [71], which showed a stronger in vitro tyramine production by *E. faecium* incubated at 37 °C than at 25 °C. According to Morandi, et al. [69], *E. faecium* metabolic activity was detected to be higher at 37 °C than at 25 °C during milk fermentation. The previous studies have indicated that lower temperatures may reduce both metabolic activity and tyramine production of *E. faecium*. BA content during fermentation at higher temperatures may even reach dangerously high levels as reported by Kang, et al. [43]. In the same report, tyramine concentrations in *E. faecium*-inoculated *Cheonggukjang* fermented at 45 °C for

48 h increased (up to 698.67 mg/kg) during the fermentation period, and consequently exceeded the recommended limit for consumption. Jeon, et al. [19] also reported strong tyramine production by *Enterococcus* spp. during soybean fermentation at 37 °C. The report demonstrated that tyramine concentrations continued to increase as fermentation progressed. Given the results, safety precautions regarding the limitation of fermentation duration and temperature appear to be necessary as extended periods of fermentation as well as high fermentation temperatures may increase tyramine content in *Cheonggukjang* beyond the recommended safe limit for consumption. Besides, the results showing a lower tyramine content in *Cheonggukjang* during fermentation at lower temperatures coincide with the results in the previous section that displayed a reduction in enterococcal count alongside a decrease in fermentation temperature. Taken together, the present study indicates that lower fermentation temperatures inhibit enterococcal growth, thereby limiting acid production and maintaining low levels of tyramine in *Cheonggukjang*. Therefore, utilizing lower temperatures for *Cheonggukjang* fermentation may reduce the risks associated with *Enterococcus* growth and tyramine accumulation.

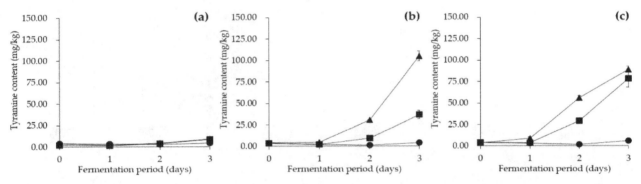

Figure 5. Effect of fermentation temperature on tyramine content in *Cheonggukjang* inoculated with (**a**) *B. subtilis* KCTC 3135, (**b**) *B. subtilis* KCTC 3135 and *E. faecium* KCCM 12118, and (**c**) *B. subtilis* KCTC 3135 and *E. faecium* CJE 216. ●: 25 °C, ■: 37 °C, ▲: 45 °C. Error bars indicate standard deviations calculated from duplicate experiments.

3.6.3. Effect of Fermentation Temperature on *tdc* Gene Expression by *E. faecium* Strains in *Cheonggukjang*

The changes in *tdc* gene expression by tyramine-producing *E. faecium* strains were detected and quantified during fermentation of *Cheonggukjang* at 25 °C, 37 °C, and 45 °C. As *Cheonggukjang* is mostly fermented at 37 °C, the *tdc* gene expression detected in *Cheonggukjang* groups fermented at 45 °C was normalized to that detected in the corresponding groups fermented at 37 °C according to the *E. faecium* strains used as inoculants. In *Cheonggukjang* fermented at 25 °C, tyramine content continuously remained at concentrations lower than 10 mg/kg, and *tdc* gene expression by *E. faecium* KCCM 12118 and *E. faecium* CJE 216 was not detected in all *Cheonggukjang* groups. In contrast, the highest *tdc* gene expression by *E. faecium* KCCM 12118 was detected in *Cheonggukjang* fermented at 45 °C and was upregulated in the range of 1.90- to 7.15-fold throughout *Cheonggukjang* fermentation, compared with that in *Cheonggukjang* fermented at 37 °C (Figure 6a–c, left). As for *Cheonggukjang* fermented at 45 °C with *E. faecium* CJE 216, downregulation of *tdc* gene expression was observed at 0.82-fold on day 1, and the expression was then upregulated in the range of 1.90- to 3.39-fold thereafter (Figure 6a–c, right). Consequently, both tyramine content and *tdc* gene expression were highest in *Cheonggukjang* groups fermented at 45 °C (viz., high-temperature group). Nonetheless, the variation in detected *tdc* gene expression levels during fermentation did not necessarily reflect the tyramine content observed for *Cheonggukjang*. After one day of fermentation, the *Cheonggukjang* group with *E. faecium* CJE 216 fermented at 37 °C contained a lower tyramine content than at 45 °C; however, *tdc* gene expression was slightly higher at 37 °C as described right above. The results showed that there may be differences between gene expression level and enzyme activity (and products thereof). Glanemann, et al. [72] reported that, in vitro, the mRNA response levels do not necessarily reflect the protein response levels or enzyme activity. As previously suggested by Ladero, et al. [73], while the correlation between BA content and gene expression is not always linear, RT-qPCR remains a reliable method to detect and quantify BA-producing bacteria in food products. Similarly, in our preliminary tests conducted under different incubation conditions, tyramine production by *E. faecium* strains in an assay medium appeared to be insignificantly related to *tdc* gene expression level (data not shown). Therefore, utilizing HPLC analysis appears to be essential for the quantification of BA content and/or bacterial BA production in food samples [31,73,74]. When utilized in conjunction, the complementary methods, that is, HPLC and RT-qPCR, sufficiently allow for the quantitative analysis of both the BA content and tyramine-producing bacteria (including enterococci) in food products [24,31]. In the present study, the results derived from both techniques indicated that the fermentation of *Cheonggukjang* at high temperatures results in increased *tdc* gene expression and tyramine production. Therefore, low-temperature fermentation appears to be necessary to minimize both *tdc* gene expression and tyramine production by *Enterococcus* spp. and thereby ensure the safety of fermented soybean products.

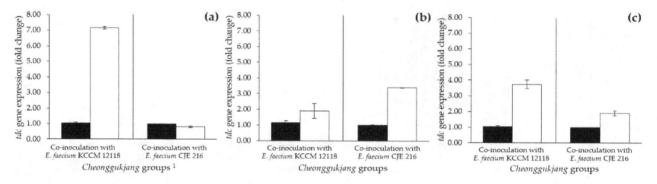

Figure 6. Effect of fermentation temperature on *tdc* expression by *E. faecium* strains in *Cheonggukjang* on (**a**) day 1, (**b**) day 2, and (**c**) day 3 of fermentation. ■: 37 °C, □: 45 °C. [1] *Cheonggukjang* groups were co-inoculated with *B. subtilis* (KCTC 3135) and *E. faecium* (KCCM 12118 or CJE 216) strains. The expression levels observed in groups fermented at 45 °C were normalized to those detected in the corresponding groups fermented at 37 °C. Expression of *tdc* gene was not detected in *Cheonggukjang* fermented at 25 °C. Error bars indicate standard deviations calculated from duplicate experiments.

4. Conclusions

The current study assessed the safety risk of tyramine in *Cheonggukjang*, diagnosed the microbial causative agent (i.e., *E. faecium*) responsible for high tyramine levels, and evaluated the impact of fermentation temperature on enterococcal growth (as well as acid production and *tdc* gene expression) and tyramine production. Of the retail *Cheonggukjang* examined, half of the products contained tyramine content that exceeded the recommended limit for safe consumption by up to a factor of approximately 14. *E. faecium* strains isolated from the retail *Cheonggukjang* products were highly capable of producing tyramine in assay media, which indicated that the species is principally, or at least partly, responsible for tyramine accumulation in the food.

During in situ fermentation at different temperatures, the tyramine content of *Cheonggukjang* groups co-inoculated with *B. subtilis* (used as an inoculant to ferment soybeans) and *E. faecium* (either isolated in this study or designated previously as the type strain) strains was highest at 45 °C, followed by 37 °C and 25 °C. On the other hand, the control group inoculated with only *B. subtilis* strain (without any *E. faecium* inoculants) had the lowest tyramine content at all fermentation temperatures, which supported the notion that *E. faecium* may be a key producer of tyramine in *Cheonggukjang*. Another implication of the results was that a lower fermentation temperature leads to a lower tyramine content below the recommended limit in *Cheonggukjang*, even though the tyramine content continually increases during fermentation. Therefore, low temperatures and a short fermentation duration may reduce the accumulation of tyramine caused by *E. faecium* growth in *Cheonggukjang*, thereby reducing the safety risks associated with consuming food with high BA concentrations.

Author Contributions: Conceptualization, Y.K.P. and J.-H.M.; Investigation, Y.K.P., Y.H.J., J.-H.L., B.Y.B., and J.L.; Formal analysis, J.L.; Writing—original draft, Y.K.P.; Writing—review and editing, Y.K.P., Y.H.J., K.C.J., and J.-H.M.; Supervision: J.-H.M. All authors have read and agreed to the published version of the manuscript.

Acknowledgments: The authors thank Jae Hoan Lee and Alixander Mattay Pawluk of Department of Food and Biotechnology at Korea University for technical assistance and English editing, respectively.

References

1. Jang, C.H.; Lim, J.K.; Kim, J.H.; Park, C.S.; Kwon, D.Y.; Kim, Y.-S.; Shin, D.H.; Kim, J.-S. Change of isoflavone content during manufacturing of *cheonggukjang*, a traditional Korean fermented soyfood. *Food Sci. Biotechnol.* **2006**, *15*, 643–646.

2. Kim, K.-J.; Ryu, M.-K.; Kim, S.-S. *Chungkook-jang Koji* fermentation with rice straw. *Korean J. Food Sci. Technol.* **1982**, *14*, 301–308.

3. Hong, S.W.; Kim, J.Y.; Lee, B.K.; Chung, K.S. The bacterial biological response modifier enriched *Chungkookjang* fermentation. *Korean J. Food Sci. Technol.* **2006**, *38*, 548–553.

4. Lee, J.-O.; Ha, S.-D.; Kim, A.-J.; Yuh, C.-S.; Bang, I.-S.; Park, S.-H. Industrial application and physiological functions of *Chongkukjang*. *Food Sci. Ind.* **2005**, *38*, 69–78.

5. Hwang, J.-S.; Kim, S.-J.; Kim, H.-B. Antioxidant and blood-pressure reduction effects of fermented soybean, Chungkookjang. *Korean J. Microbiol.* **2009**, *45*, 54–57.

6. Askar, A.; Treptow, H. *Biogene Amine in Lebensmitteln: Vorkommen, Bedeutung und Bestimmung*, 1st ed.; Eugen Ulmer: Stuttgart, Germany, 1986; pp. 21–74.

7. Silla Santos, M.H. Biogenic amines: Their importance in foods. *Int. J. Food Microbiol.* **1996**, *29*, 213–231. [CrossRef]

8. Shalaby, A.R. Significance of biogenic amines to food safety and human health. *Food Res. Int.* **1996**, *29*, 675–690. [CrossRef]

9. Ladero, V.; Calles-Enriquez, M.; Fernández, M.; Alvarez, M.A. Toxicological effects of dietary biogenic amines. *Curr. Nutr. Food Sci.* **2010**, *6*, 145–156. [CrossRef]

10. EFSA Panel on Biological Hazards (BIOHAZ). Scientific opinion on risk based control of biogenic amine formation in fermented foods. *EFSA J.* **2011**, *9*, 2393. [CrossRef]

11. Taylor, S.L.; Eitnmiller, R.R. Histamine food poisoning: Toxicology and clinical aspects. *Crit. Rev. Toxicol.* **1986**, *17*, 91–128. [CrossRef]

12. Kovacova-Hanuskova, E.; Buday, T.; Gavliakova, S.; Plevkova, J. Histamine, histamine intoxication and intolerance. *Allergol. Immunopathol.* **2015**, *43*, 498–506. [CrossRef] [PubMed]

13. Maintz, L.; Novak, N. Histamine and histamine intolerance. *Am. J. Clin. Nutr.* **2007**, *85*, 1185–1196. [CrossRef] [PubMed]

14. Smith, T.A. Amines in food. *Food Chem.* **1981**, *6*, 169–200. [CrossRef]

15. Stratton, J.E.; Hutkins, R.W.; Taylor, S.L. Biogenic amines in cheese and other fermented foods: A review. *J. Food Prot.* **1991**, *54*, 460–470. [CrossRef]

16. Linares, D.M.; Martín, M.; Ladero, V.; Alvarez, M.A.; Fernández, M. Biogenic amines in dairy products. *Crit. Rev. Food Sci. Nutr.* **2011**, *51*, 691–703. [CrossRef]

17. Schirone, M.; Tofalo, T.; Fasoli, G.; Perpetuini, G.; Corsetti, A.; Manetta, A.C.; Ciarrocchi, A.; Suzzi, G. High content of biogenic amines in Pecorino cheeses. *Food Microbiol.* **2013**, *34*, 137–144. [CrossRef]

18. Ko, Y.-J.; Son, Y.-H.; Kim, E.-J.; Seol, H.-G.; Lee, G.-R.; Kim, D.-H.; Ryu, C.-H. Quality properties of commercial *Chungkukjang* in Korea. *J. Agric. Life Sci.* **2012**, *46*, 177–187.

19. Jeon, A.R.; Lee, J.H.; Mah, J.-H. Biogenic amine formation and bacterial contribution in *Cheonggukjang*, a Korean traditional fermented soybean food. *LWT Food Sci. Technol.* **2018**, *92*, 282–289. [CrossRef]

20. Seo, M.-J.; Lee, C.-D.; Lee, J.-N.; Yang, H.-J.; Jeong, D.-Y.; Lee, G.-H. Analysis of biogenic amines and inorganic elements in *Cheonggukjang*. *Korean J. Food Preserv.* **2019**, *26*, 101–108. [CrossRef]

21. Ten Brink, B.; Damink, C.; Joosten, H.M.L.J.; Huis in 't Veld, J.H.J. Occurrence and formation of biologically active amines in foods. *Int. J. Food Microbiol.* **1990**, *11*, 73–84. [CrossRef]

22. Ibe, A.; Nishima, T.; Kasai, N. Bacteriological properties of and amine-production conditions for tyramine-and histamine-producing bacterial strains isolated from soybean paste (miso) starting materials. *Jpn. J. Toxicol. Environ. Health* **1992**, *38*, 403–409. [CrossRef]

23. Torriani, S.; Gatto, V.; Sembeni, S.; Tofalo, R.; Suzzi, G.; Belletti, N.; Gardini, F.; Bover-Cid, S. Rapid detection and quantification of tyrosine decarboxylase gene (*tdc*) and its expression in gram-positive bacteria associated with fermented foods using PCR-based methods. *J. Food Prot.* **2008**, *71*, 93–101. [CrossRef] [PubMed]

24. Ladero, V.; Fernández, M.; Cuesta, I.; Alvarez, M.A. Quantitative detection and identification of tyramine-producing enterococci and lactobacilli in cheese by multiplex qPCR. *Food Microbiol.* **2010**, *27*, 933–939. [CrossRef] [PubMed]

25. Kang, T.-M.; Park, J.-H. Isolation and antibiotic susceptibility of *Enterococcus* spp. from fermented soy paste. *J. Korean Soc. Food Sci. Nutr.* **2012**, *41*, 714–720. [CrossRef]

26. Kang, H.-R.; Lee, Y.-L.; Hwang, H.-J. Potential for application as a starter culture of tyramine-reducing strain. *J. Korean Soc. Food Sci. Nutr.* **2017**, *46*, 1561–1567. [CrossRef]

27. Svec, P.; Devriese, L.A. Enterococcus. In *Bergey's Manual of Systematics of Archaea and Bacteria*, 2nd ed.; De Vos, P., Garrity, G.M., Jones, D., Krieg, N.R., Ludwig, W., Rainey, F.A., Schleifer, K.-H., Whitman, W.B., Eds.; Springer: New York, NY, USA, 2015; Volume 3, pp. 594–607.

28. Food Information Statistics System. Available online: http://www.atfis.or.kr/article/M001050000/view.do?articleId=2452&page=5&searchKey=&searchString=&searchCategory= (accessed on 29 May 2020).

29. Kim, I.-J.; Kim, H.-K.; Chung, J.-H.; Jeong, Y.-K.; Ryu, C.-H. Study of functional *Chungkukjang* contain fibrinolytic enzyme. *Korean J. Life Sci.* **2002**, *12*, 357–362.

30. Lee, N.-R.; Go, T.-H.; Lee, S.-M.; Hong, C.-O.; Park, K.-M.; Park, G.-T.; Hwang, D.-Y.; Son, H.-J. Characteristics of Chungkookjang prepared by *Bacillus amyloliquefaciens* with different soybeans and fermentation temperatures. *Korean J. Microbiol.* **2013**, *49*, 71–77. [CrossRef]

31. Bhardwaj, A.; Gupta, H.; Iyer, R.; Kumar, N.; Malik, R.K. Tyramine-producing enterococci are equally detected on tyramine production medium, by quantification of tyramine by HPLC, or by *tdc* gene-targeted PCR. *Dairy Sci. Technol.* **2009**, *89*, 601–611. [CrossRef]

32. AOAC. *Official Methods of Analysis of AOAC International*, 18th ed.; AOAC International: Gaithersburg, MD, USA, 2005.

33. ISO 7218:2007. *Microbiology of Food and Animal Feeding Stuffs—General Requirements and Guidance for Microbiological Examinations*; ISO: Geneva, Switzerland, 2007.

34. Mareková, M.; Lauková, A.; DeVuyst, L.; Skaugen, M.; Nes, I.F. Partial characterization of bacteriocins produced by environmental strain *Enterococcus faecium* EK13. *J. Appl. Microbiol.* **2003**, *94*, 523–530. [CrossRef]

35. Ryu, M.S.; Yang, H.-J.; Kim, J.W.; Jeong, S.-J.; Jeong, S.-Y.; Eom, J.-S.; Jeong, D.-Y. Potential probiotics activity of *Bacillus* spp. from traditional soybean pastes and fermentation characteristics of *Cheonggukjang*. *Korean J. Food Preserv.* **2017**, *24*, 1168–1179. [CrossRef]

36. Lee, J.S.; Lee, M.H.; Kim, J.M. Changes in quality characteristics of *cheonggukjang* added with quinoa during fermentation period. *Korean J. Food Nutr.* **2018**, *31*, 24–32.

37. National Institute of Agricultural Sciences. Available online: https://koreanfood.rda.go.kr:2360/eng/fctFoodSrchEng/engMain (accessed on 7 May 2020).

38. Ben-Gigirey, B.; Vieites Baptista De Sousa, J.M.; Villa, T.G.; Barros-Velazquez, J. Changes in biogenic amines and microbiological analysis in albacore (*Thunnus alalunga*) muscle during frozen storage. *J. Food Prot.* **1998**, *61*, 608–615. [CrossRef] [PubMed]

39. Eerola, S.; Hinkkanen, R.; Lindfors, E.; Hirvi, T. Liquid chromatographic determination of biogenic amines in dry sausages. *J. AOAC Int.* **1993**, *76*, 575–577. [CrossRef]

40. Ben-Gigirey, B.; Vieites Baptista De Sousa, J.M.; Villa, T.G.; Barros-Velazquez, J. Histamine and cadaverine production by bacteria isolated from fresh and frozen albacore (*Thunnus alalunga*). *J. Food Prot.* **1999**, *62*, 933–939. [CrossRef] [PubMed]

41. Marcobal, Á.; Martín-Álvarez, P.J.; Moreno-Arribas, M.V.; Muñoz, R. A multifactorial design for studying factors influencing growth and tyramine production of the lactic acid bacteria *Lactobacillus brevis* CECT 4669 and *Enterococcus faecium* BIFI-58. *Res. Microbiol.* **2006**, *157*, 417–424. [CrossRef] [PubMed]

42. Yoon, H.; Park, J.H.; Choi, A.; Hwang, H.-J.; Mah, J.-H. Validation of an HPLC analytical method for determination of biogenic amines in agricultural products and monitoring of biogenic amines in Korean fermented agricultural products. *Toxicol. Res.* **2015**, *31*, 299–305. [CrossRef]

43. Kang, H.-R.; Kim, H.-S.; Mah, J.-H.; Kim, Y.-W.; Hwang, H.-J. Tyramine reduction by tyrosine decarboxylase inhibitor in *Enterococcus faecium* for tyramine controlled *cheonggukjang*. *Food Sci. Biotechnol.* **2018**, *27*, 87–93. [CrossRef]

44. Top, J.; Paganelli, F.L.; Zhang, X.; van Schaik, W.; Leavis, H.L.; Van Luit-Asbroek, M.; van der Poll, T.; Leendertse, M.; Bonten, M.J.M.; Willems, R.J.L. The *Enterococcus faecium* enterococcal biofilm regulator, EbrB, regulates the *esp* operon and is implicated in biofilm formation and intestinal colonization. *PLoS ONE* **2013**, *8*, e65224. [CrossRef]

45. Lee, E.S.; Kim, Y.S.; Ryu, M.S.; Jeong, D.Y.; Uhm, T.B.; Cho, S.H. Characterization of *Bacillus licheniformis* SCK A08 with antagonistic property against *Bacillus cereus* and degrading capacity of biogenic amines. *J. Food Hyg. Saf.* **2014**, *29*, 40–46. [CrossRef]

46. Yoo, S.-M.; Choe, J.-S.; Park, H.-J.; Hong, S.-P.; Chang, C.-M.; Kim, J.-S. Physicochemical properties of traditional *Chonggugjang* produced in different regions. *Appl. Biol. Chem.* **1998**, *41*, 377–383.

47. Kim, J.-W.; Kim, Y.-S.; Jeong, P.-H.; Kim, H.-E.; Shin, D.-H. Physicochemical characteristics of traditional fermented soybean products manufactured in folk villages of Sunchang region. *J. Food Hyg. Saf.* **2006**, *21*, 223–230.

48. Jeong, W.J.; Lee, A.R.; Chun, J.; Cha, J.; Song, Y.-S.; Kim, J.H. Properties of *cheonggukjang* fermented with *Bacillus* strains with high fibrinolytic activities. *J. Food Sci. Nutr.* **2009**, *14*, 252–259. [CrossRef]

49. Kang, S.J.; Kim, S.S.; Chung, H.Y. Comparison of physicochemical characteristics and consumer perception of *Cheongkukjang*. *J. Korean Soc. Food Sci. Nutr.* **2014**, *43*, 1104–1111. [CrossRef]

50. Oh, S.-J.; Mah, J.-H.; Kim, J.-H.; Kim, Y.-W.; Hwang, H.-J. Reduction of tyramine by addition of *Schizandra chinensis* Baillon in Cheonggukjang. *J. Med. Food* **2012**, *15*, 1109–1115. [CrossRef] [PubMed]

51. International Commission on Microbiological Specifications for Foods International Association of Microbiological Societies. Reduced water activity. In *Microbial Ecology of Foods*, 1st ed.; Silliker, J.H., Elliot, R.P., Baird-Parker, A.C., Bryan, F.L., Christian, J.H.B., Clark, D.S., Olson, J.C., Roberts, T.A., Eds.; Academic Press: New York, NY, USA, 1980; Volume 1, pp. 70–91.

52. Li, L.; Ruan, L.; Ji, A.; Wen, Z.; Chen, S.; Wang, L.; Wei, X. Biogenic amines analysis and microbial contribution in traditional fermented food of Douchi. *Sci. Rep.* **2018**, *8*, 1–10. [CrossRef] [PubMed]

53. Takebe, Y.; Takizaki, M.; Tanaka, H.; Ohta, H.; Niidome, T.; Morimura, S. Evaluation of the biogenic amine-production ability of lactic acid bacteria isolated from tofu-misozuke. *Food Sci. Technol. Res.* **2016**, *22*, 673–678. [CrossRef]

54. Giraffa, G. Enterococci from foods. *FEMS Microbiol. Rev.* **2002**, *26*, 163–171. [CrossRef]

55. Giraffa, G.; Carminati, D.; Neviani, E. Enterococci isolated from dairy products: A review of risks and potential technological use. *J. Food Prot.* **1997**, *60*, 732–738. [CrossRef]

56. Bandara, N.; Chung, S.-J.; Jeong, D.-Y.; Kim, K.-P. The use of the pathogen-specific bacteriophage BCP8-2 to develop a rice straw-derived *Bacillus cereus*-free starter culture. *Korean J. Food Sci. Technol.* **2014**, *46*, 115–120. [CrossRef]

57. Heu, J.-S.; Lee, I.-J.; Yoon, M.-H.; Choi, W.-Y. Adhesive microbial populations of rice straws and their effects on Chungkukjang fermentation. *Korean J. Agric. Sci.* **1999**, *26*, 77–83.

58. Seok, Y.-R.; Kim, Y.-H.; Kim, S.; Woo, H.-S.; Kim, T.-W.; Lee, S.-H.; Choi, C. Change of protein and amino acid composition during *Chungkook-Jang* fermentation using *Bacillus licheniformis* CN-115. *Korean J. Agic. Sci.* **1994**, *37*, 65–71.

59. Cho, T.-Y.; Han, G.-H.; Bahn, K.-N.; Son, Y.-W.; Jang, M.-R.; Lee, C.-H.; Kim, S.-H.; Kim, D.-B.; Kim, S.-B. Evaluation of biogenic amines in Korean commercial fermented foods. *Korean J. Food Sci. Technol.* **2006**, *38*, 730–737.

60. Han, G.-H.; Cho, T.-Y.; Yoo, M.-S.; Kim, C.-S.; Kim, J.-M.; Kim, H.-A.; Kim, M.-O.; Kim, S.-C.; Lee, S.-A.; Ko, Y.-S.; et al. Biogenic amines formation and content in fermented soybean paste (*cheonggukjang*). *Korean J. Food Sci. Technol.* **2007**, *39*, 541–545.

61. Rice, S.L.; Eitenmiller, R.R.; Koehler, P.E. Biologically active amines in food: A review. *J. Milk Food Technol.* **1976**, *39*, 353–358. [CrossRef]

62. Rodriguez-Jerez, J.J.; Giaccone, V.; Colavita, G.; Parisi, E. *Bacillus macerans*—A new potent histamine producing micro-organism isolated from Italian cheese. *Food Microbiol.* **1994**, *11*, 409–415. [CrossRef]

63. Novella-Rodríguez, S.; Veciana-Nogues, M.T.; Roig-Sagues, A.X.; Trujillo-Mesa, A.J.; Vidal-Carou, M.C. Evaluation of biogenic amines and microbial counts throughout the ripening of goat cheeses from pasteurized and raw milk. *J. Dairy Res.* **2004**, *71*, 245–252. [CrossRef] [PubMed]

64. Marcobal, A.; de las Rivas, B.; Moreno-Arribas, M.V.; Munoz, R. Evidence for horizontal gene transfer as

origin of putrescine production in *Oenococcus oeni* RM83. *Appl. Environ. Microbiol.* **2006**, *72*, 7954–7958. [CrossRef] [PubMed]

65. Condori, J.; Nopo-Olazabal, C.; Medrano, G.; Medina-Bolivar, F. Selection of reference genes for qPCR in hairy root cultures of peanut. *BMC Res. Notes* **2011**, *4*, 392. [CrossRef] [PubMed]

66. Loizzo, M.R.; Menichini, F.; Picci, N.; Puoci, F.; Spizzirri, U.G.; Restuccia, D. Technological aspects and analytical determination of biogenic amines in cheese. *Trends Food Sci. Technol.* **2013**, *30*, 38–55. [CrossRef]

67. Cosansu, S. Determination of biogenic amines in a fermented beverage, boza. *J. Food Agric. Environ.* **2009**, *7*, 54–58.

68. Mann, S.-Y.; Kim, E.-A.; Lee, G.-Y.; Kim, R.-U.; Hwang, D.-Y.; Son, H.-J.; Kim, D.-S. Isolation and identification of GABA-producing microorganism from *Chungkookjang*. *J. Life Sci.* **2013**, *23*, 102–109. [CrossRef]

69. Morandi, S.; Brasca, M.; Alfieri, P.; Lodi, R.; Tamburini, A. Influence of pH and temperature on the growth of *Enterococcus faecium* and *Enterococcus faecalis*. *Le Lait* **2005**, *85*, 181–192. [CrossRef]

70. Van den Berghe, E.; De Winter, T.; De Vuyst, L. Enterocin A production by *Enterococcus faecium* FAIR-E 406 is characterised by a temperature-and pH-dependent switch-off mechanism when growth is limited due to nutrient depletion. *Int. J. Food Microbiol.* **2006**, *107*, 159–170. [CrossRef] [PubMed]

71. Kalhotka, L.; Manga, I.; Přichystalová, J.; Hůlová, M.; Vyletělová, M.; Šustová, K. Decarboxylase activity test of the genus *Enterococcus* isolated from goat milk and cheese. *Acta Vet. BRNO* **2012**, *81*, 145–151. [CrossRef]

72. Glanemann, C.; Loos, A.; Gorret, N.; Willis, L.B.; O'brien, X.M.; Lessard, P.A.; Sinskey, A.J. Disparity between changes in mRNA abundance and enzyme activity in *Corynebacterium glutamicum*: Implications for DNA microarray analysis. *Appl. Microbiol. Biotechnol.* **2003**, *61*, 61–68. [CrossRef] [PubMed]

73. Ladero, V.; Linares, D.M.; Fernández, M.; Alvarez, M.A. Real time quantitative PCR detection of histamine-producing lactic acid bacteria in cheese: Relation with histamine content. *Food Res. Int.* **2008**, *41*, 1015–1019. [CrossRef]

74. Spano, G.; Russo, P.; Lonvaud-Funel, A.; Lucas, P.; Alexandre, H.; Grandvalet, C.; Coton, E.; Coton, M.; Barnavon, L.; Bach, B.; et al. Biogenic amines in fermented foods. *Eur. J. Clin. Nutr.* **2010**, *64*, S95–S100. [CrossRef]

10

Influence of Iodine Feeding on Microbiological and Physico-Chemical Characteristics and Biogenic Amines Content in a Raw Ewes' Milk Cheese

Maria Schirone *, Rosanna Tofalo, Giorgia Perpetuini, Anna Chiara Manetta, Paola Di Gianvito, Fabrizia Tittarelli, Noemi Battistelli, Aldo Corsetti, Giovanna Suzzi and Giuseppe Martino *

Faculty of Bioscience and Technology for Food, Agriculture and Environment, University of Teramo, Via R. Balzarini, 1, 64100 Teramo, Italy; rtofalo@unite.it (R.T.); giorgia.perpetuini@gmail.com (G.P.); acmanetta@unite.it (A.C.M.); digianvito.paola@gmail.com (P.D.G.); ftittarelli@unite.it (F.T.); noemi.battistelli@gmail.com (N.B.) acorsetti@unite.it (A.C.); gsuzzi@unite.it (G.S.)
* Correspondence: mschirone@unite.it (M.S.); gmartino@unite.it (G.M.)

Abstract: Iodine is an essential trace element involved in the regulation of thyroid metabolism and antioxidant status in humans and animals. The aim of this study was to evaluate the effect of ewes' dietary iodine supplementation on biogenic amines content as well as microbiological and physico-chemical characteristics in a raw milk cheese at different ripening times (milk, curd, and 2, 7, 15, 30, 60, and 90 days). Two cheese-making trials were carried out using milk from ewes fed with unifeed (Cheese A) or with the same concentrate enriched with iodine (Cheese B). The results indicated that the counts of principal microbial groups and physico-chemical characteristics were quite similar in both cheeses at day 90. Cheese B was characterized by a higher content of biogenic amines and propionic acid. Propionic bacteria were found in both cheeses mainly in Trial B in agreement with the higher content of propionic acid detected.

Keywords: raw milk cheese; biogenic amines; iodine feed; physico-chemical composition

---

1. Introduction

Milk and dairy products represent the second most important source of iodine in the European Union or in the United States [1] particularly for infants and children. Iodine deficit in the diet causes various thyroid dysfunctions and infant mortality [2]; iodine has a recommended daily intake of 150 µg for both adolescents and adults [3,4]. The concentration of this element in milk and dairy products has been reported in different papers and it can vary in terms of animal feed, the season (the higher concentration is in winter), and exposure to iodophors [2]. Changes in animal feeding have been proposed as one of the most promising approaches to modify iodine content in milk [5]. Some studies [6,7] have been carried out to evaluate the effects of dietary iodine supplementation in dairy goats and cows on milk iodine content and milk production traits. Nudda et al. [6] reported that the iodine supplementation in dairy goat diets doubled the milk iodine content when compared with the control group, even if no evident effect was observed in the gross composition of milk. On the contrary, Weiss et al. [7] found that iodine concentration increased in serum but not in milk after supplementation of this element in diets of dairy cows. In fact, very little information is available about the effects of iodine addition on ewes' milk and milk-based product composition, nor about the response of dairy product microbiota.

Pecorino Incanestrato di Castel del Monte (ICM) is an artisanal semi-hard pasta filata cheese obtained starting from ewes' raw milk without the addition of starter cultures. ICM is produced in the Abruzzo region (Central Italy) and is included in the list of typical products (PAT—Prodotti Agroalimentari Tradizionali). As other raw milk cheeses, the characteristics of the final product are influenced by several parameters such as raw milk microbiota, microorganisms deriving from equipment and from the dairy environments, and outside and inside grazing animal feeding systems [8].

In this study, the effect of dietary iodine supplementation in dairy ewes on biogenic amine (BA) content as well as microbiological and physico-chemical characteristics in ICM cheese was evaluated.

2. Materials and Methods

2.1. Cheese-Making Procedure

Cheese samples were manufactured in a small factory, located in the production area of ICM (L'Aquila, Abruzzo Region, Italy), from raw whole ewes' milk of one or two daily milking without the addition of natural or commercial starter cultures. The milk was filtered and heated at 35–40 °C for 15–25 min and coagulated with lamb rennet at 38 °C, according to routine manufacture. Afterwards, the curd was broken and fit into special baskets, the so-called *fiscelle*. The product was salted and ripened up to 3 months. The final products weighted about 2 kg. Two different cheese-making trials were carried out in triplicate using milk (100 L). In a completely randomized block design, 2 groups of 15 Sopravissana ewes were assigned to 2 diets. In the first group, ewes were fed with unifeed (hay and concentrate) (Cheese A), while in the second group ewes were fed with unifeed enriched with iodine at a final concentration of 10 mg/kg (Cheese B). This concentration of iodine was selected according to Regulation EC No. 1459/2005 [9]. The cheese yield was about 24% in both cheese-making trials. Analyses were performed in triplicate on milk, curd, and cheese samples at different ripening times: 2, 7, 15, 30, 60, and 90 days.

2.2. Microbiological Analyses

Milk and cheese samples (10 mL or g) were diluted in 90 mL of a sodium citrate (2% w/v) solution and homogenized with a Stomacher Lab-Blender 400 (Steward Medical, London, UK) for 1 min. Serial dilutions in sterile peptone water (0.1% w/v) were plated in triplicate on different media to enumerate the following microorganisms: mesophilic lactobacilli, lactococci, aerobic mesophilic bacteria (AMB), yeasts, *Enterobacteriaceae*, enterococci, and coagulase-negative staphylococci (CNS), according to Schirone et al. [10]. The presence of *Escherichia coli* O157:H7, *Salmonella* spp., and *Listeria monocytogenes* was determined according to standard methods reported in ISO [11–13].

For the detection of propionibacteria, a semi-quantitative approach was applied as described previously [14,15]. DNA was extracted using PowerSoil DNA Isolation Kit (MoBio Laboratories) according to manufacturer's protocol starting from 5 g of cheese as previously described [14]. PB1 (5′-AGTGGCGAAGGCGGTTCTCTGGA-3′) and PB2 (5′-TGGGGTCGAGTTGCAGACCCCAAT-3′) primer set was used. PCR amplification program consisted of denaturation at 94 °C for 4 min, 40 cycles of denaturation at 94 °C for 30 s, annealing at 70 °C for 15 s, and extension at 72 °C for 1 min followed by a final extension at 72 °C for 5 min.

2.3. Gross Physico-Chemical Composition

A radial slice of each cheese was randomly taken and used for physico-chemical assays. The rind of each slice was carefully removed, and the rind-less material was fully shredded. pH, water activity (a_w), dry matter, total protein, fat, and ash content were determined according to Schirone et al. [9]. Iodine concentration was evaluated using a commercial kit according to manufacture instructions (Celltech, Turin, Italy) in milk and cheese samples.

Organic acids (mg/g) were determined as reported by Tofalo et al. [15] and Bouzas et al. [16] using an HPLC 200 series (Perkin Elmer, Monza, Italy) connected to a UV VIS detector at 210 nm. ROA Organic Acid H^+ column (Phenomenex, Bologna, Italy) was used for the analyses.

All determinations were performed isocratically with a flow rate of 0.7 mL/min at 65 °C using H_2SO_4 solution 0.009 N as mobile phase.

The nitrogen fractions determined were water-soluble nitrogen (WSN, expressed in %N) [17], trichloroacetic acid-soluble nitrogen (12% TCA-SN, expressed as %N) [18], and amino acid nitrogen (AAN, expressed as mg leucine/g) [19].

2.4. BA Determination

Determination of BA (mg/kg) was carried out as described by Schirone et al. [20]. In brief, 10 g of cheese samples were extracted and derivatized with dansyl chloride (Fluka Chimica, Milan, Italy). The chromatographic system consisted of an HPLC Waters Alliance (Waters SpA, Vimodrone, Italy), equipped with a Waters 2695 separation module connected to a Waters 2996 photodiode array detector. The separation of the analytes was carried out using a Waters Spherisorb C18 S3ODS-2 column (3 μm particle size, 150 mm × 4.6 mm Inner Diameter) equipped with a Waters Spherisorb S5ODS2 guard column. A linear gradient made up of acetonitrile and ultrapure water was applied: acetonitrile 57% (v/v) for 5 min; acetonitrile 80% (v/v) for 4 min, acetonitrile 90% (v/v) for 5 min. The peaks were detected at 254 nm.

2.5. Statistical Analyses

Statistical analyses were performed using the software STATISTICA for Windows (STAT. version 8.0, StatSoft Inc., Tulsa, OK, USA). Collected data were subjected to two-way analysis of variance (ANOVA) to detect significant differences. The principal component analysis (PCA) was performed on physico-chemical and microbiological data after auto-scaling.

3. Results

3.1. Microbial Analyses

Microbial counts are shown in Table 1. Overall, mesophilic lactobacilli, lattococci, AMB, enterococci, and yeasts showed a significant increase during the first days ripening. This was partly due to both microbial growth during coagulation and the physical retention of microorganisms in curds. The count of AMB obtained from the milk was higher in Cheese A (6.9 log CFU/mL) than in Cheese B (5.5 log CFU/mL) and then increased up to 8.4 log and 8.7 log CFU/g at 90 days of ripening, respectively. These counts are common in cheeses produced from raw milk, and they agree with those obtained in different cheese varieties such as Montasio [21] or Cebreiro [22,23].

Lactic acid bacteria (LAB) dominated in ICM cheeses during all ripening. In Cheese B, a higher number of lactococci and mesophilic lactobacilli was observed than in the Cheese A during the first stages of ripening, while at the end of ripening both cheeses showed similar counts, more than 8 log CFU/g. In general, LAB dominated the adventitious microbiota prevailing in all cheeses. Overall, in the early phase of manufacture, non-starter lactic acid bacteria (NSLAB) were present at very low values, whereas during ripening they increase from approximately 2.0 to 6.0 log CFU/g in ripened cheese [24]. Enterococci counts during ripening resulted to be quite different in ICM Cheeses A and B. In Trial A, their number increased from 2.8 log CFU/mL in milk up to a maximum value of 6.5 log CFU/g at 2 days and then decreased at 5.5 log CFU/g at the end of ripening. In Trial B, the counts started from 2.5 log CFU/g in milk, increased up to 6.3 log CFU/g after 15 days, and then decreased at 3.8 log CFU/g at 90 days of ripening. Enterococci represent the major part of curd microbiota and in some cases, they are the predominant microorganisms in the fully ripened product, constituting about the 41% of the LAB population [25]. In particular, enterococci have been recognized as an essential part of the natural microbial population of many dairy products, where they can sometimes prevail over lactobacilli and lactococci [22,26,27]. High levels of enterococci observed in other cheeses have been suggested to have a relevant role during the whole ripening process, because of their proteolytic and lipolytic activities that contribute to aroma compounds production (C4 metabolites such as diacetyl acetoin or 2,3-butanediol) [28,29].

Table 1. Evolution of principal microbial groups during the ripening in the two different trials expressed as log CFU/g.

Microbial Groups	Trial	Milk	Curd	2 Days	7 Days	15 Days	30 Days	60 Days	90 Days
Mesophilic lactobacilli	A	5.2 (0.02) *	6.6 (0.006) *	7.2 (0.01) *	7.8 (0.002) *	8.3 (0.02) *	7.2 (0.009) *	8.4 (0.006) *	8.7 (0.001) *
	B	6.8 (0.003) *	7.9 (0.001) *	8.6 (0.004) *	8.5 (0.004) *	8.8 (0.002) *	8.5 (0.002) *	8.1 (0.09) *	8.7 (0.001) *
Lactococci	A	5.4 (0.006) *	6.1 (0.013) *	7.5 (0.01) *	7.6 (0.01) *	8.1 (0.02) *	7.9 (0.004) *	8.7 (0.05) *	8.5 (0.004) *
	B	6.8 (0.001) *	8.4 (0.007) *	8.6 (0.005) *	8.7 (0.001) *	8.8 (0.001) *	8.8 (0.002) *	8.5 (0.05) *	8.5 (0.007) *
AMB [a]	A	6.9 (0.001) *	6.7 (0.001) *	7.9 (0.001) *	7.4 (0.007) *	8.6 (0.004) *	8.9 (0.001) *	8.4 (0.001) *	8.4 (0.006) *
	B	5.5 (0.01) *	7.0 (0.014) *	8.7 (0.002) *	8.8 (0.004) *	8.9 (0.001) *	8.9 (0.001) *	8.9 (0.09) *	8.7 (0.006) *
Yeasts	A	-	4.3 (0.005) *	3.5 (0.007) *	4.6 (0.002) *	4.5 (0.006) *	4.4 (0.005) *	4.5 (0.004)	4.4 (0.007) *
	B	-	3.5 (0.003) *	4.7 (0.004) *	4.4 (0.004) *	5.2 (0.009) *	5.1 (0.009) *	5.3 (0.4)	5.1 (0.003) *
Enterobacteriaceae	A	3.4 (0.009) *	5.4 (0.007) *	4.7 (0.003) *	4.3 (0.009) *	4.5 (0.007) *	2.6 (0.005) *	2.4 (0.002) *	2.4 (0.004)
	B	3.4 (0.001) *	5.4 (0.005) *	5.8 (0.002) *	5.8 (0.001) *	5.8 (0.001) *	3.8 (0.002) *	3.9 (0.003) *	<1
Enterococci	A	2.8 (0.001) *	4.7 (0.004) *	6.5 (0.004) *	5.3 (0.002) *	4.5 (0.004) *	4.3 (0.004) *	5.8 (0.003) *	5.5 (0.003) *
	B	2.5 (0.006) *	5.7 (0.002) *	5.6 (0.005) *	6.0 (0.001) *	6.3 (0.002) *	6.1 (0.001) *	5.1 (0.07) *	3.8 (0.001) *
CNS [b]	A	4.9 (0.002) *	6.4 (0.003) *	5.4 (0.003) *	6.5 (0.001) *	6.5 (0.003) *	6.7 (0.001) *	3.1 (0.003) *	3.4 (0.01) *
	B	6.3 (0.005) *	6.4 (0.003) *	7.3 (0.007) *	6.8 (0.001) *	5.7 (0.002) *	5.6 (0.003) *	5 (1) *	4.3 (0.01) *

The data are reported as mean (S.D.); samples for each microbial group at the same ripening time marked with * showed statistically significant differences ($p < 0.05$). [a] aerobic mesophilic bacteria, [b] coagulase-negative staphylococci.

As regards *Enterobacteriaceae*, they are associated to the natural microbiota of many dairy products, and together with coliforms are considered indicators of the microbiological quality of cheese. These microorganisms were present in milk of both cheeses and after 15 days of ripening, ranging from about 4.5 to 5.8 log CFU/g for Cheeses A and B respectively, whereas they were not enumerable (<10 CFU/g) in Cheese B at 90 days of ripening. *Enterobacteriaceae* are generally considered as microorganisms with a high decarboxylase activity, particularly in relation to the production of cadaverine and putrescine [30] and are common in many traditional cheeses of Mediterranean area [31]. The counts of CNS were higher in Milk B (6.3 log CFU/mL) than in Milk A (4.9 log CFU/mL). These microorganisms increased during the first days of ripening and decreased at 90 days with values of 3.4 and 4.3 log CFU/g in Cheeses A and B, respectively.

Yeasts, absent in milk in both trials, were present in curd and reached values of 4.4 and 5.1 log CFU/g in Cheeses A and B, respectively, at the end of ripening. Similar data have been reported in other raw milk cheeses such as Pecorino di Farindola [32,33].

Pathogens such as *Salmonella* spp., *L. monocytogenes*, and *E. coli* O157:H7 resulted absent in all the examined samples.

3.2. Gross Physico-Chemical Composition

The physico-chemical parameters and organic acids content for the two cheese-making procedures at 90 days of ripening are reported in Table 2. After 2 days of ripening, the pH values were 5.75 and 5.44 in Cheeses A and B, respectively (data not shown), and they then slightly decreased at the end of ripening. These differences, generally attributed to the metabolic activity of different species and strains of LAB, are typical of low acidified cheese produced with ewes' raw milk [10]. The mean a_w values decreased as ripening progressed and at day 90 they were similar in both cheeses (about 0.97). A higher percentage of fat was observed in Cheese B (51.69% w/w) than in Cheese A (47.03% w/w) at the end of ripening, whereas proteins were present in the amount in Cheese A (44.12% w/w) than in Cheese B (40.92% w/w), even if no statistical differences were observed ($p < 0.05$). The average values of iodine concentration were 86.1 and 481.3 µg/100 mL in Milks A and B, respectively. At day 90, the iodine concentration was 128.7 µg/100 g in Cheeses A and 375.9 µg/100 g in Cheese B. The iodine amount in milk has been reported to reflect the dietary iodine content, and it is an indicator of the iodine status of the animal [34]. The iodine concentration in milk is directly proportional to the iodine levels in feedstuffs. Moreover, the season of milk production and fat content of milk can significantly affect its rate [35]. Manca et al. [36] found that iodine supplementation did not influence the goat milk fatty acid profile, except for some short-chain fatty acids. Milk fat and protein content did not vary between two groups of dairy sheep fed with iodine supplementation in diets at different concentrations [37].

Lactic acid was the most abundant organic acid with values of about 35 mg/g in both Cheeses A and B (Table 2). Similar concentrations of citric, acetic, and succinic acids were detected in both cheeses with values of about 0.5, 0.9, and 0.2 mg/g, respectively. Propionic acid was present in higher concentration in Cheese B. Therefore, to verify the origin of this organic acid, a genus-specific PCR was used. *Propionibacterium freudenreichii* was the only propionic bacteria detected, as demonstrated by the presence of a specific fragment of 850 bp. It was present in all samples, the only exception being Cheese A at 7 days of ripening (data not shown). Band intensities were correlated to propionic bacteria abundance. In Cheese A, the *P. freudenreichii* presence ranged from 10 to 10^3 CFU/g—in Cheese B, from 10^3 to 10^6 CFU/g. Similar results were found by Tofalo et al. [15] in a traditional Abruzzo cheese where the presence of *P. freudenreichii* has been shown to play an important role in its sensorial characteristic and aromatic quality conferring an intense flavor.

Assessment of proteolysis in Cheeses A and B, through the determination of WSN, 12% TCA-SN and AAN over three months of ripening, is reported in Figure 1. The WSN value was 9%N in Curd A and about 12%N in Curd B. During the first weeks of ripening, there were no statistically significant differences between the examined cheeses in the level of WSN and the concentrations increased with a more intense rise in Cheese B, reaching a final rate of 14.5%N at day 90. The effect of feeding system on nitrogen fractions was more marked starting from 30 days of ripening, probably due to the impact of the milk as a source of microbial enzymes. The amount of 12% TCA-SN also increased progressively in both cheeses, but it was stronger always in Cheese B. Starter LAB and non-starter LAB (NSLAB) proteinases are principally responsible for the formation of 12% TCA-SN [38], that contains small peptides (2–20 residues) and free amino acids [39]. The average content of AAN showed a general similar evolution in both cheeses, but Trial A showed a slower proteolytic activity than that in Trial B. However, the final values obtained were similar: 8.42 and 8.60 mg leucine/g, respectively, for Cheeses A and B at 90 days of ripening.

Table 2. Physico-chemical characteristics and organic acids content (mg/g) in cheeses at the end of ripening.

Parameters	Cheese A	Cheese B
Physico-chemical		
pH	5.55 ± 0.20	5.39 ± 0.10
a_W	0.98 ± 0.01	0.97 ± 0.01
% Dry matter	67.68 ± 4.80	66.15 ± 2.23
% Fat [1]	47.03 ± 3.13	51.69 ± 1.73
% Protein [1]	44.12 ± 6.68	40.92 ± 7.54
% Ash [1]	8.85 ± 0.26	7.39 ± 0.47
Organic acids		
Citric acid	0.5 ± 0.1	0.40 ± 0.07
Succinic acid	0.10 ± 0.05	0.21 ± 0.03
Lactic acid	35 ± 4	36 ± 4
Acetic acid	0.80 ± 0.05	0.90 ± 0.06
Propionic acid	0.04 ± 0.03	0.13 ± 0.06

Data are expressed as mean ± S.D.; [1] these parameters (fat, protein, and ash) are expressed in dry matter; no statistically significant differences were observed ($p < 0.05$).

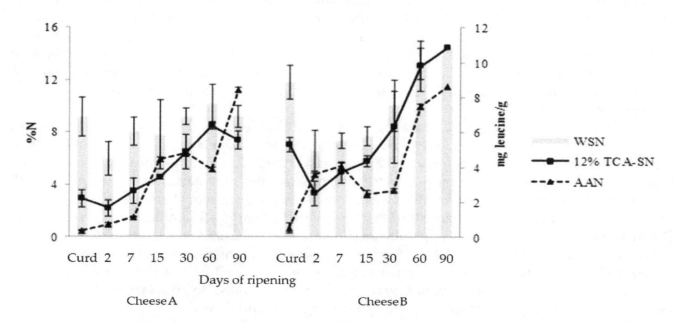

Figure 1. Evolution of nitrogen fractions during ripening in Cheeses A and B. WSN: water-soluble nitrogen; TCA-SN: trichloroacetic acid-soluble nitrogen; AAN: amino acid nitrogen.

3.3. BA Content

The high content of BA in cheese is well documented [20,40]. The accumulation of BA has been mainly ascribed to the activity of NSLAB, even if an indirect role of LAB proteolytic activity could be hypothesized providing the precursor amino acids used for BA synthesis. Moreover, some factors, such as environmental hygienic conditions, decarboxylase microorganisms, and temperature and ripening of cheese can contribute to the qualitative and quantitative BA profiles [31]. The accumulation of BA at high concentrations and the presence of BA-producing microorganisms cannot be avoided in raw milk cheeses as well as in fermented foods and beverages. Total BA content was found to be similar up to 60 days in both cheeses with an average content of about 400 mg/kg (Figure 2A,B). In raw milk A and B, only putrescine was detected at low concentrations (about 2 mg/kg). At day 60, the main amine was putrescine, followed by cadaverine, tyramine, and histamine. In Cheese B, a significant increase was observed in the total BA content at day 90 (760.7 mg/kg); in Cheese A, a decrease was detected at that time (244.30 mg/kg). The reduction was particularly evident for histamine (5.80 mg/kg) and cadaverine that disappeared. This fact could be explained by the presence of some BA-degrading strains, as reported by other authors [41–43]. Recently, Alvarez et al. [41] reported that a significant alternative to reduce BA content in fermented foods (such as cheese, wine, and sausages) was the use of BA-degrading strains, isolated from different origins. Fresno et al. [42] suggested that the addition of two strains (*Lactobacillus casei* 4a and 5b) were able to reduce BA contents in a Cabrales-like mini-cheese manufacturing model, although the exact mechanism via which this occurs remains unknown. In order to identify the pathways involved in the catabolism of these compounds, Ladero et al. [43] reported the draft genome of the *L. casei* 5b strain isolated from cheese. The use of BA-degrading strains could be particularly useful during cheeses manufacturing from raw milk in which a specific non-starter microbiota is essential for the organoleptic characteristic of the final product.

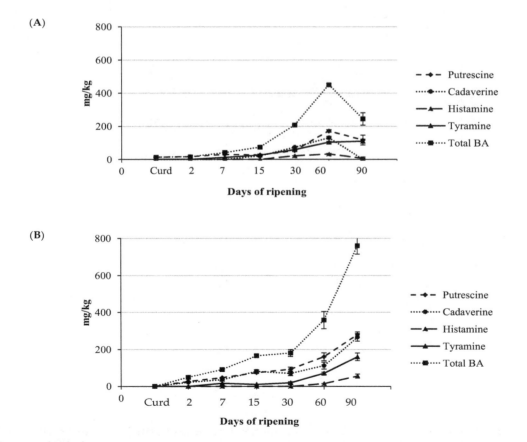

Figure 2. Biogenic amines content (mg/kg) during the ripening in Cheese A (**A**) and Cheese B (**B**). BA, biogenic amine.

In order to understand the variability between the two different cheeses, PCA was carried out using as variables physico-chemical and microbiological data. The PCA results were shown in Figure 3, the score plot (A) and the loading plot (B). The two principal components (PCs) captured 60.64% of total variance in the first two dimensions with 43.30% and 17.34% explained by Factors 1 and 2, respectively. In the score plot (A), both Cheeses A and B at days 60 and 90 of ripening clustered together in the positive section of PC 1 and were closely related to the high values of dry matter, fat, protein, and ash content as well as propionic and lactic acids; meanwhile, the negative counterpart of PC 1 was mainly associated to the succinic acid and grouped the samples of Cheese A in the first month of ripening. The different ripening times of Cheese B were discriminated over the first PC, based above all on the counts of microorganisms.

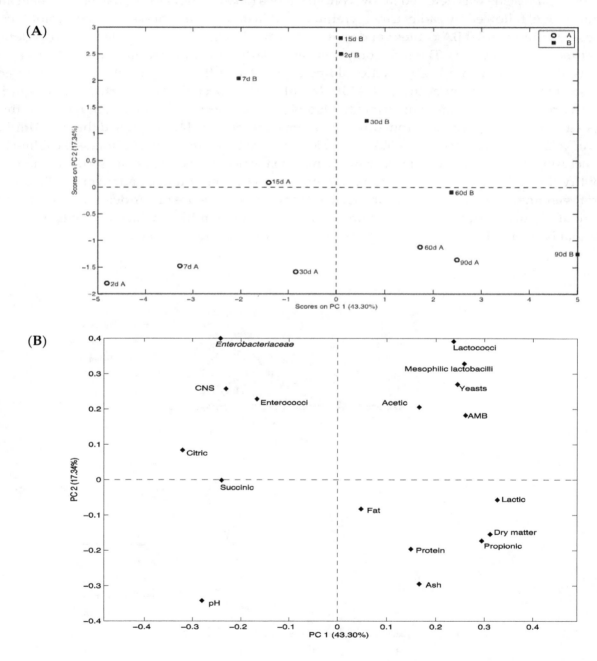

Figure 3. Score plot (**A**) and loading plot (**B**) of the first and second principal components (PCs) after PC analysis encompassing microbiological and physico-chemical parameters.

4. Conclusions

The overall management system of the farm was the same, so it is possible to hypothesize that iodine influenced the main features of Pecorino Incanestrato di Castel del Monte cheese. The physico-chemical and microbiological data highlighted a relevant effect of dietary iodine supplementation on ewes' raw milk and cheese microbiota. Even if the counts of principal microbial groups were quite similar in both cheeses, differences were found in some biochemical activities of microorganisms such as proteolytic/peptidasic activities or total BA content at day 90 of ripening. The findings call for a deep study on the selective effect of iodine on microbial populations in raw milk cheeses.

Author Contributions: Conceptualization: G.M. and R.T. Methodology: A.C.M., P.D.G., and N.B. Software: N.B., and G.P. Formal Analysis: A.C.M., P.D.G., and G.P. Investigation: A.C. and F.T. Resources: G.M. Writing—Original Draft Preparation: M.S., R.T., and G.S. Writing—Review & Editing: F.T., A.C., G.S. Supervision: R.T.

Acknowledgments: The authors thank Dr. Giuseppe Fasoli for molecular analysis.

References

1. Bader, N.; Möller, U.; Leiterer, M.; Franke, K.; Jahreis, G. Pilot study: Tendency of increasing iodine content in human milk and cow's milk. *Exp. Clin. Endocrinol. Diabetes* **2005**, *113*, 8–12. [CrossRef] [PubMed]
2. Gaucheron, F. Milk minerals, trace elements, and macroelements. In *Milk and Dairy Products in Human Nutrition: Production, Composition and Health*; Park, Y.W., Haenlein, G.F.W., Eds.; Wiley & Sons Ltd.: Hoboken, NJ, USA, 2013; pp. 172–199. ISBN 9781118534168.
3. Haldimann, M.; Alt, A.; Blank, A.; Blondeau, K. Iodine content of food groups. *J. Food Compos. Anal.* **2005**, *18*, 461–471. [CrossRef]
4. World Health Organization (WHO). *Assessment of Iodine Deficiency Disorders and Monitoring Their Elimination*, 3rd ed.; World Health Organization (WHO): Geneva, Switzerland, 2007; ISBN 9789241595827.
5. Knowles, S.O.; Grace, N.D.; Knight, T.W.; McNabba, W.C.; Lee, J. Reasons and means for manipulating the micronutrient composition of milk from grazing dairy cattle. *Anim. Feed Sci. Technol.* **2006**, *131*, 154–167. [CrossRef]
6. Nudda, A.; Battacone, G.; Decandia, M.; Acciaro, M.; Aghini-Lombardi, F.; Frigeri, M.; Pulina, G. The effect of dietary iodine supplementation in dairy goats on milk production traits and milk iodine content. *J. Dairy Sci.* **2009**, *92*, 5133–5138. [CrossRef] [PubMed]
7. Weiss, W.P.; Wyatt, D.J.; Kleinschmit, D.H.; Socha, M.T. Effect of including canola meal and supplemental iodine in diets of dairy cows on short-term changes in iodine concentrations in milk. *J. Dairy Sci.* **2015**, *98*, 4841–4849. [CrossRef] [PubMed]
8. Giello, M.; La Storia, A.; Masucci, F.; Di Francia, A.; Ercolini, D.; Villani, F. Dynamics of bacterial communities during manufacture and ripening of traditional Caciocavallo of Castelfranco cheese in relation to cows' feeding. *Food Microbiol.* **2017**, *63*, 170–177. [CrossRef] [PubMed]
9. European Commission (EC). No. 1459/2005 of 8 September 2005 amending the conditions for the authorization of a number of feed additives belonging to the group of trace elements. *Off. J. Eur. Union* **2005**, *L233*, 8–10.
10. Schirone, M.; Tofalo, R.; Mazzone, G.; Corsetti, A.; Suzzi, G. Biogenic amine content and microbiological profile of Pecorino di Farindola cheese. *Food Microbiol.* **2011**, *28*, 128–136. [CrossRef] [PubMed]
11. International Organization for Standardization (ISO). *Microbiology of Food and Animal Feeding Stuffs—Horizontal Method for the Detection and Enumeration of Escherichia coli O157:H7*; ISO 16654; International Organization for Standardization (ISO): Geneva, Switzerland, 2001.
12. International Organization for Standardization (ISO). *Microbiology of Food and Animal Feeding Stuffs—Horizontal Method for the Detection of Salmonella spp.*; ISO 6579; International Organization for Standardization (ISO): Geneva, Switzerland, 2002.
13. International Organization for Standardization (ISO). *Microbiology of Food and Animal Feeding Stuffs—Horizontal Method for the Detection and Enumeration of Listeria monocytogenes—Part 1: Enumeration Method*; ISO 11290-1; International Organization for Standardization (ISO): Geneva, Switzerland, 2005.

14. Rossi, F.; Torriani, S.; Dellaglio, F. Genus and species-specific PCR-based detection of dairy propionibacteria in environmental samples using primers targeted to the 16S rDNA. *Appl. Environ. Microbiol.* **1999**, *65*, 4241–4244. [PubMed]

15. Tofalo, R.; Schirone, M.; Fasoli, G.; Perpetuini, G.; Patrignani, F.; Manetta, A.C.; Lanciotti, R.; Corsetti, A.; Martino, G.; Suzzi, G. Influence of pig rennet on proteolysis, organic acids content and microbiota of Pecorino di Farindola, a traditional Italian ewe's raw milk cheese. *Food Chem.* **2015**, *175*, 121–127. [CrossRef] [PubMed]

16. Bouzas, J.; Kantt, C.A.; Bodyfelt, F.; Torres, J.A. Simultaneous determination of sugars and organic acids in Cheddar cheese by high-performance liquid chromatography. *J. Food Sci.* **1991**, *56*, 276–278. [CrossRef]

17. Kuchroo, C.N.; Fox, P.F. Soluble nitrogen in Cheddar cheese: Comparison of extraction procedures. *Milchwissenschaft* **1982**, *37*, 331–335.

18. Polychroniadou, A.; Michaelidou, A.; Paschaloudis, N. Effect of time, temperature and extraction method on the trichloroacetic acid-soluble nitrogen of cheese. *Int. Dairy J.* **1999**, *9*, 559–568. [CrossRef]

19. Folkertsma, B.; Fox, P.F. Use of Cd-ninhydrin reagent to assess proteolysis in cheese during ripening. *J. Dairy Res.* **1992**, *59*, 217–224. [CrossRef]

20. Schirone, M.; Tofalo, R.; Fasoli, G.; Perpetuini, G.; Corsetti, A.; Manetta, A.C.; Ciarrocchi, A.; Suzzi, G. High content of biogenic amines in Pecorino cheeses. *Food Microbiol.* **2013**, *34*, 137–144. [CrossRef] [PubMed]

21. Manzano, M.; Sarais, I.; Stecchini, M.L.; Rondinini, G. Etiology of gas defects in Montasio cheese. *Ann. Microbiol.* **1990**, *40*, 255–259.

22. Centeno, J.A.; Menéndez, S.; Rodriguez-Otero, J.L. Main microbial flora present as natural starters in Cebreiro raw cow's-milk cheese (Northwest Spain). *Int. J. Food Microbiol.* **1996**, *33*, 307–313. [CrossRef]

23. Quinto, E.; Franco, C.; Rodriguez-Otero, J.L.; Fente, C.; Cepeda, A. Microbiological quality of Cebreiro cheese from Northwest Spain. *J. Food Saf.* **1994**, *14*, 1–8. [CrossRef]

24. Berthier, F.; Beuvier, E.; Dasen, A.; Grappin, R. Origin and diversity of mesophilic lactobacilli in Comté cheese, as revealed by PCR with repetitive and species-specific primers. *Int. Dairy J.* **2001**, *11*, 293–305. [CrossRef]

25. Samelis, J.; Lianou, A.; Kakouri, A.; Delbes, C.; Rogelj, I.; Bogovič-Matijašić, B.O.J.A.N.A.; Montel, M.C. Changes in the microbial composition of raw milk induced by thermization treatments applied prior to traditional Greek hard cheese processing. *J. Food Prot.* **2009**, *72*, 783–790. [CrossRef] [PubMed]

26. Poullet, B.; Huertas, M.; Sánchez, A.; Cáceres, P.; Larriba, G. Main lactic acid bacteria isolated during ripening of Casar de Cáceres cheese. *J. Dairy Res.* **1993**, *60*, 123–127. [CrossRef]

27. Suzzi, G.; Caruso, M.; Gardini, F.; Lombardi, A.; Vannini, L.; Guerzoni, M.E.; Andrighetto, C.; Lanorte, M.T. A survey of enterococci isolated from an artisanal Italian goat's cheese (semicotto caprino). *J. Appl. Microbiol.* **2000**, *89*, 267–274. [CrossRef] [PubMed]

28. Giraffa, G. Functionality of enterococci in dairy products. *Int. J. Food Microbiol.* **2003**, *88*, 215–222. [CrossRef]

29. Martino, G.P.; Quintana, I.M.; Espariz, M.; Blancato, V.S.; Magni, C. Aroma compounds generation in citrate metabolism of *Enterococcus faecium*: Genetic characterization of type I citrate gene cluster. *Int. J. Food Microbiol.* **2016**, *218*, 27–37. [CrossRef] [PubMed]

30. Suzzi, G.; Gardini, F. Biogenic amines in dry fermented sausages: A review. *Int. J. Food Microbiol.* **2003**, *88*, 41–54. [CrossRef]

31. Schirone, M.; Tofalo, R.; Visciano, P.; Corsetti, A.; Suzzi, G. Biogenic amines in Italian Pecorino cheese. *Front. Microbiol.* **2012**, *3*, 171. [CrossRef] [PubMed]

32. Tofalo, R.; Fasoli, G.; Schirone, M.; Perpetuini, G.; Pepe, A.; Corsetti, A.; Suzzi, G. The predominance, biodiversity and biotechnological properties of *Kluyveromyces marxianus* in the production of Pecorino di Farindola cheese. *Int. J. Food Microbiol.* **2014**, *187*, 41–49. [CrossRef] [PubMed]

33. Coloretti, F.; Chiavari, C.; Luise, D.; Tofalo, R.; Fasoli, G.; Suzzi, G.; Grazia, L. Detection and identification of yeasts in natural whey starter for Parmigiano Reggiano cheese-making. *Int. Dairy J.* **2017**, *66*, 13–17. [CrossRef]

34. Moschini, M.; Battaglia, M.; Beone, G.M.; Piva, G.; Masoero, F. Iodine and selenium carry over in milk and cheese in dairy cows: Effect of diet supplementation and milk yield. *Animal* **2010**, *4*, 147–155. [CrossRef] [PubMed]

35. O'Kane, S.M.; Pourshadidi, L.K.; Mulhern, M.S.; Weir, R.R.; Hill, S.; O'Reilly, J.; Kmiotek, D.; Deitrich, C.; Mackle, E.M.; Fitzgerald, E.; et al. The effect of processing and seasonality on the iodine and selenium concentration of cow's milk produced in Northern Ireland (NI): Implications for population dietary intake. *Nutrients* **2018**, *10*, 287. [CrossRef] [PubMed]

36. Manca, M.G.; Nudda, A.; Rubattu, R.; Boe, R.; Pulina, G. Fatty acid profile of milk fat in goat supplemented with iodized salt. *Ital. J. Anim. Sci.* **2009**, *8* (Suppl. 2), 460.

37. Secchiari, P.; Mele, M.; Frigeri, M.; Serra, A.; Conte, G.; Pollicardo, A.; Formisano, G.; Alghini-Lombardi, F. Iodine supplementation in dairy ewes nutrition: Effect on milk production traits and milk iodine content. *Ital. J. Anim. Sci.* **2009**, *8* (Suppl. 2), 463.

38. Fox, P.F.; Law, J.; McSweeney, P.L.H.; Wallace, J. Biochemistry of cheese ripening. In *Cheese: Chemistry, Physics and Microbiology. General Aspects*, 3rd ed.; Fox, P.F., McSweeney, P.L.H., Cogan, T.M., Guinee, T.P., Eds.; Chapman & Hall: London, UK, 1993; pp. 389–438. ISBN 97800805000942.

39. Sousa-Gallagher, M.J.; Ardo, Y.; McSweeney, P.L.H. Advances in study of proteolysis during cheese ripening. *Int. Dairy J.* **2001**, *11*, 327–345. [CrossRef]

40. Spano, G.; Russo, P.; Lonvaud-Funel, A.; Lucas, P.; Alexandre, H.; Grandvalet, C.; Coton, E.; Coton, M.; Barnavon, L.; Bach, B.; et al. Biogenic amines in fermented foods. *Eur. J. Clin. Nutr.* **2010**, *64*, S95–S100. [CrossRef] [PubMed]

41. Alvarez, M.A.; Moreno-Arribas, M.V. The problem of biogenic amines in fermented foods and the use of potential biogenic amine-degrading microorganisms as a solution. *Trends Food Sci. Technol.* **2014**, *39*, 146–155. [CrossRef]

42. Herrero-Fresno, A.; Martínez, N.; Sánchez-Llana, E.; Díaz, M.; Fernández, M.; Martin, M.C.; Ladero, V.; Alvarez, M.A. *Lactobacillus casei* strains isolated from cheese reduce biogenic amine accumulation in an experimental model. *Int. J. Food Microbiol.* **2012**, *157*, 297–304. [CrossRef] [PubMed]

43. Ladero, V.; Herrero-Fresno, A.; Martinez, N.; del Río, B.; Linares, D.M.; Fernández, M.; Martin, M.C.; Alvarez, M.A. Genome sequence analysis of the biogenic amine-degrading strain *Lactobacillus casei* 5b. *Genome Announc.* **2014**, *2*, 1–2. [CrossRef] [PubMed]

11

Casing Contribution to Proteolytic Changes and Biogenic Amines Content in the Production of an Artisanal Naturally Fermented Dry Sausage

Annalisa Serio, Jessica Laika, Francesca Maggio, Giampiero Sacchetti, Flavio D'Alessandro, Chiara Rossi, Maria Martuscelli, Clemencia Chaves-López * and Antonello Paparella

Faculty of Bioscience and Technology for Food, Agriculture and Environment, University of Teramo, via Balzarini 1, 64100 Teramo, Italy; aserio@unite.it (A.S.); jessica.laika@hotmail.com (J.L.); fmaggio@unite.it (F.M.); gsacchetti@unite.it (G.S.); flavio_dale@live.it (F.D.); crossi@unite.it (C.R.); mmartuscelli@unite.it (M.M.); apaparella@unite.it (A.P.)

* Correspondence: cchaveslopez@unite.it

Abstract: The effect of two kinds of casings on the production and characteristics of a dry fermented sausage was investigated. In detail, an Italian product, naturally fermented at low temperatures and normally wrapped in beef casing instead of the most diffused hog one, was selected. Two different productions (one traditionally in beef casing (MCB) and another in hog casing (MCH)) were investigated over time to determine the differences particularly regarding proteolytic changes during fermentation and ripening. First of all, the product in hog casing required a longer ripening time, up to 120 days, instead of 45–50 days, because of the lower drying rate, while the microbial dynamics were not significantly modified. Conversely, the proteolysis showed a different evolution, being more pronounced, together with the biogenic amines content up to 341 mg/Kg instead of 265 mg/Kg for the traditional products. The latter products were instead characterized by higher quantities of total free amino acids, 3-methyl butanoic acid, 3-Methyl-1-butanal, and 2-Methylpropanal, enriching the final taste and aroma. The traditional product MCB also showed lower hardness and chewiness than MCH. The results highlight how the choice of casing has a relevant impact on the development of the final characteristics of fermented sausages.

Keywords: proteolysis; dry fermented sausage; casing; biogenic amines; volatile compounds; texture; low temperature

1. Introduction

The characteristics of dry fermented sausages depend on many factors, including the ingredients, the recipe, the microbiota composition, and the different production steps. These factors, combined together, determine the variety of products widespread in different European countries [1]. Many fermented meats are produced in Italy; among them, a very particular one is produced during autumn and winter in the Abruzzo Region (Central Italy), and is called "*Mortadella di Campotosto*". It is made of lean pork meat mixed with salt, pepper, and nitrite and it contains a bar of lard, previously cured with salt and spices for about two weeks. It shows a characteristic sub-ovoid shape and length about 15–20 cm, with a diameter of 8–10 cm. Among the typical characteristics, this product is handmade and wrapped (instead of being stuffed) into a natural beef casing; afterwards, it is fermented at low temperature without any starter addition. Traditionally, the product is exposed to dry cold northern winds that provide the ideal conditions for product ripening, which lasts about 40–50 days, with temperatures varying from −1 °C to about 10 °C.

As for this product, the traditional casing is made of beef middles, we decided to evaluate the effect of both the traditional casing and a more usual hog casing on the characteristics of the sausage.

In fact, although *"Mortadella di Campotosto"* sausage manufacturing does not have any particular specifications about the natural casing to use, most of the producers use beef casing, but some may use hog casing because of its larger availability.

As regards fermented sausages, besides containing the meat batter, the casing exerts mechanical protection and guarantees permeability, which is the basis of the exchange of water and oxygen necessary for adequate and homogeneous drying and ripening [2]. Many features affect the quality of natural casings, such as the portion of the intestinal tract used, the manipulation of the product [3] and the mechanical and physical characteristics such as casing elasticity, permeability to water and gases, diameter and its uniformity, adhesion, and resistance to temperature variations [2]. In addition, casing can be a relevant source of enzymes, while casing microbiota is important overall when starter cultures are not added for fermentation [4].

In light of these considerations, we hypothesized that different casings could affect ripening by modifying sausage microbiota, water diffusion, and proteolysis dynamics, with a potentially significant impact on the product quality. Therefore, in this research, we aimed to deepen the knowledge on the influence of the type of casing on the characteristics of *"Mortadella di Campotosto"*-like sausages and to highlight the effect of casing on specific quality traits of the product. For these reasons, the specific objectives were: to follow the evolution of the principal microbial groups during the production process; to evaluate the protein hydrolysis of the fermented meats soon after stuffing and during ripening; to evaluate the texture and flavor of the finished product; and to follow the production of biogenic amines during the process.

2. Materials and Methods

2.1. Samples Production

"Mortadella di Campotosto"-like sausages were manufactured in autumn–winter in Macelleria "L'Olmo", sited in Scanno (AQ), Italy. The sausages were produced according to the traditional formulation and procedure: pork lean meat (ham, shoulder, and loin) and bacon were first minced, then mixed with salt (20 g/Kg), black pepper (1.0 g/Kg), and a mixture of sucrose, glucose, ascorbic acid (E300), and potassium nitrite (E252) in a total quantity of 4.0 g/Kg. After 12 h at 4 °C, the meat mix was then minced again, and about 420 g was taken for each sausage. A stripe of back lard was previously cured at 4 °C with salt, spices (pepper, oregano, rosemary, pimento, juniper, and cilantro), and a mixture of sucrose, glucose, ascorbic acid, and potassium nitrite, for about 14 days, after which it was cut into portions of about 3 × 3 × 10 cm that were inserted into the batter. Then, the product was shaped in its typical oval form. Successively, the meat balls were hand-wrapped with the natural casing, previously washed with water and vinegar, then air-dried for about 10 h. Then, the products were left in air for 3–4 h, to favor the sealing of the casing edges; afterward, they were linked with a cotton string of medium caliber and tied in pairs. Thirty-one samples for each batch were used. The size of the samples at T0 was 17 ± 2 cm in length × 9 ± 2 cm in width and the weight was 452 ± 18 g. At the end of ripening, the size was about 10 ± 1 cm in length × 6 ± 1 cm in width and about 20 ± 1.5 in diameter and a final weight of about 270 ± 10 g.

Two different batches were produced: batch A, named MCB, wrapped with beef casing (beef middles), and batch B, named MCH, wrapped with hog casing (hog bung). All samples were transferred to a fermentation room, in which the temperature varied from −2 and 5 °C for 5 days. The samples were then moved to a drying–ripening room, where they stayed up to 120 days. In the room, natural ventilation was favored, with variable relative humidity (depending on the external weather conditions) and temperatures below 12 °C.

2.2. Microbiological Analysis

Microbiological analyses were carried out on the batter section, excluding the lard, after casing removal. Ten grams of the sausages were homogenized with 90 mL of sterile 0.1% (*w/v*) peptone water for 2 min, using sterile plastic pouches in a Stomacher Lab-Blender 400 (PBI International Milan, Italy). Serial 10-fold dilutions were prepared in sterile peptone water solution and inoculated in duplicate in appropriate culture media. The following microbial groups were determined: aerobic mesophilic bacteria on Plate Count Agar (PCA) at 30 °C for 48 h; mesophilic lactobacilli (LAB) and cocci on MRS agar and M17, respectively, at 30 °C for 48–72 h under anaerobiosis; presumptive enterococci on Slanetz and Bartley agar (S&B) at 37 °C for 48 h; total enterobacteria on Violet Red Bile Glucose Agar (VRBGA) at 37 °C for 24 h; micrococci and staphylococci on Mannitol Salt Agar (MSA) at 30 °C for 72 h; yeasts on Yeast extract-Peptone-Dextrose agar (YPD), added with 150 ppm of chloramphenicol, at 30 °C for 72 h.

Mold development on the casing was evaluated by sampling 10 cm^2 of casing and determining mold growth on YPD added with 150 ppm of chloramphenicol. All the culture media were from Oxoid SpA (Rodano, Italy).

2.3. Physical Analyses

Measurement of water activity (a_w) was performed by the Aqualab instrument CX/2 (Series 3, Decagon Devices, Inc., Pullman, WA, USA). Samples (10 g) were randomly obtained from the sausage (batter section). Moisture (g water/100 g sample) was measured by drying a 3 g sample at 100 °C to constant weight [5], and the pH values were obtained using a MP 220 pH meter (Mettler-Toledo, Columbus, OH, USA).

The weight loss of "*Mortadella di Campotosto*"-like samples during drying was gravimetrically determined and calculated as shown in the following Equation (1):

$$\text{Weight loss (\%)} = [(m0 - mt)/m0] \times 100 \tag{1}$$

where m0 is the weight of sausage obtained after filling and mt is the weight of the sausages after a specific processing time (0, 6, 11, 18, 30, 45, and 120 days).

The measurements of moisture were performed by air oven drying [6], while for the NaCl content, the method of Volhard (ISO 1841-2: 1996) was used [7].

2.4. Chemical Determinations

Total nitrogen content (TN, % w/w) was determined by the Kjeldahl method, while proteins were obtained by multiplying TN × 6.25 [8]. Non-protein nitrogen content (NPN, % w/w) was measured by the precipitation of proteins with trichloroacetic acid, followed by determination of the nitrogen in the extract by the Kjeldahl method. Proteolysis Index (PI, %) was calculated as the ratio between NPN and TN (PI % = 100 × NPN × TN^{-1}), as previously reported [9].

To evaluate the intensity of the primary proteolytic changes during the process, sarcoplasmic and myofibrillar proteins were extracted [10]. The protein concentrations were determined using Bradford reagent (Sigma-Aldrich, Milan, Italy) and bovine serum albumin (BSA, Sigma-Aldrich) as standard reference, according to Bradford (1976) [11]. Sodium dodecyl sulphate-polyacrylamide gel electrophoresis (SDS-PAGE) was used to analyze proteins [12] by Mini Protean III electrophoresis equipment (Bio-Rad, Segrate, Italy), as previously described [10]. The GS-800™ Calibrated Densitometer (Bio-Rad, Segrate, Italy) was used to quantify the relative abundance of each protein band.

Total amino acids were extracted and measured on 2 g of each sample, using the Cadmium-ninhydrin method [13]. For the extraction of the free amino acids, the method proposed by Berardo and colleagues [14] was followed and concentrations were determined by Reverse-phase high performance liquid chromatography (RP-HPLC), as previously reported [15], by using the Waters AccQ Tag method (Millipore Co-Operative, Milford, MA, USA). Amino acids were converted to stable fluorescent derivatives by reaction with AccQ·Fluor reagent (6-Aminoquinolyl-N-hydroxysuccinimidyl carbamate). RP-HPLC was performed using a Waters liquid chromatography system consisting of a Waters™ 626 pump, Waters™ 600 S controller, and Waters™ 717 S autosampler (Millipore Co-operative, Milford, MA, USA), by means of a Nova-Pak™ C18 column (4 μm, 3.9 × 4.6 mm), heated to 37 °C in a Shimadzu model

CTO-10AC column oven. Elution was performed in a gradient of solvent A (Waters AccQ·Tag eluent A), solvent B (acetonitrile: Aldrich Chemical Co., Milan, Italy), and solvent C (20% methanol in Milli-Q water), prepared as follows: initial eluent 100% A; 99% A and 1% B at 0.5 min; 95% A and 5% B at 18 min; 91% A and 9% B at 19 min; 83% A and 17% B at 29.5 min; 60% B and 40% C at 33 min and held under these conditions for 20 min before returning to 100% A. The concentration of A was maintained at 100% up to 65 min, after which the gradient was changed to 60% B and 40% C for 35 min, before returning to the starting conditions. The single amino acids were identified by comparing their retention times with calibration standards. Peak areas were processed using Millennium 32 software v.4.0 (Waters, Milford, MA, USA).

2.5. Determination of Volatile Compounds

Volatile compounds were determined by solid phase micro-extraction coupled with gas chromatography mass spectrometry (SPME/GC-MS) [16] on 5 g of MCB or MCH at the selected sampling times. Volatile peaks identification was carried out by computer matching of mass spectral data with those of the compounds contained in the Agilent Hewlett-Packard NIST 98 and Wiley v. 6 mass spectral database. The volatile compounds content was expressed as relative percentage area.

2.6. Determination of Biogenic Amines

The following eight biogenic amines were detected, identified, and quantified: tryptamine (TRP), β-phenylethylamine (β-PHE), putrescine (PUT), cadaverine (CAD), histamine (HIS), tyramine (TYR), spermidine (SPD), and spermine (SPM).

The procedure of amines extraction and derivatization was carried out as described by Martuscelli et al. [17]: an aliquot of 2 g was homogenized (in Stomacher Lab blender 400, International PBI, Milan, Italy) with 10 mL of 5% trichloroacetic acid (TCA) and centrifuged (Hettich Zentrifugen, Tuttlingen, Germany) at a relative centrifugal force of $2325 \times g$ for 10 min; the supernatant was recovered and the extraction was performed with 5% TCA acid. The two acid extracts were mixed and made up to 50 mL with 5% TCA acid; the final acid extract was filtered through Whatman 54 paper (Carlo Erba, Milan, Italy). For derivatization of the samples, an aliquot of each acid extract (0.5 mL) was mixed with 150 µL of a saturated $NaHCO_3$ solution and the pH was adjusted to 11.5 with about 150 µL NaOH 1.0 M. Dansyl chloride (Fluka, Milan, Italy) solution (2 mL of 10 mg/mL dansyl chloride/acetone) was added to the alkaline amine extract. Derivatized extracts were transferred to an incubator and kept for 60 min at 40 °C under agitation (195 stokes) (Dubnoff Bath-BSD/D, International PBI, Milano, Italy). The residual dansyl chloride was removed by adding 200 mL of 300 g/L ammonia solution (Carlo Erba). After 30 min at 20 ± 1 °C and protected from light, each sample was brought up to 5 mL with acetonitrile (Carlo Erba) and filtered through a 0.22 µm PTFE filter (Alltech, Sedriano, Italy).

Biogenic amines were determined, after extraction and derivatization, by high-performance liquid chromatography (HPLC) using an Agilent 1200 Series (Agilent Technologies, Milano, Italy). In a Spherisorb S30ODS Waters C18-2 column (3 µm, 150 mm × 4.6 mm ID), 10 µL of sample was injected with gradient elution, acetonitrile (solvent A), and water (solvent B) as follows: 0–1 min 35% B isocratic; 1–5 min, 35%–20% B linear; 5–6 min, 20%–10% linear B; 6–15 min, 10% B isocratic; 15–18 min, 35% linear B; 18–20 min, 35% B isocratic. Identification of the biogenic amines (BAs) was based on their retention times and BAs content was reported as mg/kg of product.

2.7. Texture Analysis

Textural properties were evaluated at room temperature (22 ± 2 °C) using an Instron Universal Testing Machine (mod. 5422, Instron LTD, Wycombe, UK) equipped with a 500 N load cell.

Slices (1 cm thick) were transversally cut from the central part of the sausage. Cubic samples (1 × 1 × 1 cm) were cut from the inner part of the slices placed between the lard and the casing. Samples were compressed by a plunger with a plane circular surface (35 mm diameter) using a crosshead speed of 0.5 mm/s. Two different tests were carried out for the textural characterization:

- Compression–relaxation test: samples were compressed by 30% of their initial height, then the run was stopped and the plunger was maintained at the maximum compression extension for 2 min, after which the load was removed. The maximum peak force in compression (N) was taken as a hardness index and the relaxation load was used to study the elastic behavior.

- Texture Profile Analysis (TPA) test: samples were submitted to a two-cycle compression test to 30% of their initial height in the first compression. After the first cycle, samples were left for 1 min to recover their deformation and then, the second cycle was started. Hardness (N), cohesiveness ($J \, J^{-1}$), springiness (mm), and chewiness (N mm) were determined as previously described [18].

Each test was carried out on 10 samples of each batch (MCB and MCH). Since, in both tests, the experimental conditions in the first compression stage were the same, 20 samples were used for hardness calculation.

2.8. Experimental Design and Statistical Analysis

Two batches of sausages characterized by different casings were analyzed over time. In detail, three sausages per batch were randomly taken and analyzed at each ripening time (0, 6, 11, 18, 30, and 45 days). Time zero was considered as the time in which the batter was just wrapped in the casings, thus samples of batter (which was the same for the two batches) were taken just before being wrapped. All the data were subjected to two-way analysis of variance (ANOVA) to test the significance of individual (casing, ripening time) and interactive (casing × ripening time) effect. The model used for the two-way ANOVA is presented in Equation (2):

$$Y_{ijk} = \mu + \alpha_i + \beta_j + \gamma_{ij} + \varepsilon_{ijk} \tag{2}$$

where μ is the intercept; α the casing factor; β the time factor; γ the interaction; ε the error; i and j are the level of the first and second factor; k the number of within group replicates. The significance of the effects was tested by Fisher's F value and the associated p value.

Since the MCH batch was still not ripened after 45 days, an additional sampling was carried out only for this batch at 120 days. As the definitive experimental design was incomplete, data were further analyzed by the two-way nested ANOVA and the model used is presented in Equation (3):

$$Y_{ijk} = \mu + \alpha_i + \beta_{j(i)} + \varepsilon_{ijk} \tag{3}$$

where β is the time factor nested with the casing factor. Post hoc mean comparison was carried out on the time nested with casing effect using the Tukey's HSD test.

The mold load and textural properties of the fully ripened sausages (MCB at 45 days and MCH at 120 days) were eventually compared among them using the Student's t-test for independent samples in order to test significant differences between the two groups of samples.

Statistical analyses were performed by using Statistica v. 6.1 (Statsoft Europe, Hamburg, Germany).

3. Results and Discussion

3.1. Effect of the Different Casings on Microbial Growth

Differently from other Mediterranean dry fermented sausages, in which the fermentation time is 1–2 days at 18–24 °C or 1 week at relatively low temperatures (10–12 °C), the fermentation of "*Mortadella di Campotosto*" sausages is carried out at very low temperatures, below 4 °C. These conditions, together with the absence of starters, cause a slow growth of lactic acid bacteria and therefore, an extension of the fermentation time.

Table 1 depicts the behavior of the different microbial groups during time. As evidenced, during the fermentation phase (up to 11 days) and the first week of ripening (day 18), no statistically significant differences were noticed in the growth dynamics of all microbial groups between MCB and MCH products. As expected, Enterobacteriaceae were not detected after 6 days of production, probably as a consequence of the progressive pH reduction. With the extension of the ripening time (from 30 days), statistically significant differences were observed in yeast and CNS counts that were lower in MCB. In detail, the yeast count increased in MCH samples at 30 days, probably because of a succession of different species. It has to be underlined that at 30 days, the a_w values of MCB and MCH samples were significantly different and the higher MCH a_w allowed a greater microbial growth with more abundant cells loads. After that, the number of cells progressively decreased during the ripening time.

In addition, in this study, we observed that the type of casing used can affect the colonization of molds, which reached values of 5.60 Log CFU/cm^2 in MCB and < 2.0 Log CFU/cm^2 in MCH at the end of the process.

3.2. Effect of the Different Casings on Physicochemical Parameters

In Table 1, the changes in pH values during fermentation and ripening of the samples in beef casing (MCB) and in hog casing (MCH) are reported. A significant ($p < 0.05$) pH decrease was detected in both cases up to day 11, which could thus be presumably considered as the end of fermentation. The pH decrease was concurrent with the increasing number of presumptive LAB that reached levels of 7.66 and 7.85 Log CFU/g for MCH and MCB, respectively. After that, pH slowly increased, due to the typical phenomena of ripening, starting with proteolysis in both batches, but with different rates throughout ripening. The end of ripening (45 days for MCB and 120 days for MCH, respectively) was first evaluated by professional manufacturers, who tested product hardness, as perceived by digital pressure, and flavor sniffing, and then, confirmed by textural analysis before final sampling. At the end of ripening, the pH reached levels of about 5.67 for MCB and of 5.90 for MCH, respectively.

The type of casing exerted a significant effect on the drying rate, as highlighted by moisture (Figure 1) and a_w (Table 1) data; no water losses were observed during fermentation but, during ripening, the moisture dramatically differed between the two batches. After 45 days of ripening (the end of the ripening for MCB products), MCB batches reached moisture (31.2%) and a_w (0.841) values significantly lower than MCH samples, in which the values of moisture and a_w were 53.7% and 0.923, respectively. These differences can be attributed to the physical characteristics of the two types of casing, such as the degree of casing permeability, which influences the level of exchange between the filling and the external environment. In fact, hog bung casings had greater thickness (about 3-fold) than MCB, leading to a lower water vapor transmission rate and higher a_w [4]. The degree of casing permeability to water, gas, and light affects water loss, fat hydrolysis, fat oxidation, as well as pH and a_w [2,4].

As regards NaCl content, given as g 100/g total solids, slightly significant differences were observed at the end of the ripening time in both types of sausages, which showed values of 4.77 ± 0.01% and 4.97 ± 0.16% for MCB and MCH, respectively.

Table 1. Physical-chemical and microbiological characteristics of *"Mortadella di Campotosto"*-like sausages produced with beef (MCB) and hog (MCH) casings over time.

Batch	Time (Days)	pH	a_w	MAB	Yeasts	Microbial Groups (Log CFU/g)					
						LAB	LAC	ENTC	CNS	PSE	ENTB
					Complete design (CD)						
MCB	0	5.96 ± 0.04 [a]	0.957 ± 0.003 [a]	7.24 ± 0.08 [b]	7.14 ± 0.00 [a]	6.37 ± 0.11 [cd]	7.07 ± 0.06 [bc]	6.52 ± 0.38 [cd]	5.87 ± 0.36 [abc]	5.56 ± 0.38 [a]	3.09 ± 0.20 [a]
MCH	0	5.86 ± 0.12 [a]	0.959 ± 0.004 [a]	7.12 ± 0.18 [b]	7.02 ± 0.10 [a]	6.60 ± 0.08 [cd]	7.02 ± 0.10 [bc]	6.92 ± 0.10 [cd]	5.91 ± 0.48 [abc]	5.71 ± 0.46 [a]	3.42 ± 0.08 [a]
MCB	6	5.50 ± 0.09 [b]	0.956 ± 0.002 [a]	7.55 ± 0.21 [ab]	7.63 ± 0.13 [a]	7.53 ± 0.18 [abc]	7.80 ± 0.04 [abc]	7.29 ± 0.13 [cd]	6.98 ± 0.14 [d]	4.72 ± 0.39 [a]	0.95 ± 0.75 [b]
MCH	6	5.47 ± 0.17 [b]	0.956 ± 0.002 [a]	7.17 ± 0.07 [b]	7.63 ± 0.09 [a]	7.14 ± 0.13 [abcd]	8.13 ± 0.33 [ab]	7.84 ± 0.04 [b]	7.18 ± 0.13 [d]	4.41 ± 0.61 [a]	1.11 ± 0.31 [b]
MCB	11	5.48 ± 0.07 [b]	0.960 ± 0.006 [a]	7.91 ± 0.09 [ab]	7.80 ± 0.26 [a]	7.85 ± 0.04 [abc]	7.93 ± 0.09 [ab]	7.79 ± 0.04 [b]	6.29 ± 0.41 [ab]	4.58 ± 0.49 [a]	<1.00 [b]
MCH	11	5.44 ± 0.08 [b]	0.944 ± 0.009 [a]	8.15 ± 0.33 [ab]	7.26 ± 0.08 [a]	7.66 ± 0.31 [abcd]	8.28 ± 0.22 [a]	8.09 ± 0.04 [b]	4.47 ± 0.31 [c]	4.45 ± 0.62 [a]	<1.00 [b]
MCB	18	5.62 ± 0.20 [ab]	0.960 ± 0.006 [a]	7.85 ± 0.21 [ab]	4.00 ± 0.04 [cd]	6.23 ± 0.27 [d]	7.83 ± 0.20 [abc]	7.64 ± 0.20 [bc]	5.11 ± 0.15 [bcd]	5.64 ± 0.01 [ab]	<1.00 [b]
MCH	18	5.55 ± 0.13 [ab]	0.952 ± 0.006 [a]	7.79 ± 0.12 [ab]	4.00 ± 0.16 [cd]	6.23 ± 0.27 [d]	7.83 ± 0.26 [abc]	7.64 ± 0.04 [bc]	5.18 ± 0.15 [bcd]	5.64 ± 0.27 [ab]	<1.00 [b]
MCB	30	5.65 ± 0.14 [ab]	0.890 ± 0.008 [b]	8.26 ± 0.15 [ab]	4.80 ± 0.39 [bc]	6.38 ± 0.06 [d]	8.14 ± 0.13 [ab]	8.03 ± 0.20 [b]	6.22 ± 0.29 [ab]	4.48 ± 0.95 [ab]	<1.00 [b]
MCH	30	5.56 ± 0.12 [ab]	0.940 ± 0.012 [a]	8.44 ± 0.20 [a]	5.36 ± 0.17 [b]	8.24 ± 0.20 [a]	8.48 ± 0.39 [a]	8.18 ± 0.08 [b]	6.78 ± 0.29 [bc]	3.95 ± 0.10 [bc]	<1.00 [b]
MCB	45	5.67 ± 0.11 [ab]	0.841 ± 0.006 [c]	8.08 ± 0.03 [ab]	2.47 ± 0.20 [e]	7.96 ± 0.22 [ab]	8.54 ± 0.29 [a]	6.31 ± 0.29 [d]	2.00 ± 0.20 [e]	5.11 ± 0.12 [ab]	<1.00 [b]
MCH	45	5.59 ± 0.09 [ab]	0.923 ± 0.008 [b]	7.97 ± 0.07 [ab]	3.54 ± 0.22 [e]	7.10 ± 0.14 [abcd]	7.72 ± 0.39 [abc]	7.76 ± 0.04 [bc]	3.98 ± 0.23 [d]	3.69 ± 0.10 [bc]	<1.00 [b]
F(C)		0.46	51.5 ***	0.20	93.5 ***	0.26	0.07	0.15	26.4 ***	2.90	0.06
F(t)		3.51 *	39.5 ***	9.10 ***	257 ***	1.7 ***	7.93 ***	12.3 ***	4.38 **	5.38 **	65.9 ***
F(C × t)		0.04	15.4 ***	0.66	25.6 ***	7.59 ***	1.68	5.16 **	34.5 ***	0.96	0.06
					Incomplete Design (CD + MCH 120 d)						
MCH	120	5.90 ± 0.05 [c]	0.812 ± 0.006 [c]	7.55 ± 0.14 [c]		3.50 ± 0.14 [d]	6.55 ± 0.36 [b]	6.60 ± 0.34 [c]	9.33 ± 0.11 [a]	2.00 ± 0.20 [c]	<1.00
F(C)		0.01	135 ***	0.47		151 ***	0.01	1.17	3.42	12.3 **	0.26
F(tt(C))		2.21 *	45.6 ***	4.60 ***		134 ***	7.77 ***	6.51 ***	13.3 ***	6.41 ***	33.3 ***

Means with different letters in the same column are significantly different ($p < 0.05$). Log CFU/g—Log Colony Forming Unit/gram of product a_w—water activity MAB—Mesophilic Aerobic Bacteria, LAB—Lactic acid bacteria; LAC—Lactococci; ENTC—Enterococci; PSE—Pseudomonas spp.; CNS—Coagulase negative Staphylococci; ENTB—Enterobacteriaceae. Fisher's F value of casing (F(C)), time (F(t)), combined casing and time (F(C × t)), and time nested with casing (F(t(C))) factors calculated for each analytical determination. ANOVA significant differences were indicated by F values. * $p < 0.05$, ** $p < 0.01$. *** $p < 0.005$.

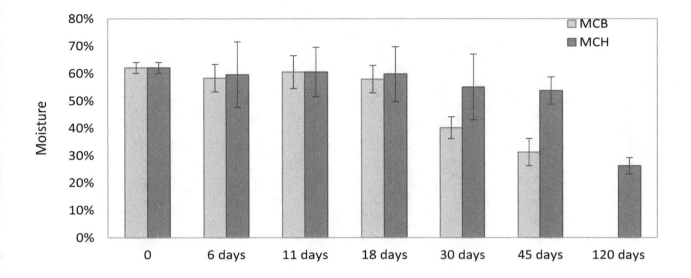

Figure 1. Evolution of the relative humidity of *"Mortadella di Campotosto"*-like sausages produced with beef (MCB) and hog (MCH) casings over time.

3.3. Effect on Proteolysis

Protein hydrolysis was evaluated by gel electrophoresis, as well as by measuring the content of total free amino acids, volatiles, and amines that greatly influence the texture, flavor, and safety of dry fermented sausages [19]. The hydrolysis of sarcoplasmic and myofibrillar proteins, determined via SDS-PAGE analysis, was influenced by the ripening time and the type of casing (Figure 2).

3.3.1. Sarcoplasmic Fraction

The electrophoretic separation of sarcoplasmic proteins of MCB and MCH at different processing stages is illustrated in Figure 2a. Proteolysis took place during fermentation, as revealed by the slight changes from the first fermentation phase; the bands most susceptible to degradation were those of about 61 and 56 kDa, followed by a huge band of about 48 kDa, and the most intensive degradation was observed in MCH samples. In addition, two fragments were generated at 36 and 35 KDa, which were assumed to be glyceraldehyde-3-phosphate dehydrogenase [20], and 18 KDa in both types of samples.

In addition, during ripening, starting from day 18, a more intense hydrolysis was observed in MCH samples, as indicated by the intensity decrease in the band at 45 kDa (data at 120 days not shown), which is assumed to be creatine kinase [21]. This band completely disappeared after 45 days, while the intensity of the bands of about 74, 37, 36, 18, and 12 KDa increased overall in the MCB samples. The appearance and increase in polypeptides in the range of 14–100 kDa have been observed also by other authors [10,22].

As a_w strongly affects the activities of all endogenous proteinases [23], the differences between MCB and MCH could be ascribable to the higher a_w values of MCH samples (Table 1). Thus, it could be possible that the highest a_w values in MCH batches could have favored the activity of cathepsins B, which are able to break down sarcoplasmic proteins [24]. However, the contribution of bacterial enzymes to protein degradation needs to be taken into account, since LAB counts at day 30 reached higher values in MCH samples (8.24 Log CFU/g) than in MCB samples (6.48 Log CFU/g). In this context, in addition to LAB microbial enzymes, also *Staphylococcus carnosus* and *Staphylococcus simulans* proteases are capable to hydrolyze sarcoplasmic proteins [25].

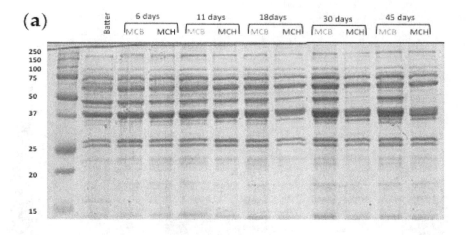

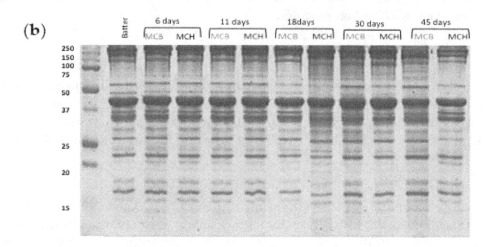

Figure 2. SDS-PAGE electrophoretic profiles of "*Mortadella di Campotosto*"-like sausages produced with beef (MCB) and hog (MCH) casings over time. Panel (**a**) sarcoplasmic fraction; panel (**b**) myofibrillar fraction. Batter: 0 days. Values on the ordinates refer to the marker and express the bands dimension in KDa.

3.3.2. Myofibrillar Fraction

Myofibrillar proteins play the most critical role during meat processing, as they are responsible for the cohesive structure and firm texture of meat products [26]. As evidenced, this protein fraction was less susceptible to degradation and the hydrolysis dynamics in MCB and MCH showed very similar profiles. Recently, Berardo et al. [14] reported that actin (45 kDa) is highly broken down during fermentation; nevertheless, we did not evidence any change of this protein during fermentation in both MCB and MCH samples, in agreement with other studies [27,28]. On the contrary, the generation of polypeptides and large peptides with molecular weight from 50 to 100 kDa was more evident at day 11.

During ripening, when the endogenous enzymes are affected by the a_w decrease, having particular impact on cathepsins and alanyl- and pyroglutamyl-amino-peptidases [29], an important degradation of the band of about 48 kDa was detected (day 18); afterwards, it disappeared in both batches. In the meantime, a band of about 33 KDa, probably corresponding to β-tropomyosin, appeared in MCH samples at day 18, while in MCB samples, it was detected at 30 days of ripening [30].

Differences between the two products were clear from the 30th day of the process, in which the a_w values were 0.890 ± 0.008 and 0.940 ± 0.012 for MCB and MCH, respectively.

As regards Myosin Heavy Chain (220 kDa), an important degradation for at least 45% in MCH samples was detected at day 45, in contrast with MCB samples, in which it remained almost unchanged during ripening. The hydrolysis of actin (45 kDa) was less severe than that of myosin (25 kDa), and it was clear from day 30 in MCB, in which the reduction was about 15%. These results are in accordance with other authors [31], who reported a lower degradation of actin in fermented sausages with higher pH, suggesting that this might be due to the low optimum pH of cathepsin D-like muscle enzymes, playing a major role in actin hydrolysis.

Additionally, changes in tropomyosin (35 kDa) were more intense in MCB samples that presented also hydrolysis of the bands at 48 and 54 kDa, probably corresponding to desmin. Moreover, the myosin II short-chain (about 18 kDa) showed an intensity increase of approximatively 10% in MCH samples; muscle proteinases predominate in proteolysis evolution along the dry fermented sausage ripening, while those from bacteria mainly act during fermentation [32]. Nevertheless, the major proteolysis of the myofibrillar fractions in MCH (presenting a_w values of 0.890 at day 30) could be attributed to particular bacteria or yeast species present in the meat or in the casing, well adapted to the particular environment of this sausage and probably dominating during fermentation and ripening. Moreover, on the MCB casing surface, the molds, reaching loads of 5.60 Log CFU/g, could have promoted the greater proteolysis [33]. In fact, during ripening, when a_w decreases, the molds, and especially strains of *Aspergillus* and *Penicillium* genera, tend to dominate due to their capability to overcome xerophilic and halophilic conditions. *Penicillium chrysogenum* and *P. nalgiovense* contribute to proteolytic activities [34] and *Penicillium chrysogenum* Pg222 proteolytic enzymes show activity on the principal myofibrillar proteins, including actin, myosin, tropomyosin, and troponin [35].

3.4. Total Amino Acids Content

The generation of free amino acids (FAA) is the final outcome of proteolysis, and it contributes to the specific taste and also to the generation of volatile compounds, which provide the flavor in fermented sausages. The FAA content, expressed as mM of leucine, of both kinds of samples, was analyzed during the experimental period. As expected, the low temperatures applied in the production of the samples resulted in a limited generation of FAA, and their content was significantly different ($p < 0.05$) depending on the type of casing.

The quantification of total free amino acids (TFAA) is reported in Table 2. As evidenced, the initial batter contained about 361.25 ± 13.32 mg amino acids/100 g of dry-matter, and during the process, this concentration changed over time to a final concentration of 84.35 ± 19.18 and 235.59 ± 6.59 mg/100 g of dry-matter for MCB and MCH, respectively, at 45 and 120 days, with the major contribution of arginine (Arg) and alanine (Ala), followed by leucine (Leu) and valine (Val). The observed fluctuations in the content of each individual amino acid could be ascribed to the balance between FAA produced by protein breakdown and microbial activity. Among the bacteria, coagulase-negative staphylococci (CNS), *Lactobacillus sakei*, *Lactobacillus curvatus*, and some yeasts such as *Saccharomyces cerevisiae* have been reported to be directly involved in meat proteolysis and in free amino acids generation [15]. At the same time, many of these microorganisms use free amino acids as substrate for further metabolic reactions (deamination, dehydrogenation, and transamination), which are related to the development of aroma and flavor that characterize the final fermented sausage [36,37]. Moreover, arginine reduction in MCH samples could also be correlated to decarboxylation reactions, with the consequent production of putrescine, starting from day 45.

Table 2. Free amino acids content (mg/100 g meat) in *"Mortadella di Campotosto"*-like sausages produced with beef (MCB) and hog (MCH) casings over time.

Batch	Time (Days)	Arg	Asp	Glu	His	Ser	Ala	Gly	Ile	Leu	Met	Phe	Val	TOTAL FAA
Complete Design														
MCB	0	254 ± 2 a	0.85 ± 0.04 ab	10.1 ± 0.24 b	n.d.	0.13 ± 0.08 bcde	21.0 ± 2.25 d	n.d.	2.36 ± 0.04 de	7.80 ± 0.57 de	59.4 ± 7.7 ab	2.36 ± 0.20 f	3.25 ± 0.20 bc	361.25 ± 13.32 aa
MCH	0	247 ± 2 a	0.70 ± 0.16 ab	11.8 ± 1.59 b	n.d.	0.13 ± 0.08 bcde	18.9 ± 0.53 d	n.d.	2.20 ± 0.17 de	6.70 ± 0.33 de	58.4 ± 6.9 ab	2.46 ± 0.13 f	2.90 ± 0.08 bc	351.19 ± 11.97 a
MCB	6	192 ± 7 bcd	0.69 ± 0.05 ab	n.d.	n.d.	0.23 ± 0.03 e	4.50 ± 0.37 e	n.d.	2.1 ± 0.17 de	7.26 ± 0.30 de	55.7 ± 0.5 b	5.31 ± 3.10 bcd	2.27 ± 0.11 c	349.40 ± 10.10 a
MCH	6	158 ± 23 d	0.45 ± 0.07 bc	n.d.	37.9 ± 0.87 b	0.07 ± 0.01 cde	6.81 ± 1.47 e	37.9 ± 0.92 e	1.31 ± 0.27 e	5.46 ± 0.52 ef	55.3 ± 3.2 b	4.23 ± 0.39 de	2.24 ± 0.18 c	318.34 ± 15.66 ab
MCB	11	236 ± 7 ab	0.70 ± 0.05 ab	1.73 ± 0.23 c	n.d.	0.16 ± 0.01 bcde	32.7 ± 2.5 c	n.d.	2.46 ± 0.20 cd	7.00 ± 0.15 de	62.4 ± 1.8 ab	3.56 ± 0.26 ef	2.96 ± 0.37 bc	270.34 ± 19.75 b
MCH	11	220 ± 16 abc	0.63 ± 0.30 ab	2.80 ± 0.45 c	2.50 ± 0.08 d	0.09 ± 0.01 bcde	17.8 ± 1.5 d	3.95 ± 0.92 d	1.53 ± 0.27 d	6.26 ± 0.52 de	55.6 ± 1.8 b	3.73 ± 0.26 ef	2.70 ± 0.38 bc	310.07 ± 37.12 b
MCB	18	177 ± 4 cd	0.87 ± 0.08 ab	n.d.	15.5 ± 1.16 c	0.24 ± 0.02 bc	136 ± 2 a	15.5 ± 1.15 bc	2.40 ± 0.15 cd	8.44 ± 0.32 cde	73.6 ± 0.2 a	6.94 ± 0.17 ab	2.57 ± 0.21 c	438.81 ± 6.61 c
MCH	18	13.8 ± 0.9 e	0.67 ± 0.07 ab	n.d.	35.6 ± 1.74 b	0.05 ± 0.01 de	123 ± 4 b	35.6 ± 2.57 a	3.22 ± 0.10 bc	14.3 ± 0.30 b	7.20 ± 0.21 c	3.42 ± 0.20 e	3.11 ± 0.11 b	237.78 ± 1.72 d
MCB	30	186 ± 10 bcd	0.95 ± 0.12 a	n.d.	13.1 ± 1.02 c	0.26 ± 0.01 b	131 ± 1 ab	13.1 ± 0.41 c	3.35 ± 0.23 bc	9.23 ± 0.37 cd	21.8 ± 0.6 c	8.42 ± 0.23 a	2.56 ± 0.17 bc	389.53 ± 20.13 a
MCH	30	4.15 ± 0.90 f	0.67 ± 0.06 ab	n.d.	38.4 ± 1.24 ab	1.07 ± 0.07 a	128 ± 2 ab	38.4 ± 2.02 a	3.49 ± 0.25 ab	13.4 ± 1.02 b	7.07 ± 1.02 c	4.31 ± 0.30 cde	3.47 ± 0.18 b	241.73 ± 9.05 b
MCB	45	63.5 ± 15 e	0.05 ± 0.11 cd	n.d.	2.03 ± 0.21 d	n.d.	1.61 ± 1.20 e	2.03 ± 1.20 ab	3.61 ± 0.20 ab	2.74 ± 0.18 f	6.51 ± 0.88 c	2.13 ± 0.20 f	n.r.	84.35 ± 19.18 e
MCH	45	8.43 ± 0.20 f	0.97 ± 0.12 a	15.9 ± 1.65 a	45.9 ± 4.83 a	0.14 ± 0.02 bcd	135 ± 2 a	32.9 ± 3.06 ab	4.49 ± 0.19 a	21.0 ± 1.52 a	8.30 ± 0.41 c	5.82 ± 0.12 bc	5.16 ± 0.32 bc	287.32 ± 30.97 d
F (C)		154 ***	0.12	48.8 ***	580 ***	34.5 ***	268 ***	109 ***	0.090	124 ***	65.3 ***	26.3 ***	76.9 ***	
F (t)		138 ***	3.58 *	96.2 ***	121 ***	143 ***	1808 ***	22.7 ***	45.5 ***	30.7 ***	119 ***	45.5 ***	4.87 **	
F (C × t)		27.8 ***	13.8 ***	41.9 ***	66.6 ***	98.6 ***	485 ***	10.7 ***	8.56 ***	63.0 ***	36.4 ***	43.3 ***	47.2 ***	
Incomplete design (CD + MCH 120d)														
MCH	120	1.61 ± 0.08 f	n.r.	14.0 ± 0.70 ab	33.0 ± 1.5 a	1.10 ± 0.08 a	123 ± 20 ab	33.0 ± 2.5 a	3.88 ± 0.20 ab	11.4 ± 0.7 bc	6.70 ± 0.67 c	3.31 ± 0.26 ef	4.75 ± 0.29 a	235.59 ± 6.6 d
F (C)		263 ***	0.06	112 ***	690 ***	125 ***	596 ***	133 ***	3.60	131 ***	120 ***	34.3 ***	111 ***	
F (tt(C))		90.7 ***	0.29 ***	75.8 ***	87.6 ***	113 ***	1130 ***	16.0 ***	27.1 ***	42.1 ***	81.7 ***	41.3 ***	25.6 ***	

Arg—arginine; Asp—aspartic acid; Glu—glutamine; His—histidine; Ser—serine; Ala—alanine; Gly—glycine; Ile—isoleucine; Leu—leucine; Met—methionine Phe—phenylalanine; Val—valine; FAA—free amino acids. Means with different letters in the same column are significantly different ($p < 0.05$). Fisher's F value of casing (F(C)), time (F(t)), combined casing and time (F(C × t)), and time nested with casing (F(t(C))) factors calculated for each analytical determination. ANOVA significant differences were indicated by F values. ** $p <0.01$. *** $p < 0.005$. n.d.—not detectable.

The major differences in FAA were observed during ripening, in which hydrophobic amino acids were accumulated. In addition, significant differences ($p < 0.05$) were observed for the greater amounts of Arg and Ala in MCB samples up to 30 days. This concentration appeared dramatically reduced at day 45, particularly for Arg in MCH samples and for Ala in MCB ones. In this respect, two hypotheses can be proposed: (1) the different environmental conditions present in MCB after 30 days could have selected microorganisms with a highly efficient arginine-converting machinery, with the aim of obtaining energy from arginine in the absence of glucose, as reported for some *Lactobacillus*, CNS and *Pseudomonas* species [38–40], and this might be reflected by the presence of *Pseudomonadaceae* at values of 5.11 ± 0.2 Log CFU/g in MCB samples at day 45; or (2) a possible oxidation of this amino acid could have happened by means of free radicals generated by lipolysis, leading to the formation of carbonyl groups [41]. In fact, amino acids such as lysine, threonine, arginine, and proline are easily attacked by these radicals.

In the case of MCH, the major FAA at the end of ripening was Ala with amounts of 123.20 mg/100 g of dry-matter.

3.5. Volatile Compounds Derived from Amino Acids

FAA are very important in fermented sausages, both for their contribution to the specific taste and for their involvement in degradation reactions that generate volatile compounds, which provide the flavor in this type of product. It is documented that the pH rise during fermentation is due to the microbial degradation of FAA by decarboxylation and deamination [42]. The transamination and decarboxylation of valine, isoleucine, and leucine, which are branched amino acids, produce the respective branched aldehydes, alcohols, and/or acids. Additionally, amino acids such as Phe, Thr, Try, Tyr, etc., are transformed into their respective aldehydes, such as phenylacetaldehyde from phenylalanine, and indole compounds from tryptophan, while the degradation of the sulfur amino acids cysteine and methionine produces sulfur volatile compounds [43]. In particular, we analyzed the accumulation dynamics of compounds derived from branched amino acids (branched aldehydes, alcohols, and carboxylic acids), such as 2-methylpropanal derived from valine (Val), 2-methylbutanal from isoleucine (Ile), 3-methylbutanal from leucine (Leu), and phenylacetaldehyde, benzaldehyde, phenylethyl alcohol, and ethyl benzoate ester that derived from phenylalanine (Phe) and the results are shown in Table 3.

In general, a greater relative abundance of branched chain aldehydes, alcohols, and acids from the catabolism of branched chain amino acids was detected in MCB samples. In particular, 3-Methyl-1-butanal was the most abundant compound in both types of samples, being more present during fermentation (up to 11 days) in MCH and during ripening in MCB samples. In this respect, other authors [44] suggested that 3-Methyl-butanoic acid and 3-Methyl-1-butanal are markers of the CNS activity in fermented meats.

On the other hand, the decrease in 3-Methyl butanol in MCB samples could be ascribed to its conversion into the corresponding 3-Methyl-butanoic acid (Table 3). MCB samples were characterized also by a major presence of 2-Methylpropanal and 2-Methylpropanoic acid, which increased over time and were not detected in MCH samples during the entire production period. On the contrary, phenyl ethyl alcohol concentration increased only in MCH batches.

Table 3. Relative abundance of branched chain compounds derived from branched chain amino acids in "*Mortadella di Campotosto*"-like sausages produced with beef (MCB) and hog (MCH) casings over time. Results are expressed as relative abundance $\times 10^6$.

	Time (Days)	2-MPA (%)	3-MBA (%)	3-MB (%)	PEA (%)	3-M1-B (%)	2-MP (%)
				Complete Design (CD)			
MCB	0	0.26 ± 0.01 [de]	1.18 ± 0.02 [de]	1.29 ± 0.07 [c]	1.63 ± 0.50 [bc]	7.93 ± 1.23 [bc]	0.49 ± 0.01 [d]
MCH	0	0.24 ± 0.01 [de]	1.18 ± 0.02 [de]	1.09 ± 0.09 [c]	1.49 ± 0.62 [bc]	7.93 ± 1.25 [bc]	0.52 ± 0.01 [d]
MCB	6	0.43 ± 0.03 [cd]	2.90 ± 0.16 [abcd]	7.37 ± 1.14 [a]	0.15 ± 0.01 [c]	2.64 ± 0.37 [e]	0.2 ± 0.03 [d]
MCH	6	n.d.	1.20 ± 0.20 [cde]	2.81 ± 0.48 [bc]	1.29 ± 0.26 [bc]	7.89 ± 0.77 [bcd]	n.d.
MCB	11	0.71 ± 0.07 [bc]	3.6 ± 0.49 [ab]	5.50 ± 1.17 [ab]	0.19 ± 0.04 [c]	5.42 ± 0.49 [cde]	0.7 ± 0.05 [d]
MCH	11	n.d.	1.50 ± 0.34 [bcde]	2.40 ± 0.62 [bc]	1.30 ± 0.29 [bc]	9.2 ± 0.68 [abc]	n.d.
MCB	18	0.88 ± 0.06 [ab]	4.75 ± 0.95 [a]	5.23 ± 1.76 [ab]	0.23 ± 0.04 [c]	12.7 ± 0.85 [a]	1.7 ± 0.34 [c]
MCH	18	0.26 ± 0.02 [de]	0.60 ± 0.10 [e]	0.2 ± 0.06 [c]	1.28 ± 0.23 [bc]	5.43 ± 1.09 [cde]	n.d.
MCB	30	n.d.	3.88 ± 0.10 [a]	n.d.	0.26 ± 0.03 [c]	10.94 ± 0.84 [ab]	2.8 ± 0.17 [b]
MCH	30	1.23 ± 0.13 [a]	0.65 ± 0.35 [e]	n.d.	1.32 ± 0.43 [bc]	7.63 ± 0.88 [bcd]	n.d.
MCB	45	n.d.	3.83 ± 0.12 [a]	n.d.	0.29 ± 0.05 [c]	7.32 ± 1.02 [bcd]	4.2 ± 0.53 [a]
MCH	45	1.01 ± 0.16 [ab]	1.07 ± 0.80 [de]	n.d.	4.33 ± 0.83 [a]	6.6 ± 0.31 [bcde]	n.d.d
F (C)		4.28 *	88.7 ***	25.5 ***	48.7 ***	0.45	206 ***
F (t)		14.9 ***	9.08 ***	17.0 ***	7.11 ***	6.06 ***	32.2 ***
F (C × t)		97.6 ***	5.54 **	5.48 **	8.21 ***	14.5 ***	35.9 ***
				Incomplete Design (CD + MCH 120 d)			
MCH	120	n.d.	3.44 ± 0.11 [abc]	n.d.	2.15 ± 0.20 [b]	3.7 ± 0.34 [de]	n.d.
F (C)		0.05	67.7 ***	34.3 ***	58.3 ***	3.68	242 ***
F (t(C))		60.2 ***	9.45 ***	11.2 ***	7.39 ***	11.7 ***	33.5 ***

2-MPA-2-Methylpropanoic acid; 3-MBA-3-Methylbutanoic acid; 3-MB-3-Methylbutanol; PEA-phenyl ethyl alcohol; 3-M1-B-3-Methyl-1-butanal; 2-MP-2-Methylpropanal. Means with different letters in the same column are significantly different ($p < 0.05$). Fisher's F value of casing (F(C)), time (F(t)), combined casing and time (F(C × t)), and time nested with casing (F(t(C))) factors calculated for each analytical determination. ANOVA significant differences were indicated by F values. * $p < 0.05$, ** $p < 0.01$, *** $p < 0.005$; n.d.—not detectable.

3.6. Texture Analysis

The textural properties of *"Mortadella di Campotosto"*-like sausages were studied both by stress relaxation and TPA (Texture Profile) analysis. TPA parameters of the two types of samples at the end of ripening are shown in Table 4. At the end of ripening, corresponding to 45 days for MCB and 120 days for MCH, the two types of samples showed very similar textural profiles, except for small differences in springiness and chewiness.

Table 4. Results of texture profile analysis and compression–relaxation test of ripened *"Mortadella di Campotosto"*-like sausages produced with beef (MCB) and hog (MCH) casings.

Product	Hardness (N)	Springiness (mm)	Cohesiveness ($J \times J^{-1}$)	Chewiness (N mm)	Relaxation Load (%)
MCB (45 days)	16.0 ± 3.4 [a]	1.91 ± 0.19 [b]	0.50 ± 0.04 [a]	15.2 ± 2.9 [b]	63.68 ± 0.63 [a]
MCH (120 days)	16.5 ± 1.7 [a]	2.06 ± 0.12 [a]	0.52 ± 0.04 [a]	17.6 ± 1.4 [a]	62.70 ± 0.82 [a]

Data in the same column with different letters are significantly different at a $p < 0.05$ level.

No statistically significant differences were found in hardness, despite MCB showing lower a_w and moisture values and a slightly higher proteolysis index (12.92% vs. 12.13%) than MCH. In general, protein breakdown during fermented sausages ripening contributes to hardness decrease [45]. However, beside proteolysis, drying is a major factor affecting the binding and textural properties of fermented meat products. In most cases, the effect of dehydration, which increases hardness by promoting the elastic behavior, could counteract and even overcome the effect of proteolysis [28,46].

MCH samples showed a more elastic physical behavior and a consequently higher chewiness. Chewiness resulted positively affected by proteolysis in many studies, independently from positive or negative hardness changes [28,47,48]. Since in TPA analysis, chewiness is the product of hardness × cohesiveness × springiness, the higher chewiness of the MCH product observed in this study is due to its higher springiness, since no differences were found in hardness and cohesiveness. Springiness, which is a measure of elasticity [18], is depleted by moisture content, since water acts as a plasticizer and promotes viscous behavior. Despite differences in springiness, observed by measuring the recovery of the deformation after uniaxial compression, no significant differences in elastic behavior were observed when a stress–relaxation test was applied, as the force dissipated by viscous flow was identical in the two samples (Table 4).

In this section, the effect of proteolysis and dehydration were discussed since they are the main factors affecting sausage texture, but it should be considered that also lipolysis may contribute to the final texture.

3.7. Biogenic Amines (BA) Content

High quantities of proteins, associated with the proteolytic activity of endogenous enzymes and decarboxylase activity of wild microbiota, can support the accumulation of biogenic amines in fermented sausages [49,50], although the final balance depends on the equilibrium between BAs formation and degradation [51]. Figure 3 depicts the BAs content in *"Mortadella di Campotosto"*-like samples up to the end of ripening (45 days for MCB, panel a; 120 days for MCH, panel b).

In general, tryptamine, phenylethylamine, and spermine were not detected, while tyramine (TYR) and polyamines such as cadaverine (CAD), putrescine (PUT), and spermidine (SPD) were found during the entire production process, although with differences between the two products. Histamine was not detected in MCH, while it was found in MCB at low concentration (up to 17 mg/Kg, after the drying step). In addition, during fermentation (up to 11 days), TYR and SPD were the most abundant amines in MCB sausages and were detected in similar concentrations in MCB and MCH samples at up to 45 days of ripening. Tyramine production has been associated with the presence of LAB and enterococci that usually possess high amino acid decarboxylase activity [52]. This characteristic is strain-dependent and could be expressed during drying and ripening [53].

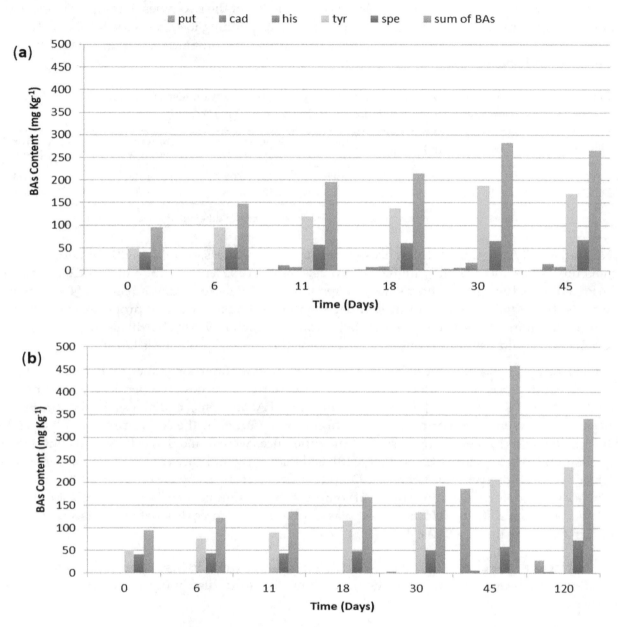

Figure 3. Putrescine (put), cadaverine (cad), histamine (his), tyramine (tyr), spermidine (spd), and sum of biogenic amines (as mg/Kg) content in "*Mortadella di Campotosto*"-like sausages produced with beef (panel (**a**), MCB) and hog (panel (**b**), MCH) casings, over time.

The sum of BAs content at the end of ripening resulted significantly different between the two types of products (265 ± 6 and 341 ± 23 mg/Kg, respectively, for MCB and MCH). These BAs levels are commonly found in other types of dry sausages produced by natural fermentation. Again, the differences in BAs content could be attributed to the higher water activity in MCH, as well as to the lower NaCl concentration due to scarce water loss, optimal for microbial development and BAs accumulation, in particular of tyramine and putrescine [54]. The significant reduction in total BAs at the end of ripening was associated with PUT decrease (more than 80%) in MCH samples, and with the decline of TYR, close to 40%, in MBH samples. Biogenic amines degradation was probably due to microorganisms possessing amino-oxidase enzymes, such as particular strains of the genera

Lactobacillus, Pediococcus, Micrococcus, as well as *Staphylococcus carnosus* [55]. The activity of these enzymes is pH-dependent and particularly active at pH values close to neutrality [56].

In the absence of a legal limit for biogenic amines content in dry fermented sausages, European Food Safety Authority (EFSA) stated that up to 50 mg of histamine and 60 mg of tyramine can be considered safe for healthy individuals; however, these limits fall dramatically if an individual takes anti-MAO drugs or is particularly sensitive to these amines [57]. Suzzi and Gardini [53] identified a sum of 200 mg/Kg of vasoactive amines (tyramine, histamine, tryptamine, and 2-phenylethylamine), as an indicator of good manufacturing procedure for fermented sausages. Among the investigated samples, only in MCH this limit was reached 45 days after the start of the manufacturing process and was still exceeded (although reduced) at the end of ripening.

4. Conclusions

The effect of different casing on some characteristics of dry fermented sausages produced at very low temperatures was investigated. This study demonstrated that *"Mortadella di Campotosto"*-like sausages did not show an intense proteolysis during fermentation and ripening, probably because production is carried out at very low temperatures. Nevertheless, the type of casing had a strong effect on ripening time, proteolysis, and production of some volatile compounds. On the other hand, although the presence of biogenic amines is considered as unavoidable in fermented meat products, the results highlighted that in *"Mortadella di Campotosto"*-like sausages processed with the traditional beef casing, the risk associated with the presence of bioactive amines is low. Furthermore, in addition to the better texture performance, when beef casing is used for this type of sausage, the process is significantly shorter in comparison to hog casing, with important positive effects on production costs for ripening and storage. For all these considerations, and despite its lower availability on the market, beef casings, traditionally used for this kind of product, are determinant for the final characteristics of this type of product.

Author Contributions: Conceptualization, A.P. and C.C.-L.; methodology, C.C.-L., A.S.; software, C.C.-L., G.S.; validation, A.S., C.C.-L. and A.P.; formal analysis, J.L., F.M., F.D., C.R.; investigation, J.L., F.M., F.D., C.R., M.M.; data curation, A.S., C.C.-L., A.P., M.M., G.S.; writing—original draft preparation, C.C.-L., A.S.; writing—review and editing, C.C.-L., A.S., A.P., G.S.; supervision, A.P.; project administration, A.P.; funding acquisition, A.P. All authors have read and agreed to the published version of the manuscript.

Acknowledgments: The authors gratefully thank Macelleria "L'Olmo", sited in Scanno (AQ), Italy for the helpfulness, and for having provided all the experimental samples.

References

1.   Toldrá, F. Characterization of proteolysis. In *Dry-Cured Meat Products*; Nip, W.K., Ed.; Wiley-Blackwell: Middletown, CT, USA, 2002; pp. 113–134. [CrossRef]

2.   Djordjevic, J.; Pecanac, B.; Todorovic, M.; Dokmanovic, M.; Glamoclija, N.; Tadic, V.; Baltic, M.Z. Fermented sausage casings. *Procedia Food Sci.* **2015,** *5,* 69–72. [CrossRef]

3.   Wu, Y.-C.; Chi, S.-P.; Christieans, S. Casings. In *Handbook of Fermented Meat and Poultry*; Toldrà, F., Ed.; Wiley Blackwell Publishing: Ames, IA, USA, 2010; pp. 86–96. [CrossRef]

4.   Pisacane, V.; Callegari, M.L.; Puglisi, E.; Dallolio, G.; Rebecchi, A. Microbial analyses of traditional Italian salami reveal microorganisms transfer from the natural casing to the meat matrix. *Int. J. Food Microbiol.* **2015,** *207,* 57–65. [CrossRef] [PubMed]

5.   AOAC. *Official Methods of Analysis of AOAC International*, 16th ed.; Association of Official Analytical Chemists: Gaithersburg, MD, USA, 1999.

6.   Park, Y.W. Moisture and water activity. In *Handbook of Processed Meats and Poultry Analysis*; Nollet, L.M.L., Toldrà, F., Eds.; CRC Press: Boca Raton, FL, USA, 2009; pp. 36–67.

7.   ISO 1841-2:1996. *Meat and Meat Products–Determination of Chloride Content. Part 1 Volhard Method*; ISO: Geneva, Switzerland, 1996.

8.   AOAC. *Official Method of Analysis of AOAC International*, 17th ed.; Association of Official Analytical Chemists: Gaithersburg, MD, USA, 2002.

9. Martuscelli, M.; Lupieri, L.; Chaves-Lopez, C.; Mastrocola, D.; Pittia, P. Technological approach to reduce NaCl content of traditional smoked dry-cured hams: Effect on quality properties and stability. *J. Food Sci. Technol.* **2015**, *52*, 7771–7782. [CrossRef] [PubMed]

10. Martín-Sánchez, A.M.; Chaves-López, C.; Sendra, E.; Sayas, E.; Fenández-López, J.; Pérez-Álvarez, J.Á. Lipolysis, proteolysis, and sensory characteristics of a Spanish fermented dry-cured meat product (salchichón) with oregano essential oil used as surface mold inhibitor. *Meat Sci.* **2011**, *89*, 35–44. [CrossRef] [PubMed]

11. Bradford, M.M. A rapid and sensitive method for the quantification of microgram quantities of protein utilizing the principle of protein-dye binding. *Anal. Biochem.* **1976**, *72*, 248–254. [CrossRef]

12. Laemmli, U.K. Cleavage of structural proteins during the assembly of the head of bacteriophage T4. *Nature* **1970**, *227*, 68–685. [CrossRef] [PubMed]

13. Folkertsma, B.; Fox, P.F. Use of the Cd-ninhydrin reagent to assess proteolysis in cheese during ripening. *J. Dairy Res.* **1992**, *59*, 217–224. [CrossRef]

14. Berardo, A.; Devreese, B.; De Maerè, H.; Stavropoulou, D.A.; Van Royen, G.; Leroy, F.; De Smet, S. Actin proteolysis during ripening of dry fermented sausages at different pH values. *Food Chem.* **2017**, *221*, 1322–1332. [CrossRef]

15. Chaves-López, C.; Paparella, A.; Tofalo, R.; Suzzi, G. Proteolytic activity of *Saccharomyces cerevisiae* strains associated with Italian dry-fermented sausages in a model system. *Int. J. Food Microbiol.* **2011**, *150*, 50–58. [CrossRef]

16. Chaves-López, C.; Serio, A.; Mazzarrino, G.; Martuscelli, M.; Scarpone, E.; Paparella, A. Control of household mycoflora in fermented sausages using phenolic fractions from olive mill wastewaters. *Int. J. Food Microbiol.* **2015**, *207*, 49–56. [CrossRef]

17. Martuscelli, M.; Pittia, P.; Casamassima, L.M.; Lupieri, L.; Neri, L. Effect of intensity of smoking treatment on the free amino acids and biogenic amines occurrence in dry cured ham. *Food Chem.* **2009**, *116*, 955–962. [CrossRef]

18. Bourne, M.C. Texture Profile Analysis. *Food Technol.* **1978**, *32*, 62–66.

19. Khan, M.I.; Arshad, M.S.; Anjum, F.M.; Sameen, A.; Aneeq ur, R.; Gill, W.T. Meat as a functional food with special reference to probiotic sausages. *Food Res. Int.* **2011**, *44*, 3125–3133. [CrossRef]

20. Soriano, A.; Cruz, B.; Gómez, L.; Mariscal, C.; García Ruiz, A. Proteolysis, physicochemical characteristics and free fatty acids composition of dry sausages made with deer (*Cervus elaphus*) or wild boar (*Sus scrofa*) meat: A preliminary study. *Food Chem.* **2006**, *96*, 173–184. [CrossRef]

21. Nakagawa, T.; Watabe, S.; Hashimoto, K. Identification of the three major components in fish sarcoplasmic proteins. *Nippon Suisan Gakk* **1988**, *54*, 999–1004. [CrossRef]

22. Dalmış, Ü.; Soyer, A. Effect of processing methods and starter culture (*Staphylococcus xylosus* and *Pediococcus pentosaceus*) on proteolytic changes in Turkish sausages (sucuk) during ripening and storage. *Meat Sci.* **2008**, *80*, 345–354. [CrossRef] [PubMed]

23. Toldrà, F. Proteolysis and lipolysis in flavour development of dry-cured meat products. *Meat Sci.* **1998**, *49*, S101–S110. [CrossRef]

24. Ladrat, C.; Verrez-Bagnis, V.; Noël, J.; Fleurence, J. In vitro proteolysis of myofibrillar and sarcoplasmic proteins of white muscle of sea bass (*Dicentrarchus labrax* L.): Effects of cathepsins B, D and L. *Food Chem.* **2003**, *81*, 517–525. [CrossRef]

25. Casaburi, A.; Blaiotta, G.; Mauriello, G.; Pepe, O.; Villani, F. Technological activities of *Staphylococcus carnosus* and *Staphylococcus simulans* strains isolated from fermented sausages. *Meat Sci.* **2005**, *71*, 643–650. [CrossRef]

26. Xiong, Y.L.; Blanchard, S.P. Dynamic gelling properties of myofibrillar protein from skeletal muscles of different chicken parts. *J. Agric. Food Chem.* **1994**, *42*, 670–674. [CrossRef]

27. Fadda, S.; López, C.; Vignolo, G. Role of lactic acid bacteria during meat conditioning and fermentation: Peptides generated as sensorial and hygienic biomarkers. *Meat Sci.* **2010**, *86*, 66–79. [CrossRef] [PubMed]

28. Saccani, G.; Fornelli, G.; Zanardi, E. Characterization of textural properties and changes of myofibrillar and sarcoplasmic proteins in salame Felino during ripening. *Int. J. Food Prop.* **2013**, *16*, 1460–1471. [CrossRef]

29. Toldrá, F. The role of muscle enzymes in dry-cured meat products with different drying conditions. *Trends Food Sci. Technol.* **2006**, *17*, 164–168. [CrossRef]

30. Claeys, E.; Uytterhaegen, L.; Buts, B.; Demeyer, D. Quantification of myofibrillar proteins by SDS-PAGE. *Meat Sci.* **1995**, *39*, 177–193. [CrossRef]

31. Demeyer, D.; Raemaekers, M.; Rizzo, A.; Holck, A.; De Smedt, A.; ten Brink, B.; Hagen, B.; Montel, C.;

Zanardi, E.; Murbrekk, E.; et al. Control of bioflavour and safety in fermented sausages: First results of a European Project. *Food Res. Int.* **2000**, *33*, 171–180. [CrossRef]

32. Roseiro, L.C.; Gomes, A.; Gonçalves, H.; Sol, M.; Cercas, R.; Santos, C. Effect of processing on proteolysis and biogenic amines formation in a Portuguese traditional dry-fermented ripened sausage "Chouriço Grosso de Estremoz e Borba PGI". *Meat Sci.* **2010**, *84*, 172–179. [CrossRef]

33. Di Cagno, R.; Chaves López, C.; Tofalo, R.; Gallo, G.; De Angelis, M.; Paparella, A.; Hammes, W.P.; Gobbetti, M. Comparison of the compositional, microbiological, biochemical and volatile profile characteristics of three Italian PDO fermented sausages. *Meat Sci.* **2008**, *79*, 224–235. [CrossRef]

34. Magistà, D.; Susca, A.; Ferrara, M.; Logrieco, A.F.; Perrone, G. *Penicillium* species: Crossroad between quality and safety of cured meat production. *Curr. Opin. Food Sci.* **2017**, *17*, 36–40. [CrossRef]

35. Benito, M.J.; Córdoba, J.J.; Alonso, M.; Asensio, M.A.; Nuñez, F. Hydrolytic activity of *Penicillium chrysogenum* Pg222 on pork myofibrillar proteins. *Int. J. Food Microbiol.* **2003**, *89*, 155–161. [CrossRef]

36. Beck, H.C.; Hansen, A.M.; Lauritsen, F.R. Catabolism of leucine to branched-chain fatty acids in *Staphylococcus xylosus*. *J. Appl. Microbiol.* **2004**, *96*, 1185–1193. [CrossRef]

37. Flores, M.; Olivares, A. Flavor. In *Handbook of Fermented Meat and Poultry*, 2nd ed.; Toldrà, F., Hui, Y.H., Astiasarán, I., Sebranek, J.G., Talon, R., Eds.; Wiley Blackwell: Hoboken, NJ, USA, 2014; pp. 217–225. [CrossRef]

38. Janssens, M.; Van der Mijnsbrugge, A.; Sánchez Mainar, M.; Balzarini, T.; De Vuyst, L.; Leroy, F. The use of nucleosides and arginine as alternative energy sources by coagulase-negative staphylococci in view of meat fermentation. *Food Microbiol.* **2014**, *39*, 53–60. [CrossRef] [PubMed]

39. Rimaux, T.; Vrancken, G.; Pothakos, V.; Maes, D.; De Vuyst, L.; Leroy, F. The kinetics of the arginine deiminase pathway in the meat starter culture *Lactobacillus sakei* CTC 494 are pH-dependent. *Food Microbiol.* **2011**, *28*, 597–604. [CrossRef] [PubMed]

40. Stalon, V.; Mercenier, A. L-arginine utilization by *Pseudomonas* species. *J. Gen. Microbiol.* **1984**, *130*, 69–76. [CrossRef] [PubMed]

41. Chen, Q.; Kong, B.; Han, Q.; Liu, Q.; Xu, L. The role of bacterial fermentation in the hydrolysis and oxidation of sarcoplasmic and myofibrillar proteins in Harbin dry sausages. *Meat Sci.* **2016**, *121*, 196–206. [CrossRef]

42. Toldrà, F.; Sanz, Y.; Flores, M. Meat fermentation technology. In *Meat Science and Application*; Hui, Y.H., Nip, W.K., Rogers, R., Eds.; CRC Press: Boca Raton, FL, USA, 2001. [CrossRef]

43. Ardö, Y. Flavour formation by amino acid catabolism. *Biotechnol. Adv.* **2006**, *24*, 238–242. [CrossRef]

44. Janssens, M.; Myter, N.; De Vuyst, L.; Leroy, F. Community dynamics of coagulase-negative staphylococci during spontaneous artisan-type meat fermentations differ between smoking and moulding treatments. *Int. J. Food Microbiol.* **2013**, *166*, 168–175. [CrossRef]

45. Barbut, S. Texture. In *Handbook of Fermented Meat and Poultry*; Toldrá, F., Hui, Y.H., Astiasarán, I., Nip, W.K., Sebranek, J.G., Silveira, E.T.F., Stahnke, L.H., Talon, R., Eds.; Blackwell Publishing: Ames, IA, USA, 2007; pp. 217–226. [CrossRef]

46. Ikonić, P.; Jokanović, M.; Petrović, L.; Tasić, T.; Škaljac, S.; Šojić, B.; Džinić, N.; Tomović, V.; Tomić, J.; Danilović, B.; et al. Effect of starter culture addition and processing method on proteolysis and texture profile of traditional dry-fermented sausage Petrovská klobása. *Int. J. Food Prop.* **2016**, *19*, 1924–1937. [CrossRef]

47. Feng, L.; Qiao, Y.; Zou, Y.; Huang, M.; Kang, Z.; Zhou, G. Effect of Flavourzyme on proteolysis, antioxidant capacity and sensory attributes of Chinese sausage. *Meat Sci.* **2014**, *98*, 34–40. [CrossRef]

48. Benito, M.J.; Rodríguez, M.; Cordoba, M.G.; Andrade, M.J.; Córdoba, J.J. Effect of the fungal protease EPg222 on proteolysis and texture in the dry fermented sausage 'salchichón'. *J. Sci. Food Agric.* **2005**, *85*, 273–280. [CrossRef]

49. Loizzo, M.R.; Spizzirri, U.G.; Bonesi, M.; Tundis, R.; Picci, N.; Restuccia, D. Influence of packaging conditions on biogenic amines and fatty acids evolution during 15 months storage of a typical spreadable salami ('Nduja). *Food Chem.* **2016**, *213*, 115–122. [CrossRef]

50. Paparella, A.; Tofalo, R. Fermented sausages: A potential source of biogenic amines. In *Biogenic Amines in Food: Analysis, Occurrence and Toxicity*; Saal, B., Tofalo, R., Eds.; The Royal Society of Chemistry: London, UK, 2020; pp. 103–118. [CrossRef]

51. Gardini, F.; Martuscelli, M.; Crudele, M.A.; Paparella, A.; Suzzi, G. Use of *Staphylococcus xylosus* as a starter culture in dried sausages: Effect on the biogenic amine content. *Meat Sci.* **2002**, *61*, 275–283. [CrossRef]

52. Serio, A.; Paparella, A.; Chaves López, C.; Corsetti, A.; Suzzi, G. *Enterococcus* populations in Pecorino Abruzzese cheese: Biodiversity and safety aspects. *J. Food Prot.* **2007**, *70*, 1561–1568. [CrossRef] [PubMed]

53.   Suzzi, G.; Gardini, F. Biogenic amines in dry fermented sausages: A review. *Int. J. Food Microbiol.* **2003**, *88*, 41–54. [CrossRef]

54.   Anastasio, A.; Draisci, R.; Pepe, T.; Mercogliano, R.; Delli Quadri, F.; Luppi, G.; Cortesi, M.L. Development of biogenic amines during the ripening of Italian dry sausages. *J. Food Prot.* **2010**, *73*, 114–118. [CrossRef] [PubMed]

55.   Lorenzo, J.M.; Sichetti Munekata, P.E.; Dominiguez, R. Role of autochthonous starter cultures in the reduction of biogenic amines in traditional meat products. *Curr. Opin. Food Sci.* **2017**, *14*, 61–65. [CrossRef]

56.   Martuscelli, M.; Crudele, M.A.; Gardini, F.; Suzzi, G. Biogenic amine formation and oxidation by *Staphylococcus xylosus* strains from artisanal fermented sausages. *Lett. Appl. Microbiol.* **2002**, *31*, 228–232. [CrossRef]

57.   EFSA. Scientific opinion on risk based control of biogenic amine formation in fermented foods. *EFSA J.* **2001**, *9*, 2393. [CrossRef]

12

The Occurrence of Biogenic Amines and Determination of Biogenic Amine-Producing Lactic Acid Bacteria in *Kkakdugi* and *Chonggak* Kimchi

Young Hun Jin, Jae Hoan Lee, Young Kyung Park, Jun-Hee Lee and Jae-Hyung Mah *

Department of Food and Biotechnology, Korea University, 2511 Sejong-ro, Sejong 30019, Korea;
younghoonjin3090@korea.ac.kr (Y.H.J.); jae-lee@korea.ac.kr (J.H.L.); eskimo@korea.ac.kr (Y.K.P.);
bory92@korea.ac.kr (J.-H.L.)
* Correspondence: nextbio@korea.ac.kr

Abstract: In this study, biogenic amine content in two types of fermented radish kimchi (*Kkakdugi* and *Chonggak* kimchi) was determined by high performance liquid chromatography (HPLC). While most samples had low levels of biogenic amines, some samples contained histamine content over the toxicity limit. Additionally, significant amounts of total biogenic amines were detected in certain samples due to high levels of putrefactive amines. As one of the significant factors influencing biogenic amine content in both radish kimchi, *Myeolchi-aekjoet* appeared to be important source of histamine. Besides, tyramine-producing strains of lactic acid bacteria existed in both radish kimchi. Through 16s rRNA sequencing analysis, the dominant species of tyramine-producing strains was identified as *Lactobacillus brevis*, which suggests that the species is responsible for tyramine formation in both radish kimchi. During fermentation, a higher tyramine accumulation was observed in both radish kimchi when *L. brevis* strains were used as inocula. The addition of *Myeolchi-aekjeot* affected the initial concentrations of histamine and cadaverine in both radish kimchi. Therefore, this study suggests that reducing the ratio of *Myeolchi-aekjeot* to other ingredients (and/or using *Myeolchi-aekjeot* with low biogenic amine content) and using starter cultures with ability to degrade and/or inability to produce biogenic amines would be effective in reducing biogenic amine content in *Kkakdugi* and *Chonggak* kimchi.

Keywords: kimchi; *Kkakdugi*; *Chonggak* kimchi; radish kimchi; biogenic amines; tyramine; lactic acid bacteria; *Lactobacillus brevis*

1. Introduction

Biogenic amines (BA) have been considered to be toxic compounds in foods. Several authors have proposed the maximum tolerable limits of some toxicologically important BA in foods as follows: histamine, 100 mg/kg; tyramine, 100–800 mg/kg; β-phenylethylamine, 30 mg/kg; total BA, 1000 mg/kg [1,2]. In addition, polyamines such as putrescine and cadaverine have been known to potentiate the toxicity of BA, especially histamine and tyramine, in foods, although they are less toxic [1]. Consumption of foods containing excessive BA may cause symptoms such as migraines, sweating, nausea, hypotension, and hypertension, unless human intestinal amine oxidases—such as monoamine oxidase (MAO), diamine oxidase (DAO), and polyamine oxidase (PAO)—quickly metabolize and detoxify BA [3]. Thus, it is important to know that, although relatively low levels of BA naturally exist in common foods, microbial decarboxylation of amino acids may sometimes lead to a significant increment of BA in fermented or contaminated foods [2]. In lactic acid fermented foods such as cheese and fermented sausage, some species of lactic acid bacteria (LAB) have been considered as producers of BA, particularly tyramine [4]. On the other hand, several reports have indicated that use of LAB starter cultures unable to produce BA may reduce BA accumulation during fermentation and storage [5,6].

Kimchi is a generic term of Korean traditional lactic fermented vegetables. According to Codex standard [7], for preparation of kimchi, salted Chinese cabbage (as a main ingredient) is mixed with seasoning paste consisting of red pepper powder, radish, garlic, green onion, and ginger, and then fermented properly, however, which, in reality, refers to *Baechu* kimchi. Alongside the Chinese cabbage, various vegetables such as radish, ponytail radish, cucumber, and green onion are also used as main ingredients of kimchi depending on kimchi varieties in Korea [8]. Among numerous kimchi varieties prepared with different vegetables, *Baechu* kimchi, *Kkakdugi* (diced radish kimchi), and *Chonggak* kimchi (ponytail radish kimchi) are the most popular varieties of kimchi in Korea [9]. In the meantime, for improving sensory quality of kimchi, various types of salted and fermented seafood (*Jeotgal*) and sauces thereof (*Aekjeot*) are usually used for kimchi preparation in Korea [10]. Particularly, *Myeolchi-jeotgal* (salted and fermented anchovy), *Saeu-jeotgal* (salted and fermented shrimp), *Myeolchi-aekjeot* (a sauce prepared from *Myeolchi-jeotgal*) are commonly used *Jeotgal* and *Aekjeot* [11]. As *Jeotgal* and *Aekjeot* contain high levels of proteins and amino acids, when kimchi is prepared with them, BA accumulation may occur during kimchi fermentation [12]. Hence, several authors have intensively investigated BA content and BA-producing LAB in *Baechu* kimchi [13–15]. On the other hand, there is a lack of study on BA content and BA-producing LAB in *Kkakdugi* and *Chonggak* kimchi, although the two types of radish kimchi are as popular as *Baechu* kimchi in Korea.

In this study, therefore, BA content in *Kkakdugi* and *Chonggak* kimchi was determined to evaluate BA-related risks. Several possible contributing factors to BA content, including physicochemical properties and microbial BA production, were also investigated in the study. Finally, fermentation of both radish kimchi was carried out to determine the most important bacterial species contributing to BA formation in the radish kimchi, employing LAB strains with distinguishable BA-producing activities as fermenting microorganisms. This is the first study describing that *Lactobacillus brevis* is the species responsible for tyramine formation in kimchi variety throughout fermentation period.

2. Materials and Methods

2.1. Sampling

Two types of radish kimchi (*Kkakdugi* and *Chonggak* kimchi) samples of five popular kimchi manufacturers made within 30 days were obtained from the retail markets. After arrival, samples were stored at 4 °C or immediately analyzed for BA content, physicochemical parameters, and microbial measurement.

2.2. Physicochemical Measurements

pH, acidity, salinity, and water activity of *Kkakdugi* and *Chonggak* kimchi samples were determined. The pH of the samples was determined by Orion 3-star Benchtop pH meter (Thermo Scientific, Waltham, MA, USA). Acidity and salinity were measured according to the AOAC method [16]. The water activity was determined by water activity meter (AquaLab Pre; Meter Group, Inc., Pullman, WA, USA).

2.3. Microbial Measurement, Isolation, and Identification of Strains

Lactic acid bacterial counts and total aerobic bacterial counts were determined on de Man, Rogosa, and Sharpe (MRS, Laboratorios Conda Co., Madrid, Spain) agar and Plate Count Agar (PCA, Difco, Becton Dickinson, Sparks, MD, USA). According to manufacturer's instructions, MRS agar was incubated at 37 °C for 48–72 h, and PCA at 37 °C for 24 h. After incubation, enumeration was carried out on plates with 30–300 colonies.

LAB strains were isolated on MRS agar. Individual colonies on MRS agar were randomly selected and streaked on the same media. The single colonies were transferred to MRS broth at 37 °C for 48–72 h. Then, the cultured broth was stored in the presence of 20% glycerol (*v/v*) at −80 °C. In *Kkakdugi* and

Chonggak kimchi samples, 130 and 120 LAB strains were isolated, respectively. The strains were identified by 16s rRNA gene sequence analysis with the universal bacterial primer pair (518F and 805R, Solgent Co., Daejeon, Korea).

2.4. BA Extraction from Samples and Bacterial Cultures for HPLC Analysis

BA extraction from *Kkakdugi* and *Chonggak* kimchi samples was conducted by the methods developed by Eerola et al. [17], with minor modification. The sample broth (5 g) was mixed with 20 mL of perchloric acid (0.4 M). The mixture was incubated at 4 °C for 2 h and centrifuged at $3000 \times g$ at 4 °C for 10 min. After collecting the supernatant, the pellet was extracted again with equal volumes of perchloric acid under the same conditions. The total volume of supernatant was adjusted to 50 mL with perchloric acid. The extract was filtered using Whatman paper no. 1 and stored before analysis.

BA extraction from bacterial cultures was carried out based on the procedures described by Ben-Gigirey et al. [18,19], with minor modification. A loopful of a strain was inoculated in 5 mL of BA production assay medium. The compositions of BA production assay medium are as follows: MRS broth with 0.5% of L-ornithine monohydrochloride, L-lysine monohydrochloride, L-histidine monohydrochloride monohydrate, and L-tyrosine disodium salt hydrate (all Sigma-Aldrich Chemical Co., St. Louis, MO, USA); 0.0005% of pyridoxal-HCl (Sigma-Aldrich); pH of the broth was adjusted to 5.8 by adding hydrochloride solution (2 M). After incubating the strain at 37 °C for 48 h, 100 μL of the culture was inoculated into the same broth and incubated under the same conditions. Subsequently, after being mixed with 0.4 M perchloric acid at a volume ratio of 1:9, the mixture was incubated at 4 °C for 2 h and stored before analysis.

2.5. Preparation of Standard Solutions for HPLC Analysis

Tryptamine, β-phenylethylamine hydrochloride, putrescine dihydrochloride, cadaverine dihydrochloride, histamine dihydrochloride, tyramine hydrochloride, spermidine trihydrochloride, and spermine tetrahydrochloride (all Sigma-Aldrich) were used for standard solutions, and 1,7-diaminoheptane (Sigma-Aldrich) was applied for an internal standard. The concentrations of all standard solutions were adjusted to 0, 10, 50, 100, and 1000 ppm.

2.6. Derivatization of Extracts and Standards

The procedures of derivatization of BA in the extract were carried out by the method developed by Eerola et al. [17]. Briefly, 200 μL of 2 M sodium hydroxide and 300 μL of saturated sodium bicarbonate were added to 1 mL of the extract/standard solutions. Then, 2 mL of 1% dansyl chloride solution (dissolved in acetone) was mixed with the solution and then incubated for 45 min at 40 °C in dark room. The incubated solution was mixed with 100 μL of 25% ammonium hydroxide and reacted for 30 min at room temperature. The volume of the sample solution was adjusted to 5 mL by adding acetonitrile. The sample solution was centrifuged at $3000 \times g$ for 5 min, and the supernatant was filtered by using a 0.2 μm-pore-size filter (Millipore Co., Bedford, MA, USA).

2.7. HPLC Analysis

HPLC analysis was carried out according to the procedure developed by Eerola et al. [17] and modified by Ben-Gigirey et al. [18]. YL9100 HPLC system equipped with YL9120 UV–vis detector (all Younglin, Anyang, Korea) was employed and the data were analyzed with Autochro-3000 data system (Younglin). For the gradient HPLC method, 0.1 M ammonium acetate (solvent A; Sigma-Aldrich) and HPLC-grade acetonitrile (solvent B; SK chemicals, Ulsan, Korea) were used as the mobile phases. The chromatographic separation was carried out using Nova-Pak C18 column (4 μm, 4.6×150 mm; Waters, Milford, MA, USA) held in 40 °C at a flow rate of 1 mL/min. The gradient elution mode was as follows; 50:50 (A:B) to 10:90 for 19 min, 50:50 at 20 min, isocratic with 50:50 before next analysis. The analysis was conducted at 254 nm, and 10 μL of the sample solution was injected.

The detection limits were within the range of 0.01 to 0.10 mg/kg for food matrices [20]. The validation parameters, including detection limits, of the analytical procedure used in the study were reported in our earlier study [20]. Figure S1 illustrates the procedure, from extraction to HPLC analysis, for BA analysis.

2.8. Fermentation of Two Types of Radish Kimchi: Kkakdugi and Chonggak Kimchi

For preparation of *Kkakdugi* and *Chonggak* kimchi, diced white radish (2 × 2 × 2 cm³) or halved ponytail radish were soaked in 10% w/v salt brine for 30 min, respectively. Then, each salted radish was rinsed with tap water three times and drained for 3 h. *Kkakdugi* and *Chonggak* kimchi samples were prepared in triplicate, as shown in Table 1, according to the standard recipes developed by the National Institute of Agricultural Sciences [21]. The salinity of all samples was adjusted to 2.5%. The *Kkakdugi* and *Chonggak* kimchi samples were divided into five experimental groups, respectively, based on the presence or absence of *Myeolchi-aekjeot* and *Saeu-jeotgal* and LAB inoculum. The experimental groups designed for the present study were B group ("Blank" samples prepared with neither *Myeolchi-aekjeot* and *Saeu-jeotgal* nor inoculum), C group ("Control" samples prepared with *Myeolchi-aekjeot* and *Saeu-jeotgal*, but without inoculum), PC group ("Positive Control" samples prepared with *Myeolchi-aekjeot* and *Saeu-jeotgal*, and inoculated with *L. brevis* JCM 1170 as a reference strain), LB group ("*L. brevis*" samples prepared with *Myeolchi-aekjeot* and *Saeu-jeotgal*, and inoculated with tyramine-producing *L. brevis* strains, i.e., KD3M5 strain for *Kkakdugi* and CG2M15 strain for *Chonggak* kimchi, respectively), and LP group ("*L. plantarum*" samples prepared with *Myeolchi-aekjeot* and *Saeu-jeotgal*, and inoculated with *L. plantarum* strains, i.e., KD3M15 strain for *Kkakdugi* and CG3M21 strain for *Chonggak* kimchi, respectively). The samples belonging to respective experimental groups were fermented at 25 °C for three days. Changes on the physicochemical and microbial properties, and BA content were measured in triplicate during fermentation.

Table 1. Ingredients used for preparation of *Kkakdugi* and *Chonggak* kimchi.

Ingredients (g)	Salted Radish	Red Pepper Powder	Garlic	Ginger	Sesame Seed	Sugar	Glutinous Rice Paste	Myeolchi-aekjeot	Saeu-jeotgal
Kkakdugi	100	3	3	1.5	1	2	5	2	2
Chonggak kimchi	100	3.5	3	1.5	0.5	1.5	4	2	2

2.9. Statistical Analyses

Statistical analyses were performed with Minitab statistical software version 12.11 (Minitab Inc. State College, PA, USA). The data were presented as means ± standard deviations of the three independent replicates. The mean values were compared by one-way analysis of variance (ANOVA) with Tukey's honest significant difference (HSD) test and a probability (*p*) values of less than 0.05 were considered statistically significant.

3. Results and Discussion

3.1. Determination of BA Content in Radish Kimchi: Kkakdugi and Chonggak Kimchi

As shown in Table 2, BA content in *Kkakdugi* and *Chonggak* kimchi samples produced by popular manufacturers in Korea was determined, and human health risk of BA in both radish kimchi was estimated based on the suggestions of both Ten Brink et al. [1] and Silla Santos [2]. In all the samples of *Kkakdugi* and *Chonggak* kimchi, low levels of tyramine (<100 mg/kg), tryptamine, β-phenylethylamine, spermidine, and spermine (<30 mg/kg) were detected, which are within safe levels for human consumption. However, one *Kkakdugi* sample (KD2) had 127.78 ± 26.78 mg/kg of histamine, which is over the toxicity limit (100 mg/kg) suggested by Ten Brink et al. [1]. Another *Kkakdugi* sample (KD5) contained putrescine and cadaverine at concentrations of 982.32 ± 19.42 mg/kg and 124.60 ± 108.78 mg/kg, respectively, consequently exceeding the 1000 mg/kg limit for total BA which is considered to provoke toxicity [2]. In *Chonggak* kimchi samples, 131.20 ± 7.90 mg/kg of histamine was detected in one sample (CG5), which also contained 853.7 ± 36.80 mg/kg of putrescine and 112.10 ± 3.60 mg/kg of cadaverine. The amounts of histamine and total BA in the sample were found to exceed toxicity limits. Meanwhile, the BA content detected in both types of radish kimchi samples varied widely in the present study, which is similar to respective BA levels in *Baechu* kimchi reported previously [13,22]. On the other hand, Mah et al. [12] reported lower concentrations of putrescine, cadaverine, histamine, tyramine, spermidine, and spermine in both *Kkakdugi* and *Chonggak* kimchi than those detected in the same kinds of kimchi used in this study. This may be due to the differences in manufacturing methods, main ingredients, and storage conditions between kimchi samples used in the present and previous studies [9]. In the meantime, Mah et al. [12] also reported that the amounts of tyramine and other BA increased during the ripening of *Baechu* kimchi. Therefore, although tyramine was detected at low levels in all the samples of *Kkakdugi* and *Chonggak* kimchi in the present study, the significance and risk of tyramine formation in both types of radish kimchi should not be overlooked.

Table 2. BA content in two types of radish kimchi samples: *Kkakdugi* and *Chonggak* kimchi.

Samples [2]	BA Content (mg/kg) [1]							
	Trp	Phe	Put	Cad	His	Tyr	Spd	Spm
KD1	ND [3]	ND	10.85 ± 1.17 [4]	2.57 ± 0.62	18.75 ± 1.16	2.97 ± 0.33	12.27 ± 0.98	0.56 ± 0.96
KD2	ND	1.93 ± 1.69	563.59 ± 45.64	ND	127.78 ± 26.78	14.73 ± 1.96	12.66 ± 2.75	ND
KD3	ND	ND	19.00 ± 2.00	6.10 ± 0.40	24.50 ± 4.00	10.80 ± 0.40	ND	ND
KD4	ND	0.86 ± 1.49	97.45 ± 77.05	3.15 ± 5.46	40.82 ± 29.05	21.67 ± 17.81	5.30 ± 4.85	3.10 ± 2.82
KD5	ND	15.24 ± 1.87	982.32 ± 19.42	124.60 ± 108.78	67.84 ± 17.46	76.95 ± 4.25	16.76 ± 0.87	1.48 ± 0.08
Average	ND	3.61 ± 6.55	334.64 ± 427.97	27.28 ± 54.44	55.94 ± 44.45	25.42 ± 29.59	9.40 ± 6.68	1.03 ± 1.31
CG1	ND	ND	8.97 ± 2.02	2.38 ± 2.12	38.61 ± 6.03	4.85 ± 4.60	9.22 ± 2.16	20.74 ± 3.47
CG2	ND	ND	3.89 ± 1.68	2.00 ± 0.77	8.24 ± 2.09	0.79 ± 0.69	8.27 ± 2.90	2.12 ± 0.53
CG3	12.30 ± 6.30	ND	175.10 ± 7.30	55.40 ± 2.80	46.30 ± 6.70	18.70 ± 2.40	7.70 ± 5.50	ND
CG4	9.10 ± 7.10	1.10 ± 1.00	303.70 ± 20.20	148.50 ± 9.00	69.30 ± 20.90	11.10 ± 2.20	6.10 ± 3.70	8.30 ± 5.60
CG5	23.70 ± 6.10	2.80 ± 1.20	853.70 ± 36.80	112.10 ± 3.60	131.20 ± 7.90	7.00 ± 2.20	14.00 ± 5.30	ND
Average	9.02 ± 9.86	0.78 ± 1.23	269.07 ± 349.93	64.08 ± 65.51	58.73 ± 46.02	8.49 ± 6.80	9.06 ± 2.99	6.23 ± 8.79

[1] Trp: tryptamine, Phe: β-phenylethylamine, Put: putrescine, Cad: cadaverine, His: histamine, Tyr: tyramine, Spd: spermidine, Spm: spermine; [2] KD: *Kkakdugi* (diced radish kimchi), CG: *Chonggak* kimchi (ponytail radish kimchi); [3] ND: not detected (<0.1 mg/kg); [4] mean ± standard deviation.

According to Tsai et al. [13], a high level of histamine in kimchi may result from the addition of salted and fermented fish products. *Myeolchi-aekjeot* is the most widely used salted and fermented fish product for the preparation of kimchi variety, and approximately 2–4% of *Kkakdugi* (on the basis of weight percent) and 2–5% of *Chonggak* kimchi, respectively, are commonly added to main ingredients during kimchi preparation [21,23–26]. *Saeu-jeotgal* is also added, alone or together with *Myeolchi-aekjeot*, to main ingredients of kimchi, but Mah et al. [12] reported that *Myeolchi-aekjeot* contains a significantly higher level of histamine (up to 1154.7 mg/kg) than *Saeu-jeotgal*. In this study, all radish kimchi samples were prepared with both *Myeolchi-aekjeot* and *Saeu-jeotgal* as ingredients. Altogether, the excessive level of histamine in several radish kimchi samples could be due to the amount of added *Myeolchi-aekjeot* with high histamine content. Unfortunately, the food labels of the samples used in this study just provided the list of ingredients.

An overdose of histamine may provoke undesirable symptoms such as a migraine, sweating, and hypotension [3]. In addition, high levels of putrescine and cadaverine can potentiate histamine toxicity by inhibiting intestinal diamine oxidase and histamine-N-methyltransferase [27] and potentially react with nitrites to form carcinogenic N-nitrosamines [28]. Taking this into account, although most *Kkakdugi* and *Chonggak* kimchi samples seem to be safe for consumption, the fact that several samples contained relatively high levels of putrescine and cadaverine in the present study indicates that it is necessary to monitor and reduce BA content, particularly histamine, putrescine, and cadaverine.

3.2. Physicochemical and Microbial Properties of Radish Kimchi: Kkakdugi and Chonggak Kimchi

To predict possible reasons as to why some samples of two types of radish kimchi contained higher levels of BA, pH, acidity, salinity, water activity (a_w), and lactic acid bacterial and total aerobic bacterial counts of *Kkakdugi* and *Chonggak* kimchi samples were determined. In *Kkakdugi* samples, the values of the parameters were as follows: pH, 4.16 ± 0.17 (minimum to maximum range of 3.94–4.41); acidity (%), 0.86 ± 0.31 (0.51–1.27); salinity (%), 3.36 ± 1.21 (1.40–4.50); a_w, 0.983 ± 0.003 (0.977–0.988); lactic acid bacterial counts, 8.52 ± 0.61 Log CFU/mL (7.88–9.38 Log CFU/mL); total aerobic bacterial counts, 8.37 ± 0.96 Log CFU/mL (6.83–9.32 Log CFU/mL). In case of *Chonggak* kimchi samples, the measured values were as follows: pH, 4.96 ± 1.17 (3.98–6.36); acidity (%), 0.71 ± 0.43 (0.19–1.10); salinity (%), 3.83 ± 1.67 (2.15–6.48); a_w, 0.984 ± 0.004 (0.979–0.991); lactic acid bacterial counts, 7.83 ± 0.48 Log CFU/mL (7.42–8.60 Log CFU/mL); total aerobic bacterial counts, 8.18 ± 1.07 Log CFU/mL (6.88–9.48 Log CFU/mL). The values are in accordance with those of previous reports [13,29]. Linear regression analysis was performed to determine the contributors influencing BA content. Results revealed weak correlations between physiochemical parameters, as well as microbial properties, and BA content (data not shown). Nonetheless, several reports have shown that physicochemical and microbial properties may affect BA content in fermented foods [2,30,31]. Altogether, the results indicate that, besides physicochemical and microbial properties, there are complex factors affecting BA content in both radish kimchi, for instance, kinds of salted and fermented fish products used for kimchi preparation as described above.

3.3. BA Production by LAB Strains Isolated from Radish Kimchi: Kkakdugi and Chonggak Kimchi

BA production by LAB strains isolated from *Kkakdugi* and *Chonggak* kimchi samples was examined to determine BA-producing LAB species in two types of radish kimchi. All the strains showed low production (below the detection limit) of tryptamine, β-phenylethylamine, putrescine, cadaverine, histamine, spermidine, and spermine. However, 39 strains (30%) of 130 LAB isolated from *Kkakdugi* samples produced higher levels of tyramine (287.23–386.17 µg/mL) than other strains (below the detection limit). Among the 120 LAB strains isolated from *Chonggak* kimchi, 16 strains (13%) also showed a stronger tyramine production capability (260.93–339.56 µg/mL), while other strains revealed lower capability (below the detection limit). In addition, the tyramine-producing LAB strains, which were isolated from either *Kkakdugi* or *Chonggak* kimchi samples, revealed a similar ability to produce tyramine, as described right above. Meanwhile, despite the low level of tyramine detected in all the samples of *Kkakdugi* and *Chonggak* kimchi, the fact that parts of LAB strains isolated from both radish kimchi samples were highly capable of producing tyramine supports that tyramine increment may occur during the ripening of the kimchi [12].

To further determine microorganisms responsible for BA formation in radish kimchi at species level, the strains were divided into two groups: (i) 55 tyramine-producing LAB strains (39 strains from *Kkakdugi*; 16 strains from *Chonggak* kimchi) and (ii) 195 LAB strains unable to produce BA. In the two groups, several strains were randomly selected and subsequently identified based on 16s rRNA sequencing analysis. Then, the selected strains able to produce tyramine were all identified as *L. brevis*, which indicates that the species is probably responsible for tyramine formation in both types of radish kimchi. On the other hand, the selected strains unable to produce BA were identified as *Leuconostoc (Leu.) mesenteroides*, *Weissella cibaria*, *W. paramesenteroides*, *L. pentosus*, and *L. plantarum*. The results are in agreement with previous reports in which *Leuconostoc*, *Weissella*, and *Lactobacillus* spp. were suggested to be responsible for kimchi fermentation [8,32]. Meanwhile, tyramine production by *L. brevis* in various fermented foods, including wine and fermented sausage, as well as *Baechu* kimchi, has been previously reported [14,33,34]. In the reports, tyramine production by *L. brevis* isolated from wine ranged from 441.6 to 1070.0 μg/mL, which is higher than that of the present study. On the contrary, *L. brevis* isolated from fermented sausage and *Baechu* kimchi produced tyramine at the range from 138.51 to 169.47 μg/mL and from 282 to 388 μg/mL, respectively, which are similar or lower than that of this study. In addition, several authors also isolated tyramine-producing *Leu. mesenteroides*, *W. cibaria*, and *W. paramesenteroides* from *Baechu* kimchi [14,15] and *L. plantarum* from wine [35]. Interestingly, as described right above, there are somewhat disparate results between the present and previous studies, which indicates that the strains belonging to the same species may possess different ability to produce tyramine especially depending upon the kinds of foods. Thus, microbial BA production in radish kimchi is likely determined at strain level, probably adapting to the respective food ecosystems, as suggested by previous reports [36,37]. Another implication is that the strains unable to produce BA isolated in the current study have potential as starter cultures for kimchi fermentation. Further investigations are needed to use them as starter cultures, which may involve tests to examine if the strains fulfill the criteria of starter culture, including the technical properties of strains, food safety requirements, and quality expectations [38].

3.4. Changes in Tyramine and Other BA Content during Fermentation of Radish Kimchi: Kkakdugi and Chonggak Kimchi

Fermentation of *Kkakdugi* and *Chonggak* kimchi was performed to investigate the influences of *Myeolchi-aekjeot* (together with *Saeu-jeotgal*) and LAB strains (particularly *L. brevis*) on BA content (especially tyramine) of both radish kimchi. Five groups of *Kkakdugi* and *Chonggak* kimchi samples were prepared based on the presence or absence of *Myeolchi-aekjeot* and types of LAB inocula. *L. brevis* strains of KD3M5 and CG2M15 with the highest tyramine production activity among the identified tyramine-producing LAB strains were used to see if the species is practically responsible for tyramine formation during fermentation of *Kkakdugi* and *Chonggak* kimchi. On the other hand, *L. plantarum* strains of KD3M15 and CG3M21 unable to produce BA were used for two reasons. (i) *L. plantarum*, like *L. brevis*, is predominant species in kimchi [39]. (ii) Differently from *L. brevis*, *L. plantarum* has been found to be negative for tyramine production in the present and previous studies [33,40,41].

As shown in Figures 1 and 2, changes in physicochemical and microbial properties of *Kkakdugi* and *Chonggak* kimchi during the fermentation for 3 days were similar with those of several previous reports [25,29,42]. In detail, the pH of all radish kimchi groups decreased during day 1 of fermentation, and stayed constantly thereafter. On the contrary, the counts of total aerobic bacteria and lactic acid bacteria, and the acidity of all radish kimchi groups increased during day 1 and day 2, respectively, and remained constantly thereafter, which indicates that an appropriate fermentation process of *Kkakdugi* and *Chonggak* kimchi took place. It is mention worthy that the initial pH of C, PC, LB, and LP groups of both radish kimchi was slightly higher than that of B group, which might be because the neutral pH of *Saeu-jeotgal* affected the pH values of the former groups [43]. Nonetheless, the initial acidity of all groups, belonging to either *Kkakdugi* or *Chonggak* kimchi, was similar to each other. The salinity of all radish kimchi groups decreased slightly during fermentation. According to Shin, Ann, and Kim [44], osmosis between radish and broth (containing seasoning paste) occurs during fermentation, which results in a steady reduction of salinity. Regardless of the drop in salinity, water activity of all radish kimchi groups was constant during fermentation. In addition, the initial counts of total aerobic bacteria and lactic acid bacteria of PC, LB, and LP groups inoculated with any of LAB strains were higher than those of B and C groups to be fermented naturally without any inocula, as expected.

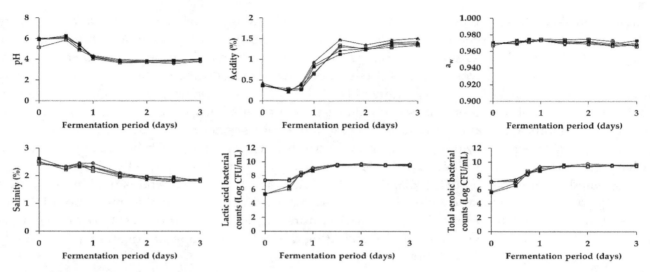

Figure 1. Changes in physicochemical and microbial properties of *Kkakdugi* during fermentation. □: B (no addition of *Myeolchi-aekjeot* and *Saeu-jeotgal*, no inoculum), ■: C (addition of *Myeolchi-aekjeot* and *Saeu-jeotgal*, no inoculum), ▲: PC (addition of *Myeolchi-aekjeot* and *Saeu-jeotgal*, *L. brevis* JCM 1170), △: LB (addition of *Myeolchi-aekjeot* and *Saeu-jeotgal*, *L. brevis* KD3M5), ○: LP (addition of *Myeolchi-aekjeot* and *Saeu-jeotgal*, *L. plantarum* KD3M15).

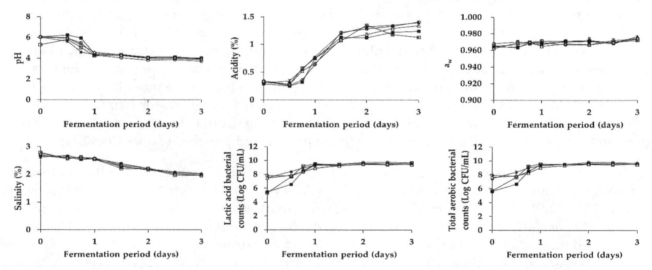

Figure 2. Changes in physicochemical and microbial properties of *Chonggak* kimchi during fermentation. □: B (no addition of *Myeolchi-aekjeot* and *Saeu-jeotgal*, no inoculum), ■: C (addition of *Myeolchi-aekjeot* and *Saeu-jeotgal*, no inoculum), ▲: PC (addition of *Myeolchi-aekjeot* and *Saeu-jeotgal*, *L. brevis* JCM 1170), △: LB (addition of *Myeolchi-aekjeot* and *Saeu-jeotgal*, *L. brevis* CG2M15), ○: LP (addition of *Myeolchi-aekjeot* and *Saeu-jeotgal*, *L. plantarum* CG3M21).

Changes in BA content (except for tryptamine and β-phenylethylamine not detected) during fermentation of *Kkakdugi* and *Chonggak* kimchi were shown in Figures 3 and 4, respectively. There appeared an increment of tyramine content in most groups (except for LP group) of both radish kimchi over the fermentation period, probably resulting from tyramine production by either inoculated or indigenous *L. brevis* strains (refer to Section 3.3). Also, the increment of tyramine content in PC and LB groups was higher than that in B and C groups of both radish kimchi (except for day 3 of *Chonggak* kimchi fermentation). This might be due to higher lactic acid bacterial counts of PC and LB groups, resulting from the inoculation of tyramine-producing *L. brevis* strains, than those of B and C groups of both radish kimchi. In the meantime, tyramine content in B and C groups of *Chonggak* kimchi steadily increased during fermentation, while that in the same groups of *Kkakdugi* increased slightly (but at a low level compared to *Chonggak* kimchi), both of which are likely associated with tyramine production by indigenous LAB strains (probably *L. brevis*). The observations are consistent with previous reports described right below. In short, Choi et al. [45] reported a dramatic increase of tyramine during natural fermentation of *Baechu* kimchi, whereas Kim et al. [46] reported that *Baechu* kimchi had a constantly low level of tyramine during natural fermentation. It is also noteworthy that, in the case of *Chonggak* kimchi, tyramine content in PC and LB groups dramatically increased during day 1 of fermentation, which was higher (and also showed a faster increment) than that in the same groups of *Kkakdugi*. The results, together with the comparison of tyramine content in B and C groups between two types of radish kimchi described above, can be explained by two speculations. The first is the difference in the ability of *L. brevis* strains to produce tyramine. The second is the distinguishable adaptation of the strains to different food ecosystems, i.e., differences in the main ingredients and/or ratio of ingredients in seasoning paste between two types of radish kimchi. Since KD3M5 strain served as an inoculum for *Kkakdugi* revealed a stronger ability to produce tyramine (377.35 ± 4.36 µg/mL) than CG2M15 strain for *Chonggak* kimchi (328.48 ± 2.61 µg/mL) when compared in vitro (refer to Section 3.3), the second speculation seems to be more probable than the first one. In addition, it is well known that bacteria produce BA to neutralize acidic environments as part of homeostatic regulation [47]. In this study, however, both radish kimchi samples of PC and LB groups showed similar patterns of acidity changes, so that the homeostatic regulation was excluded from possible reasons. Either way, there seem to be much complicated cross effects by the combinations of factors influencing the intensity of BA production by LAB during fermentation of kimchi variety. Interestingly, LP group of both radish kimchi had significantly lower levels of tyramine than the other groups. Thus, it seems that *L. plantarum* strains unable to produce BA in vitro not only have incapability of producing BA during fermentation, but also may inhibit tyramine production by indigenous LAB strains. This indicates the applicability of this species as a starter culture for reducing BA in kimchi variety.

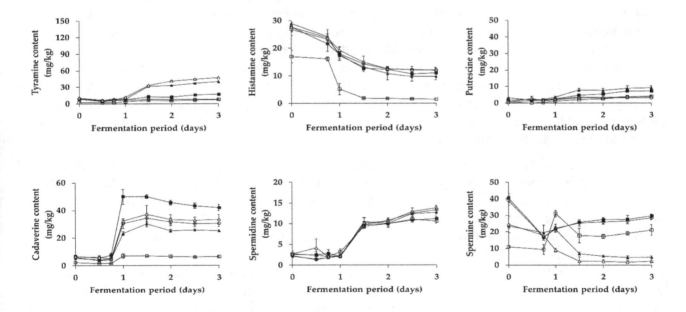

Figure 3. Changes in BA content in *Kkakdugi* during fermentation. □: B (no addition of *Myeolchi-aekjeot* and *Saeu-jeotgal*, no inoculum), ■: C (addition of *Myeolchi-aekjeot* and *Saeu-jeotgal*, no inoculum), ▲: PC (addition of *Myeolchi-aekjeot* and *Saeu-jeotgal*, *L. brevis* JCM 1170), △: LB (addition of *Myeolchi-aekjeot* and *Saeu-jeotgal*, *L. brevis* KD3M5), ○: LP (addition of *Myeolchi-aekjeot* and *Saeu-jeotgal*, *L. plantarum* KD3M15).

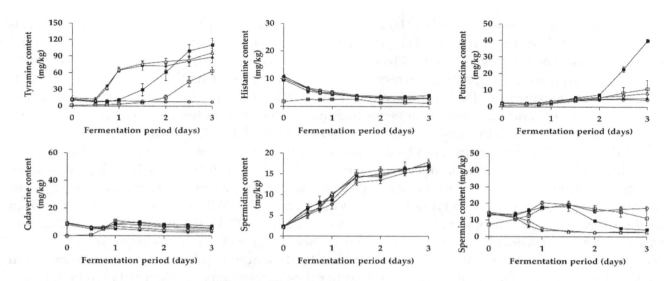

Figure 4. Changes in BA content in *Chonggak* kimchi during fermentation. □: B (no addition of *Myeolchi-aekjeot* and *Saeu-jeotgal*, no inoculum), ■: C (addition of *Myeolchi-aekjeot* and *Saeu-jeotgal*, no inoculum), ▲: PC (addition of *Myeolchi-aekjeot* and *Saeu-jeotgal*, *L. brevis* JCM 1170), △: LB (addition of *Myeolchi-aekjeot* and *Saeu-jeotgal*, *L. brevis* CG2M15), ○: LP (addition of *Myeolchi-aekjeot* and *Saeu-jeotgal*, *L. plantarum* CG3M21).

Differently from tyramine, histamine content in all groups of both radish kimchi gradually decreased during fermentation. This result might be because there were some indigenous LAB strains with histamine-degrading activity. Similarly, Kim et al. [48] reported a significant reduction of histamine content in *Baechu* kimchi inoculated with type strains of different LAB species including *L. sakei*, *L. plantarum*, *Leu. carnosum*, and *Leu. mesenteroides*, when compared with non-inoculated kimchi, suggesting that some LAB stains in kimchi are capable of degrading histamine. Meanwhile, the experimental groups of *Kkakdugi* and *Chonggak* kimchi prepared with *Myeolchi-aekjeot* (C, PC, LB, and LP groups) contained a significantly higher level of histamine than B group, which is in accordance with the suggestion of previous studies [12,22]. In the studies, the authors assumed that histamine level in *Baechu* kimchi could be affected by histamine in *Myeolchi-aekjeot*. Taking this into account, histamine content of *Kkakdugi* and *Chonggak* kimchi in the present study seems to come from *Myeolchi-aekjeot* rather than microbial histamine production during fermentation.

Putrescine and spermidine content steadily increased in all groups of *Kkakdugi* and *Chonggak* kimchi during fermentation, which is in agreement with previous reports [12,46]. There was a small and insignificant difference in putrescine and spermidine content among the groups of both radish kimchi during fermentation, which indicates that LAB strains—including *L. brevis* and *L. plantarum*—produced the polyamines during fermentation. Meanwhile, the initial concentrations putrescine and spermidine in *Kkakdugi* and *Chonggak* kimchi might be come from main ingredients, i.e., white radish and ponytail radish, respectively. In addition, a sharp increment of putrescine was observed during day 3 of fermentation, in the case of C group of *Chonggak* kimchi. To ignore the possibility of outliers, the fermentation experiment was repeatedly performed; however, the same results were observed, and the reason for such observation was not clear.

Somewhat differently from above, cadaverine content in all groups of *Kkakdugi* and *Chonggak* kimchi showed an increment during day 1 of fermentation and slight decline thereafter, although the increased cadaverine amount was mostly higher in *Kkakdugi* than in *Chonggak* kimchi. The difference in the intensity of cadaverine formation between two types of radish kimchi seems to be attributed to the complex combinations of factors described above to explain difference in the kinetics of tyramine formation between two radish kimchi. Interestingly, the initial cadaverine content in C, PC, LB, and LP groups of both radish kimchi was higher than that in B group, which might be come from *Myeolchi-aekjeot* rather than *Saeu-jeotgal*. The speculation is supported by a study by Cho et al. [22]

who reported a significantly higher level of cadaverine in *Myeolchi-aekjeot* (up to 263.6 mg/kg) than that in *Saeu-jeotgal* (up to 7.0 mg/kg). For both radish kimchi, C group contained the highest level of cadaverine, as compared to the other groups, over the fermentation period. This may be explained by a presumption that while cadaverine-producing bacteria derived from *Myeolchi-aekjeot* are probably responsible for cadaverine formation during fermentation of both radish kimchi, LAB strains (*L. brevis* and *L. plantarum*) served as inocula are probably capable of degrading cadaverine. Supporting this presumption, Mah et al. [49] reported that *Bacillus* strains isolated from *Myeolchi-jeotgal* were highly capable of producing cadaverine. Capozzi et al. [50] also reported that *L. plantarum* strains isolated from wine were capable of degrading cadaverine. At present, however, investigations on cadaverine-degrading activity of *L. brevis* are rarely found in literature.

As for change in spermine content, there appeared difference among groups of *Kkakdugi* and *Chonggak* kimchi. In PC and LB groups of both radish kimchi, a gradual decrease of spermine content was observed over the fermentation period, and the content was relatively lower than that in the other groups of both types of radish kimchi. This implies that *L. brevis* could be able to degrade spermine, although relevant reports are scarce to date. It is worth nothing that in B, C, and LP groups, spermine content decreased for day 1 of fermentation and slightly increased thereafter in *Kkakdugi*, whereas that in *Chonggak* kimchi increased for day 1 and slightly decreased thereafter. The different kinetics of spermine formation seems to result from the complex combinations of factors mentioned above. Therefore, it would be interesting in a future study to identify the factors (and combinations thereof) associated with BA formation or degradation by LAB strains during fermentation of *Kkakdugi* and *Chonggak* kimchi. The factors may involve time-related successional changes and/or interactions of microorganisms during fermentation as well as ingredients of foods and metabolic activities of strains [51]. In addition, recent studies suggested that results of in vitro BA production by food fermenting microorganisms were in disagreement with those of BA formation during fermentation of the corresponding foods [52,53]. In the present study, however, *L. brevis* was considered to be responsible for tyramine formation not only in vitro but also during practical fermentation of *Kkakdugi* and *Chonggak* kimchi.

4. Conclusions

The present study indicated that the amounts of BA in most samples of *Kkakdugi* and *Chonggak* kimchi were considered safe for consumption, but some samples contained histamine and total BA at concentrations over toxicity limits ($\geq$100 mg/kg and $\geq$1000 mg/kg, respectively). It was also found that, while *Myeolchi-aekjeot* seems to be an important source of histamine in both types of radish kimchi, *L. brevis* strains isolated from *Kkakdugi* and *Chonggak* kimchi are highly capable of producing tyramine in assay media. On the other hand, the physicochemical and microbial properties of both radish kimchi revealed weak correlations with BA content in the respective kimchi types in the present study. Through the practical fermentation of *Kkakdugi* and *Chonggak* kimchi, it turned out that *L. brevis* is responsible for tyramine formation, and *Myeolchi-aekjeot* influences histamine and cadaverine content in both radish kimchi. Consequently, this study suggests strategies for reducing BA in radish kimchi: the alteration of the ratio of ingredients used for kimchi preparation, particularly reducing ratio of *Myeolchi-aekjeot* to others, and use of starter cultures other than tyramine-producing *L. brevis* strains, especially BA-degrading LAB starter cultures. Studies on other contributing factors influencing the intensity of BA production by LAB are also required to understand complex kinetics of BA formation in the kimchi.

Author Contributions: Conceptualization: J.-H.M.; Investigation: Y.H.J., J.H.L., Y.K.P., and J.-H.L.; Writing—original draft: Y.H.J. and J.-H.M.; Writing—review and editing: Y.H.J. and J.-H.M.; Supervision: J.-H.M.

Acknowledgments: The authors thank Junsu Lee of Department of Food and Biotechnology at Korea University for technical assistance.

References

1.   Ten Brink, B.; Damink, C.; Joosten, H.M.L.J.; Huis In't Veld, J.H.J. Occurrence and formation of biologically active amines in foods. *Int. J. Food Microbiol.* **1990**, *11*, 73–84. [CrossRef]

2.   Silla Santos, M.H. Biogenic amines: Their importance in foods. *Int. J. Food Microbiol.* **1996**, *29*, 213–231. [CrossRef]

3.   Ladero, V.; Calles-Enríquez, M.; Fernández, M.; Alvarez, M.A. Toxicological effects of dietary biogenic amines. *Curr. Nutr. Food Sci.* **2010**, *6*, 145–156. [CrossRef]

4.   Marcobal, A.; De Las Rivas, B.; Landete, J.M.; Tabera, L.; Muñoz, R. Tyramine and phenylethylamine biosynthesis by food bacteria. *Crit. Rev. Food Sci. Nutr.* **2012**, *52*, 448–467. [CrossRef]

5.   Bover-Cid, S.; Hugas, M.; Izquierdo-Pulido, M.; Vidal-Carou, M.C. Reduction of biogenic amine formation using a negative amino acid-decarboxylase starter culture for fermentation of *Fuet* sausages. *J. Food Prot.* **2000**, *63*, 237–243. [CrossRef]

6.   Bover-Cid, S.; Hugas, M.; Izquierdo-Pulido, M.; Vidal-Carou, M.C. Effect of the interaction between a low tyramine-producing *Lactobacillus* and proteolytic *staphylococci* on biogenic amine production during ripening and storage of dry sausages. *Int. J. Food Microbiol.* **2001**, *65*, 113–123. [CrossRef]

7.   Codex Alimentarius Commission. *Codex Standard for kimchi, Codex Stan 223-2001*; Food and Agriculture Organization of the United Nations: Rome, Italy, 2001.

8.   Park, E.-J.; Chun, J.; Cha, C.-J.; Park, W.-S.; Jeon, C.O.; Bae, J.-W. Bacterial community analysis during fermentation of ten representative kinds of kimchi with barcoded pyrosequencing. *Food Microbiol.* **2012**, *30*, 197–204. [CrossRef]

9.   Cheigh, H.-S.; Park, K.-Y. Biochemical, microbiological, and nutritional aspects of kimchi (Korean fermented vegetable products). *Crit. Rev. Food Sci. Nutr.* **1994**, *34*, 175–203. [CrossRef]

10. Jang, K.-S.; Kim, M.-J.; Oh, Y.-A.; Kim, I.-D.; No, H.-K.; Kim, S.-D. Effects of various sub-ingredients on sensory quality of Korean cabbage kimchi. *J. Korean Soc. Food Nutr.* **1991**, *20*, 233–240.

11. Park, D.-C.; Park, J.-H.; Gu, Y.-S.; Han, J.-H.; Byun, D.-S.; Kim, E.-M.; Kim, Y.-M.; Kim, S.-B. Effects of salted-fermented fish products and their alternatives on angiotensin converting enzyme inhibitory activity of *Kimchi* during fermentation. *Korean J. Food Sci. Technol.* **2000**, *32*, 920–927.

12. Mah, J.-H.; Kim, Y.J.; No, H.-K.; Hwang, H.-J. Determination of biogenic amines in *kimchi*, Korean traditional fermented vegetable products. *Food Sci. Biotechnol.* **2004**, *13*, 826–829.

13. Tsai, Y.-H.; Kung, H.-F.; Lin, Q.-L.; Hwang, J.-H.; Cheng, S.-H.; Wei, C.-I.; Hwang, D.-F. Occurrence of histamine and histamine-forming bacteria in kimchi products in Taiwan. *Food Chem.* **2005**, *90*, 635–641. [CrossRef]

14. Kim, M.-J.; Kim, K.-S. Tyramine production among lactic acid bacteria and other species isolated from kimchi. *LWT-Food Sci. Technol.* **2014**, *56*, 406–413. [CrossRef]

15. Jeong, D.-W.; Lee, J.-H. Antibiotic resistance, hemolysis and biogenic amine production assessments of *Leuconostoc* and *Weissella* isolates for kimchi starter development. *LWT-Food Sci. Technol.* **2015**, *64*, 1078–1084. [CrossRef]

16. AOAC. *Official Methods of Analysis of AOAC International*, 17th ed.; AOAC International: Gaithersburg, MD, USA, 2000.

17. Eerola, S.; Hinkkanen, R.; Lindfors, E.; Hirvi, T. Liquid chromatographic determination of biogenic amines in dry sausages. *J. AOAC Int.* **1993**, *76*, 575–577. [PubMed]

18. Ben-Gigirey, B.; De Sousa, J.M.V.B.; Villa, T.G.; Barros-Velazquez, J. Changes in biogenic amines and microbiological analysis in albacore (*Thunnus alalunga*) muscle during frozen storage. *J. Food Prot.* **1998**, *61*, 608–615. [CrossRef]

19. Ben-Gigirey, B.; De Sousa, J.M.V.B.; Villa, T.G.; Barros-Velazquez, J. Histamine and cadaverine production by bacteria isolated from fresh and frozen albacore (*Thunnus alalunga*). *J. Food Prot.* **1999**, *62*, 933–939. [CrossRef]

20. Yoon, H.; Park, J.H.; Choi, A.; Hwang, H.-J.; Mah, J.-H. Validation of an HPLC analytical method for determination of biogenic amines in agricultural products and monitoring of biogenic amines in Korean fermented agricultural products. *Toxicol. Res.* **2015**, *31*, 299–305. [CrossRef]

21. National Institute of Agricultural Sciences. Available online: http://koreanfood.rda.go.kr/kfi/kimchi/kimchi_01 (accessed on 2 February 2019).

22. Cho, T.-Y.; Han, G.-H.; Bahn, K.-N.; Son, Y.-W.; Jang, M.-R.; Lee, C.-H.; Kim, S.-H.; Kim, D.-B.; Kim, S.-B.

Evaluation of biogenic amines in Korean commercial fermented foods. *Korean J. Food Sci. Technol.* **2006**, *38*, 730–737.

23. Kim, M.R.; Oh, Y.; Oh, S. Physicochemical and sensory properties of Kagdugi prepared with fermentation northern sand sauce during fermentation. *Korean J. Soc. Food Sci.* **2000**, *16*, 602–608.

24. Park, S.-O.; Kim, W.-K.; Park, D.-J.; Lee, S.-J. Effect of blanching time on the quality characteristics of elderly-friendly *kkakdugi*. *Food Sci. Biotechnol.* **2017**, *26*, 419–425. [CrossRef] [PubMed]

25. Kang, J.-H.; Kang, S.-H.; Ahn, E.-S.; Chung, H.-J. Quality properties of *Chonggak* kimchi fermented at different combination of temperature and time. *J. Korean Soc. Food Cult.* **2003**, *18*, 551–561.

26. Kim, Y.-J.; Jin, Y.-Y.; Song, K.-B. Study of quality change in Chonggak-kimchi during storage, for development of a freshness indicator. *Korean J. Food Preserv.* **2008**, *15*, 491–496.

27. Stratton, J.E.; Hutkins, R.W.; Taylor, S.L. Biogenic amines in cheese and other fermented foods: A review. *J. Food Prot.* **1991**, *54*, 460–470. [CrossRef]

28. Warthesen, J.J.; Scanlan, R.A.; Bills, D.D.; Libbey, L.M. Formation of heterocyclic *N*-nitrosamines from the reaction of nitrite and selected primary diamines and amino acids. *J. Agric. Food Chem.* **1975**, *23*, 898–902. [CrossRef] [PubMed]

29. Mheen, T.-I.; Kwon, T.-W. Effect of temperature and salt concentration on *Kimchi* fermentation. *Korean J. Food Sci. Technol.* **1984**, *16*, 443–450.

30. Lu, S.; Xu, X.; Shu, R.; Zhou, G.; Meng, Y.; Sun, Y.; Chen, Y.; Wang, P. Characterization of biogenic amines and factors influencing their formation in traditional Chinese sausages. *J. Food Sci.* **2010**, *75*, M366–M372. [CrossRef]

31. Özdestan, Ö.; Üren, A. Biogenic amine content of tarhana: A traditional fermented food. *Int. J. Food Prop.* **2013**, *16*, 416–428. [CrossRef]

32. Kim, M.; Chun, J. Bacterial community structure in kimchi, a Korean fermented vegetable food, as revealed by 16S rRNA gene analysis. *Int. J. Food Microbiol.* **2005**, *103*, 91–96. [CrossRef]

33. Landete, J.M.; Ferrer, S.; Pardo, I. Biogenic amine production by lactic acid bacteria, acetic bacteria and yeast isolated from wine. *Food Control* **2007**, *18*, 1569–1574. [CrossRef]

34. Latorre-Moratalla, M.L.; Bover-Cid, S.; Talon, R.; Garriga, M.; Zanardi, E.; Ianieri, A.; Fraqueza, M.J.; Elias, M.; Drosinos, E.H. Vidal-Carou, M.C. Strategies to reduce biogenic amine accumulation in traditional sausage manufacturing. *LWT-Food Sci. Technol.* **2010**, *43*, 20–25. [CrossRef]

35. Arena, M.E.; Fiocco, D.; Manca de Nadra, M.C.; Pardo, I.; Spano, G. Characterization of a *Lactobacillus plantarum* strain able to produce tyramine and partial cloning of a putative tyrosine decarboxylase gene. *Curr. Microbiol.* **2007**, *55*, 205–210. [CrossRef] [PubMed]

36. Bover-Cid, S.; Holzapfel, W.H. Improved screening procedure for biogenic amine production by lactic acid bacteria. *Int. J. Food Microbiol.* **1999**, *53*, 33–41. [CrossRef]

37. Landete, J.M.; Ferrer, S.; Polo, L.; Pardo, I. Biogenic amines in wines from three Spanish regions. *J. Agric. Food Chem.* **2005**, *53*, 1119–1124. [CrossRef] [PubMed]

38. Holzapfel, W.H. Appropriate starter culture technologies for small-scale fermentation in developing countries. *Int. J. Food Microbiol.* **2002**, *75*, 197–212. [CrossRef]

39. Lee, C.-H. Lactic acid fermented foods and their benefits in Asia. *Food Control* **1997**, *8*, 259–269. [CrossRef]

40. Moreno-Arribas, M.V.; Polo, M.C.; Jorganes, F.; Muñoz, R. Screening of biogenic amine production by lactic acid bacteria isolated from grape must and wine. *Int. J. Food Microbiol.* **2003**, *84*, 117–123. [CrossRef]

41. Park, S.; Ji, Y.; Park, H.; Lee, K.; Park, H.; Beck, B.R.; Shin, H.; Holzapfel, W.H. Evaluation of functional properties of lactobacilli isolated from Korean white kimchi. *Food Control* **2016**, *69*, 5–12. [CrossRef]

42. Kim, S.-D.; Jang, M.-S. Effects of fermentation temperature on the sensory, physicochemical and microbiological properties of *Kakdugi*. *J. Kor. Soc. Food Sci. Nutr.* **1997**, *26*, 800–806.

43. Um, M.-N.; Lee, C.-H. Isolation and identification of *Staphylococcus* sp. from Korean fermented fish products. *J. Microbiol. Biotechnol.* **1996**, *6*, 340–346.

44. Shin, Y.-H.; Ann, G.-J.; Kim, J.-E. The changes of hardness and microstructure of Dongchimi according to different kinds of water. *Korean J. Food Cookery Sci.* **2004**, *20*, 86–94.

45. Choi, Y.-J.; Jang, M.-S.; Lee, M.-A. Physicochemical changes in kimchi containing skate (*Raja kenojei*) pretreated with organic acids during fermentation. *Food Sci. Biotechnol.* **2016**, *25*, 1369–1377. [CrossRef] [PubMed]

46. Kim, S.-H.; Kang, K.H.; Kim, S.H.; Lee, S.; Lee, S.-H.; Ha, E.-S.; Sung, N.-J.; Kim, J.G.; Chung, M.J. Lactic

47. Arena, M.E.; Manca de Nadra, M.C. Biogenic amine production by *Lactobacillus*. *J. Appl. Microbiol.* **2001**, *90*, 158–162. [CrossRef]

48. Kim, S.-H.; Kim, S.H.; Kang, K.H.; Lee, S.; Kim, S.J.; Kim, J.G.; Chung, M.J. Kimchi probiotic bacteria contribute to reduced amounts of N-nitrosodimethylamine in lactic acid bacteria-fortified kimchi. *LWT-Food Sci. Technol.* **2017**, *84*, 196–203. [CrossRef]

49. Mah, J.-H.; Ahn, J.-B.; Park, J.-H.; Sung, H.-C.; Hwang, H.-J. Characterization of biogenic amine-producing microorganisms isolated from Myeolchi-Jeot, Korean salted and fermented anchovy. *J. Microbiol. Biotechnol.* **2003**, *13*, 692–699.

50. Capozzi, V.; Russo, P.; Ladero, V.; Fernández, M.; Fiocco, D.; Alvarez, M.A.; Grieco, F.; Spano, G. Biogenic amines degradation by *Lactobacillus plantarum*: Toward a potential application in wine. *Front. Microbiol.* **2012**, *3*, 122. [CrossRef]

51. Yılmaz, C.; Gökmen, V. Formation of tyramine in yoghurt during fermentation—Interaction between yoghurt starter bacteria and *Lactobacillus plantarum*. *Food Res. Int.* **2017**, *97*, 288–295. [CrossRef]

52. Nie, X.; Zhang, Q.; Lin, S. Biogenic amine accumulation in silver carp sausage inoculated with *Lactobacillus plantarum* plus *Saccharomyces cerevisiae*. *Food Chem.* **2014**, *153*, 432–436. [CrossRef]

53. Jeon, A.R.; Lee, J.H.; Mah, J.-H. Biogenic amine formation and bacterial contribution in *Cheonggukjang*, a Korean traditional fermented soybean food. *LWT-Food Sci. Technol.* **2018**, *92*, 282–289. [CrossRef]
acid bacteria directly degrade N-nitrosodimethylamine and increase the nitrite-scavenging ability in kimchi. *Food Control* **2017**, *71*, 101–109. [CrossRef]

13

Occurrence and Reduction of Biogenic Amines in Kimchi and Korean Fermented Seafood Products

Young Kyoung Park, Jae Hoan Lee and Jae-Hyung Mah *

Department of Food and Biotechnology, Korea University, 2511 Sejong-ro, Sejong 30019, Korea;
eskimo@korea.ac.kr (Y.K.P.); jae-lee@korea.ac.kr (J.H.L.)
* Correspondence: nextbio@korea.ac.kr

Abstract: Biogenic amines produced during fermentation may be harmful when ingested in high concentrations. As current regulations remain insufficient to ensure the safety of fermented vegetable products, the current study determined the risks associated with the consumption of kimchi by evaluating the biogenic amine concentrations reported by various studies. Upon evaluation, some kimchi products were found to contain histamine and tyramine at potentially hazardous concentrations exceeding the recommended limit of 100 mg/kg for both histamine and tyramine. The biogenic amines may have originated primarily from metabolic activity by microorganisms during fermentation, as well as from *Jeotgal* (Korean fermented seafood) and *Aekjeot* (Korean fermented fish sauce) products commonly used as ingredients for kimchi production. Many studies have suggested that *Jeotgal* and *Aekjeot* may contribute to the histamine and tyramine content in kimchi. Microorganisms isolated from kimchi and *Jeotgal* have been reported to produce both histamine and tyramine. Despite the potential toxicological risks, limited research has been conducted on reducing the biogenic amine content of kimchi and *Jeotgal* products. The regulation and active monitoring of biogenic amine content during kimchi production appear to be necessary to ensure the safety of the fermented vegetable products.

Keywords: kimchi; *Jeotgal*; *Aekjeot*; *Myeolchi-jeot*; *Myeolchi-aekjeot*; biogenic amines; recommended limits; occurrence; reduction; starter cultures

1. Introduction

Kimchi refers to a group of traditional Korean fermented vegetable products consumed worldwide [1]. Dating back to the 12th century during the Three Kingdoms period of ancient Korea, salted and fermented vegetable products represent the earliest form of kimchi, however, the addition of several ingredients such as the introduction of red peppers in the 16th century was eventually adopted for kimchi production [2]. The availability of local ingredients across different provinces in Korea led to the development of many regional kimchi varieties [3] (Figure 1). Currently, there are over 200 varieties of kimchi with over 100 different ingredients used for kimchi production [4]. Each kimchi variety is categorized according to the ingredients selected for production [5]. Kimchi in its current form has been recognized globally through international standardization as well [6]. Kimchi is prepared by trimming Napa cabbage, followed by salting, rinsing, and then draining excess water. The seasoning ingredients include red pepper powder, garlic, ginger, radish, glutinous rice paste, sugar, *Jeotgal*, and *Aekjeot*. The salted Napa cabbage is then mixed with the seasoning and stored at low temperatures (typically 0–10 °C in Korea [5]) to ferment until ripened [6]. While the production method described by the Codex only describes *Baechu* kimchi (Napa cabbage kimchi), slight variations are used to produce other kimchi varieties.

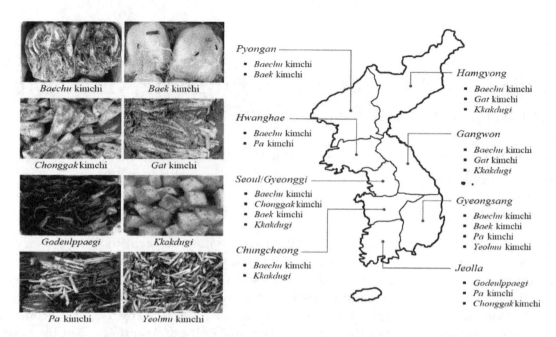

Figure 1. Kimchi varieties available across different provinces in Korea. *Baechu* kimchi: Napa cabbage kimchi; *Baek* kimchi: Napa cabbage kimchi prepared without red pepper powder; *Chonggak* kimchi: ponytail radish kimchi; *Gat* kimchi: mustard leaf kimchi; *Godeulppaegi*: Korean lettuce kimchi; *Kkakdugi*: diced radish kimchi; *Pa* kimchi: green onion kimchi; *Yeolmu* kimchi: young radish kimchi.

Nonetheless, nearly every kimchi variety benefits from preliminary brining, which inhibits the growth of pathogenic bacteria while selecting for lactic acid bacteria (LAB) known for promoting beneficial effects such as gastrointestinal regulation and prevention of colon cancer [7,8]. The LAB such as *Leuconostoc*, *Lactobacillus*, and *Weissella* species as well as the enzymes present in the ingredients are responsible for kimchi fermentation [9,10]. Consumption of kimchi is reported to provide numerous health benefits such as anti-oxidative, anti-carcinogenic, anti-mutagenic, and anti-aging effects [8,11,12].

Despite the numerous beneficial functional qualities, fermented foods such as kimchi may contain potentially harmful substances known as biogenic amines (BA). The nitrogenous compounds are mostly produced by microorganisms during fermentation through enzymatic decarboxylation of amino acids, as well as transamination of ketones and aldehydes [13]. BA are often categorized as aliphatic: putrescine, cadaverine, spermidine, spermine; aromatic: β-phenylethylamine, tyramine; heterocyclic: tryptamine, histamine [14,15]. The intake of BA at high concentrations as well as amine oxidase inhibition and deficiency may lead to toxic effects [16]. Recently, histamine, tyramine, putrescine, cadaverine, spermidine, and spermine were found to be cytotoxic toward human intestinal cells [17–19]. Furthermore, BA may also be converted to potentially carcinogenic N-nitrosamines in the presence of nitrites [20,21]. Excessive intake of foods containing high concentrations of histamine may potentially induce "scombroid poisoning" with symptoms such as headaches, hives, diarrhea, dyspnea, and hypotension [22]. Similarly, ingestion of foods with excessive tyramine content may cause a "cheese crisis" with symptoms that include severe headaches, hemorrhages, hypertensive effects or even heart failure [23]. As a result, many countries have implemented regulations on the production of histamine-rich seafood products, however many other food products are not currently regulated [24]. Several studies have suggested limits for BA content in food products of 100 mg/kg for histamine, 100–800 mg/kg for tyramine, 30 mg/kg for β-phenylethylamine, and 1000 mg/kg for total BA content [14,15]. The concentrations of BA in many fermented food products such as fermented meats and cheese have been widely reported to exceed limits for safe consumption. Similarly, BA have been detected in kimchi products, the most widely consumed traditional Korean food. High concentrations of BA have also been detected in kimchi ingredients *Jeotgal* (Korean fermented seafood) and *Aekjeot* (Korean fermented fish sauce), which contribute to the overall BA content in kimchi [25]. In addition, microorganisms isolated from kimchi as well as the fermented seafood products *Jeotgal* and *Aekjeot* have been reported to produce BA. Current regulations remain insufficient to address the potential health risks associated with the consumption of kimchi with high concentrations of BA. Therefore, the current article evaluated the risks associated with the BA content of kimchi products according to intake limits for β-phenylethylamine (30 mg/kg), histamine (100 mg/kg), and tyramine (100 mg/kg) as recommended by Ten Brink et al. [15], and reviewed potential sources of BA, and methods for reducing BA content.

2. Biogenic Amine Content in Kimchi Products

Table 1 displays the BA content of kimchi products as reported by various studies. The BA content of *Baechu* kimchi (Napa cabbage kimchi), the most popular kimchi variety consumed worldwide, has been reported by several studies. Cho et al. [25] reported histamine and tyramine concentrations in *Baechu* kimchi that exceeded recommended limits. Another study also showed that tyramine content in *Baechu* kimchi exceeded the recommended limit [26]. Tsai et al. [27] notably reported the highest histamine content which exceeded the recommended limit by a factor of 53. Tsai et al. [27] suggested that the high concentration of histamine in kimchi might be due to ingredients such as fish sauce or shrimp paste used in the kimchi production process. Shin et al. [28] reported β-phenylethylamine, histamine, and tyramine content at safe concentrations below 30 mg/kg. Similarly, Mah et al. [29] reported both histamine and tyramine content at safe concentrations below 30 mg/kg. In ripened *Baechu* kimchi, Kang et al. [26] reported tyramine content at concentrations that nearly reached the recommended limit.

Aside from *Baechu* kimchi, several studies have also reported the BA content of other kimchi varieties as well. *Chonggak* kimchi (ponytail radish kimchi) as reported by Jin et al. [30] contained histamine concentrations that exceeded the recommended limit. Tyramine content in *Chonggak* kimchi as reported by Kang et al. [26] were at safe concentrations, while Mah et al. [29] reported safe concentrations of both histamine and tyramine below recommended limits. As for *Gat* kimchi (mustard leaf kimchi), Lee et al. [31] reported histamine concentrations which exceeded the recommended limit by a factor of 2, while tyramine content slightly exceeded the limit. In contrast, Mah et al. [29] reported that *Gat* kimchi did not contain histamine and tyramine at detectable levels. *Kkakdugi* (diced radish kimchi) as reported by Jin et al. [30] contained tyramine at safe concentrations below the recommended limit, however, histamine concentrations exceeded the recommended limit. In contrast, Mah et al. [29] reported that histamine was not detected in *Kkakdugi*, and tyramine content was at safe concentrations below recommended limits. Similarly, Kang et al. [26] also reported tyramine concentrations in *Kkakdugi* below the recommended limit. As for *Pa* kimchi (green onion kimchi), Lee et al. [31] reported histamine and tyramine concentrations exceeded recommended limits by a factor of 4 and 2, respectively. In contrast, Mah et al. [29] reported histamine and tyramine in *Pa* kimchi at safe concentrations as tyramine content was not detected while histamine content remained below 30 mg/kg. Other kimchi varieties such as *Baek* kimchi (Napa cabbage kimchi prepared without red pepper powder), *Godeulppaegi* (Korean lettuce kimchi), and *Yeolmu* kimchi (young radish kimchi) were reported to contain histamine and tyramine at safe concentrations below 100 mg/kg [29].

Nonetheless, as the vast majority of studies are primarily focused upon *Baechu* kimchi, further research on the BA content of other kimchi varieties remains necessary. Currently, the severity of the risks associated with the BA content of kimchi remains difficult to thoroughly assess as limited research has been conducted. Though various BA have been detected in kimchi products, several studies have reported histamine and tyramine content at concentrations that exceeded the recommended intake limits of 100 mg/kg. Furthermore, the risk of nitrosamine formation entails the need for continuous monitoring of BA content during fermentation, especially as putrescine and cadaverine were detected at particularly high concentrations. Due to the toxicological risks associated with the consumption of BA, the content in kimchi necessitates regulation and control to ensure its safety.

Table 1. Biogenic amine content of Korean fermented vegetable products.

Korean Fermented Vegetable Products	N[1]	Biogenic Amines (mg/kg)[2]								Ref.
		TRP	PHE	PUT	CAD	HIS	TYR	SPD	SPM	
Baechu kimchi (Napa cabbage kimchi)	3	NT[3]	NT	11.2–89.0[4]	ND[5]–151.8	ND–5.1	ND–28.2	ND	ND	[29]
	20	2.3–22.6	ND–6.8	15.1–240.4	3.6–44.9	0.6–142.3	9.7–118.2	7.7–16.5	ND–3.7	[25]
	37	ND–114	ND	ND–73	ND–1550	ND–5350	ND–42	ND–88	ND–121	[27]
	18	ND–43.9	NT	ND–245.9	ND–63.3	NT	ND–103.6	ND–74.8	NT	[26]
	20	ND–74.8	ND–2.0	2.3–148.6	0.9–39.8	ND–21.8	1.1–27.9	ND–6.7	ND–5.1	[28]
Baek kimchi (Napa cabbage kimchi prepared without red pepper powder)	3	NT	NT	ND–54.7	ND–94.8	ND	ND	ND	ND	[29]
	3	tr[6]	NT	1.9–39.6	11.5–25.6	NT	7.8–64.9	ND–1.7	NT	[26]
Chonggak kimchi (ponytail radish kimchi)	3	2.3–15.2	NT	ND–11.2	ND–70.7	ND	20.2–58.1	ND	ND	[29]
	3	NT	NT	ND–20.3	ND–85.7	NT		ND	NT	[26]
	5	ND–23.70	ND–2.80	3.89–853.70	2.00–148.50	8.24–131.20	0.79–18.70	6.10–14.00	ND–20.74	[30]
Gat kimchi (mustard leaf kimchi)	3	NT	NT	ND–10.4	ND–11.6	ND	ND	ND	ND	[29]
	13	ND–26.74	ND–15.75	1.89–720.82	2.12–52.43	3.30–232.10	1.28–149.77	12.26–32.62	ND–61.94	[31]
Godeulppaegi (Korean lettuce kimchi)	3	NT	NT	ND–6.4	ND–26.7	ND	ND	ND	ND	[29]
Kkakdugi (diced radish kimchi)	3	NT	NT	ND–15.4	ND–55.1	ND	ND–9.0	ND	ND	[29]
	5	5.5–18.6	NT	ND–51.6	ND–56.2	NT	ND–10.8	ND–21.8	NT	[26]
	5	ND	ND–15.24	10.85–982.32	ND–124.60	18.75–127.78	2.97–76.95	ND–16.76	ND–3.10	[30]
Pa kimchi (green onion kimchi)	3	NT	NT	ND–7.8	ND–15.9	ND–21.7	ND	ND	ND	[29]
	13	ND–15.95	ND–5.97	ND–254.47	ND–123.29	8.67–386.03	ND–181.10	2.32–18.74	ND–33.84	[31]
Yeolmu kimchi (young radish kimchi)	3	NT	NT	ND	ND	ND	ND	ND	ND	[29]

[1] N: Number of samples examined; [2] TRP: tryptamine, PHE: β-phenylethylamine, PUT: putrescine, CAD: cadaverine, HIS: histamine, TYR: tyramine, SPD: spermidine, SPM: spermine; [3] NT: not tested; [4] Values are the minimum and maximum concentrations reported. The same number of digits is used after the decimal point in the values, as was presented in the corresponding references; [5] ND: not detected; [6] tr: trace.

3. Biogenic Amine Content of Other Vegetable Products

Research has also been conducted on the BA content of vegetable products originating from other countries (Table S1). The popular fermented food sauerkraut is produced through lactic acid fermentation of white cabbage [32,33]. Among European fermented food products, sauerkraut most closely resembles Korean kimchi [34]. Despite its popularity, Taylor et al. [35] reported that sauerkraut contained histamine concentrations that exceeded recommended limits. Ten Brink et al. [15] also reported that histamine and tyramine in sauerkraut exceeded recommended limits by a factor of 1 and 2, respectively. Many varieties of Japanese *Tsukemono* are preserved vegetables produced utilizing methods such as fermentation, salting, and pickling [36]. *Tsukemono* are differentiated based on ingredients, pickling method, and microorganisms responsible for fermentation [5]. Handa et al. [37] reported that histamine and tyramine in *Tsukemono* exceeded recommended limits by a factor of 3 and 4, respectively. As an important part of the Taiwanese diet, mustard pickle is prepared using mustard greens submerged in 14% NaCl brine for 4 months [38]. Kung et al. [38] reported that mustard pickles contained histamine and tyramine at safe concentrations below 100 mg/kg. Though fermented vegetable products are consumed worldwide, limited research has been conducted on the BA content of vegetable-based fermented foods. The few studies available had reported a wide range of BA content, including concentrations that exceeded recommended limits. Therefore, as the risks associated with the consumption of fermented vegetables remains largely undetermined, additional research is necessary to ensure the safe consumption of fermented foods.

4. Determinants for Biogenic Amine Content in Kimchi

4.1. Biogenic Amine Content of Kimchi Ingredients: Jeotgal and Aekjeot

Kimchi production involves the use of many ingredients including the fermented seafood products *Jeotgal* and *Aekjeot*. Used as seasoning ingredients during the production of kimchi [39], *Jeotgal* and *Aekjeot* contain flavor compounds that contribute greatly to the ripening process during kimchi fermentation [40]. Reports of the fermented seafood products as kimchi ingredients date back to the 16th century during the age of the *Chosun* dynasty of Korea [41]. Though *Jeotgal* and *Aekjeot* used during modern kimchi production vary by region, the most commonly used varieties include *Myeolchi-jeot* (salted and fermented anchovy), *Myeolchi-aekjeot* (salted and fermented anchovy sauce), *Saeu-jeot* (salted and fermented shrimp), and *Kkanari-aekjeot* (salted and fermented sand lance sauce) [42]. *Jeotgal* production typically involves submersion of seafood in brine with 20% salinity for 2–3 months at room temperature, and results in the final product resembling the initial seafood ingredient [43]. Some *Jeotgal* products undergo additional seasoning for consumption as side dishes rather than as ingredients during kimchi production [44,45]. Similarly, *Aekjeot* production involves the submersion of seafood in brine with salinity ranging from 20 to 30% for 1–2 years, however solid particles are removed through filtration for the final product [46]. In both *Jeotgal* and *Aekjeot*, the salt content inhibits putrefactive bacteria, and the enzymatic activity partially breaks down the proteins to develop a rich flavor [41]. Also, the addition of *Jeotgal* contributes to the protein, amino acid, and mineral content of kimchi, further reinforcing the nutritional value of kimchi products [5].

Despite the benefits described above, *Jeotgal* and *Aekjeot* have been reported to contain high concentrations of potentially hazardous BA such as histamine and tyramine [29]. Table 2 displays the BA content of the fermented seafood products. The reported BA content of *Aekjeot* and *Jeotgal* were evaluated according to recommended limits for intake. *Myeolchi-jeot* was reported to contain histamine and tyramine concentrations which exceeded recommended limits by a factor of approximately 6 and 2, respectively [47]. *Myeolchi-aekjeot* reportedly contained histamine and tyramine at concentrations that exceeded recommended limits by a factor of approximately 12 and 4, respectively [29]. The BA content of *Myeolchi-aekjeot* as studied by Cho et al. [25] showed β-phenylethylamine, histamine, and tyramine content at concentrations that exceeded recommended limits by a factor of about 2, 11, and 6, respectively. Moon et al. [48] also studied the BA content of *Myeolchi-aekjeot* by reporting β-phenylethylamine, histamine, and tyramine content at concentrations that exceeded recommended limits by a factor of approximately 3, 12, and 4, respectively. Similarly, Shin et al. [28] reported that *Myeolchi-aekjeot* contained β-phenylethylamine, histamine, and tyramine at concentrations that exceeded recommended limits by a factor of approximately 1, 4, and 4, respectively. Cho et al. [49] and Joung and Min [50] reported histamine concentrations in *Myeolchi-aekjeot* which greatly exceeded recommended limits by a factor of about 21 and 11, respectively.

As for *Kkanari-aekjeot*, histamine and tyramine content were reported at concentrations that exceeded recommended limits by a factor of approximately 10 and 2, respectively [29]. Cho et al. [25] also reported the β-phenylethylamine, histamine, and tyramine content in *Kkanari-aekjeot* at concentrations which exceeded recommended limits by a factor of 2, 11, and 6, respectively. Moon et al. [48] reported histamine and tyramine content at concentrations that exceeded recommended limits by a factor of about 7 and 3, respectively. Similarly, Shin et al. [28] reported that *Kkanari-aekjeot* contained β-phenylethylamine, histamine, and tyramine at concentrations that exceeded recommended limits by a factor of approximately 1, 10, and 3, respectively. Notably, the highest histamine content in *Kkanari-aekjeot* was reported by Cho et al. [49] as concentrations greatly exceeded the recommended limit by a factor of approximately 18.

As for *Saeu-jeot*, Mah et al. [47], Cho et al. [25], Moon et al. [48], and Shin et al. [28] reported BA content at safe concentrations below recommended limits for β-phenylethylamine, histamine, and tyramine, respectively.

Overall, the considerably high BA concentrations, especially histamine, reported for both retail *Jeotgal* and *Aekjeot* products may be potentially hazardous. All *Kkanari-aekjeot* and *Myeolchi-aekjeot* products contained histamine concentrations which exceeded 100 mg/kg indicating that safety regulations are necessary. According to Mah et al. [29], the high BA content may be due to the considerably long fermentation duration for the production of the fermented seafood. Furthermore, the results of the research conducted by Moon et al. [48] suggested that total BA content increased alongside crude protein concentrations for both *Jeotgal* and *Aekjeot*. After all, the high concentrations of BA reported for kimchi appears to originate partly from fish sauce such as *Myeolchi-aekjeot* and *Kkanari-aekjeot* [25]. Given the high concentrations of BA detected in kimchi and fermented seafood products, safety regulation and standardization of the manufacturing process appears to be necessary.

High BA concentrations were not limited to *Jeotgal* and *Aekjeot* products as the similar observations were reported for fermented seafood products originating from other countries (Table S1). Saaid et al. [51] studied the BA content of Malaysian seafood. The study showed that *Cincalok* (salted and fermented shrimp) contained histamine and tyramine at high concentrations that exceeded the recommended limits by a factor of approximately 3 and 7, respectively. *Budu* (salted and fermented anchovy) also contained high histamine and tyramine concentrations that exceeded recommended limits by a factor of 4 and 9, respectively. Similarly, research conducted by Rosma et al. [52] revealed histamine concentrations in *Budu* exceeding the recommended limit by a factor of 11.

The reported results indicated that fermented seafood products tended to contain high concentrations of BA, especially histamine. As the BA content exceeded well beyond recommended limits, consumption of the fermented seafood products may lead to adverse effects on human health. Due to the potential toxicological risks, expansion of current regulations regarding the BA content of seafood appears to be necessary to cover the aforementioned fermented seafood products as well as to include other amines such as tyramine and β-phenylethylamine.

Table 2. Biogenic amine content of Korean fermented seafood products.

Korean Fermented Seafood Products	N[1]	Biogenic Amines (mg/kg)[2]								Ref.
		TRP	PHE	PUT	CAD	HIS	TYR	SPD	SPM	
Myeolchi-jeot (salted and fermented anchovy)	3	NT[3]	NT	92–241[4]	ND[5]–665	155–579	63–244	ND–43	ND–77	[47]
Myeolchi-aekjeot (salted and fermented anchovy sauce)	4	NT	NT	86.1–178.9	ND	684.6–1154.7	222.6–383.1	ND–358.6	ND	[29]
	8	60.1–296.8	9.3–54.1	33.8–182.1	81.6–263.6	352.5–1127.6	93.9–611.3	4.7–27.1	1.9–12.2	[25]
	15	ND–382.2	ND–85.3	ND–680.0	ND–126.1	684.5–1205.0	77.5–381.1	ND	ND	[48]
	10	NT	NT	NT	NT	584.59–2070.58	NT	NT	NT	[49]
	12	NT	NT	NT	NT	150–1112	NT	NT	NT	[50]
	5	35.0–193.5	20.0–36.9	41.8–173.3	100.0–253.0	196.0–393.2	211.4–446.0	0.8–6.7	1.4–4.1	[28]
Kkanari-aekjeot (salted and fermented sand lance sauce)	4	NT	NT	55.5–136.4	ND–2.8	308.2–959.7	131.0–203.1	ND–30.9	ND	[29]
	8	62.0–187.2	10.4–51.7	1.6–311.6	52.1–314.8	215.4–1124.1	142.7–583.0	4.0–23.4	2.2–12.8	[25]
	16	ND–410.0	ND–17.9	ND–674.3	ND–96.9	308–732.2	112.3–328.0	ND	ND	[48]
	10	NT	NT	NT	NT	194.01–1839.68	NT	NT	NT	[49]
	5	122.5–242.5	18.3–32.5	30.8–43.8	52.5–168.3	183.4–1038.9	155.7–252.4	3.4–6.4	1.2–5.6	[28]
Saeu-jeot (salted and fermented shrimp)	2	NT	NT	ND	ND	ND	ND	ND	33–62	[47]
	5	5.3–10.6	ND–1.9	2.0–5.2	6.7–8.5	28.6–33.0	11.2–15.2	1.6–2.9	ND–0.7	[25]
	8	11.8–14.5	5.3–10.6	5.2–12.4	6.7–8.5	28.6–32.0	45.5	ND–45.5	ND	[48]
	5	3.3–8.1	ND–4.2	2.8–5.4	ND–1.5	2.3–12.7	1.4–7.4	ND–0.8	0.4–9.6	[28]

[1] N: Number of samples examined; [2] TRP: tryptamine, PHE: β-phenylethylamine, PUT: putrescine, CAD: cadaverine, HIS: histamine, TYR: tyramine, SPD: spermidine, SPM: spermine; [3] NT: not tested; [4] Values are the minimum and maximum concentrations reported. The same number of digits is used after the decimal point in the values, as was presented in the corresponding references; [5] ND: not detected.

4.2. Biogenic Amine Production by Bacterial Strains from Kimchi and Fermented Seafood Products

Microorganisms play a major role in the production of BA during fermentation through the decarboxylation of free amino acids. LAB responsible for fermentation have been reported to produce putrescine, cadaverine, histamine, and tyramine [15]. Table 3 displays the BA production by bacterial strains isolated from various kimchi and fermented seafood products. Tsai et al. [27] reported that LAB strains isolated from kimchi products purchased from Taiwanese markets were capable of producing histamine and other BA. The isolated strains identified as *Lactobacillus paracasei* subsp. *paracasei*, *Lb. brevis*, and *Brevibacillus brevis* were tested for β-phenylethylamine, putrescine, cadaverine, histamine, and spermine production in assay media. The reported results showed that *Lb. paracasei* subsp. *paracasei*, *Lb. brevis*, and *Bb. brevis* produced histamine at concentrations of 15.1, 13.6, and 16.3–43.1 μg/mL, respectively. Other BA were detected at concentrations lower than 15 μg/mL. Kim and Kim [53] isolated LAB strains from kimchi identified as *Lb. brevis*, *Lb. curvatus*, *Leuconostoc mesenteroides*, and *Staphylococcus hominis* that demonstrated tyramine production capabilities at over 200 μg/mL in assay media. Jeong and Lee [54] reported on putrescine, cadaverine, histamine, and tyramine production in assay media by LAB isolated from kimchi including *Leu. citreum*, *Leu. lactis*, *Leu. mesenteroides*, *Weissella cibaria*, *W. confusa*, and *W. paramesenteroides*. The results revealed that *Leuconostoc* spp. did not produce histamine and tyramine, however, putrescine and cadaverine were produced at concentrations lower than 20 μg/mL. *Weissella* spp. also produced putrescine and cadaverine at concentrations lower than 20 μg/mL, however, some strains produced histamine and tyramine at concentrations higher than 50 μg/mL. Compared to *Leuconostoc* spp., *Weissella* spp. produced a wider variety of BA at higher concentrations, prompting recommendations for stricter safety guidelines for screening starter *Weissella* strains suitable for kimchi fermentation [54].

Other varieties of kimchi were also reported to contain microorganisms capable of BA production. While the majority of the LAB strains isolated from *Chonggak* kimchi and *Kkakdugi* did not produce BA at detectable levels, some isolated LAB strains reportedly produced tyramine in the ranges of 260.93–339.56 μg/mL and 287.23–386.17 μg/mL, respectively, in BA production assay media [30]. Aside from tyramine, other BA were not detected in the same assay media. Although the study did not specify the bacterial species capable of producing BA, *Lb. brevis* was suggested as a strong producer of BA. Lee et al. [31] reported the BA production in assay media by LAB strains isolated from *Gat* kimchi and *Pa* kimchi. From *Gat* kimchi, *Enterococcus faecium*, *Lb. brevis*, and *Leu. mesenteroides* produced the highest concentrations of tyramine in the ranges of 259.10–269.57 μg/mL, ND–365.96 μg/mL, and 145.14–301.67 μg/mL, respectively. *Lb. brevis* strains also produced putrescine ranging from ND to 320.42 μg/mL. From *Pa* kimchi, the isolated LAB strains identified as *Lb. brevis* and *Lb. sakei* produced the highest concentration of BA such as tyramine in the ranges of ND-301.52 μg/mL and 113.98–131.36 μg/mL, respectively. Also, a *Lb. brevis* strain produced putrescine at 362.44 μg/mL. Aside from putrescine and tyramine, other BA produced by LAB strains were reported at concentrations lower than 60 μg/mL. Based on the reported BA production capabilities of isolated strains, LAB appear to contribute to the BA content in kimchi, especially tyramine which were produced at the highest concentrations.

Aside from LAB, other bacterial species isolated from *Jeotgal* products were reported to have BA production capabilities. *S. equorum* strains isolated from *Saeu-jeot* and *Myeolchi-jeot* were reported to be capable of producing putrescine, cadaverine, histamine, and tyramine in assay media [55,56]. The reported results showed that all BA were detected at concentrations below 50 μg/mL. Lim [57] isolated bacterial strains from *Myeolchi-jeot* which were identified as *Bacillus licheniformis*, *Serratia marcescens*, *S. xylosus*, *Aeromonas hydrophila*, and *Morganella morganii*, and the strains were capable of producing high concentrations of histamine in assay media at 1699.3 ± 35.6 μg/mL, 1987.2 ± 27.8 μg/mL, 2257 ± 30.7 μg/mL, 1655.5 ± 41.2 μg/mL, and 2869.4 ± 49.0 μg/mL, respectively. Mah et al. [58] suggested that *Bacillus* species, especially *B. licheniformis*, contributed towards BA content as the isolated strains isolated from *Myeolchi-aekjeot* were capable of producing putrescine, cadaverine, histamine, and tyramine. Thus, the isolated bacterial strains appear to contribute to the high histamine content of fermented seafood products, which in turn contribute to the BA content of kimchi.

Table 3. Biogenic amine production by bacterial strains isolated from Korean fermented vegetable and seafood products.

Korean Fermented Vegetable and Seafood Products	Strains	N [1]	Biogenic Amines (µg/mL) [2]								Ref.
			TRP	PHE	PUT	CAD	HIS	TYR	SPD	SPM	
	Lactobacillus paracasei subsp. *paracasei*	1	NT [3]	ND [4]	ND	0.3	15.1	NT	NT	4.5	[27]
	Lactobacillus brevis	1	NT	4.3	0.2	0.8	13.6	NT	NT	5.6	
	Brevibacillus brevis	2	NT	ND-3.8 [5]	ND-0.1	ND-11.2	16.3-43.1	NT	NT	6.8-8.8	
Baechu kimchi (Napa cabbage kimchi)	*Lactobacillus brevis*	6	NT	NT	NT	NT	NT	287-372	NT	NT	[53]
	Lactobacillus curvatus	4	NT	NT	NT	NT	NT	333-388	NT	NT	
	Leuconostoc mesenteroides	2	NT	NT	NT	NT	NT	282-322	NT	NT	
	Staphylococcus hominis	2	NT	NT	NT	NT	NT	287-296	NT	NT	
	Leuconostoc citreum	2	NT	NT	ND	18.1-18.2	ND	ND	NT	NT	[54]
	Leuconostoc lactis	4	NT	NT	15.6-16.2	ND	ND	ND	NT	NT	
	Leuconostoc mesenteroides	3	NT	NT	ND	17.6-19.1	ND	ND	NT	NT	
	Weissella cibaria	16	NT	NT	ND-17.7	ND-18.8	ND-72.9	ND-59.9	NT	NT	
	Weissella confusa	8	NT	NT	ND-17.1	ND-19.5	ND-73.3	ND-56.6	NT	NT	
	Weissella paramesenteroides	1	NT	NT	ND	ND	55.2	56.3	NT	NT	
Kkakdugi (diced radish kimchi)	Lactic acid bacteria	39	ND	ND	ND	ND	ND	287.23-386.17	ND	ND	[30]
Chonggak kimchi (ponytail radish kimchi)	Lactic acid bacteria	16	ND	ND	ND	ND	ND	260.93-339.56	ND	ND	
Pa kimchi (green onion kimchi)	*Lactobacillus brevis*	14	ND	ND-2.39	ND-362.44	ND-54.79	ND	ND-301.52	ND	ND	[31]
	Lactobacillus sakei	2	ND	1.00-3.96	ND	ND	ND	113.98-131.36	ND	ND	
Gat kimchi (mustard leaf kimchi)	*Enterococcus faecium*	2	ND	3.51-3.88	ND-320.42	ND-47.73	ND	259.10-269.57	ND	ND	
	Lactobacillus brevis	7	ND	ND-2.34	ND	ND	ND	ND-365.96	ND	ND	
	Leuconostoc mesenteroides	2	ND	1.47-1.91	ND	ND	ND	145.14-301.67	ND	ND	
Myeolchi-jeot (salted and fermented anchovy) Saeu-jeot (salted and fermented shrimp)	*Staphylococcus equorum*	39	NT	NT	ND-22.6	ND-29.6	ND-40.0	ND-29.7	NT	NT	[56]
Myeolchi-jeot (salted and fermented anchovy)	*Bacillus licheniformisr*	1	NT	NT	NT	NT	1699.3 ± 35.6 [6]	NT	NT	NT	[57]
	Serratia marcescens	1	NT	NT	NT	NT	1987.2 ± 27.8	NT	NT	NT	
	Staphylococcus xylosus	1	NT	NT	NT	NT	2257.4 ± 30.7	NT	NT	NT	
	Aeromonas hydrophila	1	NT	NT	NT	NT	1655.5 ± 41.2	NT	NT	NT	
	Morganella morganii	1	NT	NT	NT	NT	2869.4 ± 49.0	NT	NT	NT	

[1] N: Number of samples examined; [2] TRP: tryptamine, PHE: β-phenylethylamine, PUT: putrescine, CAD: cadaverine, HIS: histamine, TYR: tyramine, SPD: spermidine, SPM: spermine; [3] NT: not tested; [4] ND: not detected; [5] Values are the minimum and maximum concentrations reported. The same number of digits is used after the decimal point in the values, as was presented in the corresponding references; [6] mean ± standard deviation. The same number of digits is used after the decimal point in the values, as was presented in the corresponding references.

The aforementioned studies reported BA production by isolated strains at widely varying concentrations, even among the same species. Lee et al. [31] suggested that the BA production by LAB isolated from kimchi may be strain-dependent. Differences in BA production are widely considered to be strain-dependent, and not species-dependent [59]. The claim is further substantiated by the evidence for horizontal gene transfer for decarboxylase genes [60–62]. For example, as tyrosine decarboxylation was observed only for some strains, even belonging to the same species of LAB, tyramine production is considered strain-specific rather than species-specific [63]. Nonetheless, BA production by isolated strains indicates a risk for BA accumulation during *Jeotgal* and kimchi fermentation. Consequently, the control of BA accumulation during the production of fermented foods necessitates the reduction of microbial BA production by control of fermentation conditions, utilization of starter cultures, and sanitary practices to prevent contamination by BA-producing microorganisms.

5. Strategies to Reduce Biogenic Amine Content in Kimchi Products

Despite the risks associated with BA accumulation, limited research has been conducted on reducing the BA content of kimchi products. Instead of directly reducing BA content in kimchi, several studies have reported various methods to reduce BA concentrations in the kimchi ingredients *Jeotgal* and *Aekjeot*. Kim et al. [64] reported that kimchi produced using fermented seafood products contained BA at significantly higher concentrations. Lee et al. [65] suggested that the BA concentration of kimchi products may be reduced by limiting the quantity of the fermented seafood products used during kimchi production. For example, Kang [66] reported the histamine content of kimchi without *Myeolchi-aekjeot* at safe levels, however, the addition of *Myeolchi-aekjeot* raised histamine content to unsafe concentrations above the recommended limit by a factor of approximately 6. The study also described the effect of heat treatment of *Myeolchi-aekjeot* on the histamine content of kimchi. Histamine concentrations in kimchi produced using heat-treated *Myeolchi-aekjeot* were reported at 546.14 ± 1.33 mg/kg, while non-treated kimchi contained 592.78 ± 3.43 mg/kg. The reported results indicate that microorganisms from *Myeolchi-aekjeot* contributed towards the production of histamine during kimchi fermentation. Also, as research shows that histamine is heat-stable [67], the lower BA content in kimchi produced using the heat-treated *Myeolchi-aekjeot* may be due to the sterilization of histamine-producing microorganisms [66]. In addition to the contribution of BA content in kimchi by *Myeolchi-aekjeot*, Lee et al. [31] suggested that microorganisms from *Myeolchi-aekjeot* may produce BA during kimchi fermentation. Utilizing substitute ingredients in lieu of *Myeolchi-aekjeot* and *Kkanari-aekjeot* may also be effective in reducing BA content in kimchi. As other *Jeotgal* products including *Ojingeo-jeot* (salted and fermented sliced squid), *Toha-jeot* (salted and fermented *toha* shrimp), *Jogae-jeot* (salted and fermented clam), *Baendaengi-jeot* (salted and fermented big-eyed herring), and *Eorigul-jeot* (salted and fermented oysters) have been found to contain individual BA content below 100 mg/kg [47], utilization of the fermented seafood products with low BA content for kimchi production is expected to reduce the overall BA content of kimchi products [29].

Research on using additives to reduce the BA content of fermented seafood products has also been reported. Mah et al. [68] conducted research to reduce BA production by microorganisms isolated from *Myeolchi-jeot*, introducing additives into assay media and *Myeolchi-jeot*. The results confirmed that compared to the control, garlic extract was the most effective inhibitor of bacterial growth and BA production by yielding lower in vitro production of putrescine, cadaverine, histamine, tyramine, and spermidine by 11.2%, 18.4%, 11.7%, 30.9%, and 17.4%, respectively. Further results revealed that compared to *Myeolchi-jeot* samples treated with ethanol (control), the addition of 5% garlic extract to *Myeolchi-jeot* (treatment) inhibited bacterial growth and consequently reduced overall BA production by up to 8.7%. In another study by Mah and Hwang [69], other additives were also used for the reduction of BA production by *Myeolchi-jeot* microorganisms in assay media and *Myeolchi-jeot*. Among the additives tested in assay media, glycine most effectively inhibited in vitro BA production by bacterial strains. In comparison to the control without additives, the addition of 10% glycine in assay media resulted in reductions in putrescine, cadaverine, histamine, tyramine, and spermidine production by

32.6%, 78.4%, 93.2%, 100.0%, and 100.0%, respectively. Compared to the *Myeolchi-jeot* samples salted at 20% NaCl, additional supplementation of 5% glycine reportedly reduced overall BA content by 73.4%. The results suggest that the addition of glycine as well as salt may improve the safety of fermented seafood products. It is noteworthy that despite the results showing effective BA reduction, the use of garlic extract or glycine may affect the flavor of the final product.

Aside from additives, other studies have utilized starter cultures to reduce BA content in *Jeotgal*. In a study by Mah and Hwang [70], some bacterial strains isolated from *Myeolchi-jeot* were found to reduce BA content in *Myeolchi-jeot*. The reported results showed that, of the 7 starter candidate strains, *S. xylosus* exhibited the highest histamine degradation capability as well as the ability to slightly degrade tyramine in assay media. In comparison to the uninoculated *Myeolchi-jeot* control, the addition of the starter culture reduced the production of putrescine, cadaverine, histamine, tyramine, and spermidine by 16.5%, 10.8%, 18.0%, 38.9%, and 45.6%, respectively. Jeong et al. [56] isolated strains from *Jeotgal* for use as potential starters and found that *S. equorum* strain KS1039 did not produce putrescine, cadaverine, histamine, and tyramine in vitro.

A limited number of studies have even attempted to directly reduce the BA content of kimchi through the inoculation of bacterial strains. Kim et al. [71] reported reductions in tryptamine, putrescine, cadaverine, histamine, and tyramine levels in *Baechu* kimchi fortified with *Leu. carnosum*, *Leu. mesenteroides*, *Lb. plantarum*, and *Lb. sakei* strains. Similarly, Jin et al. [30] reported that *Kkakdugi* and *Chonggak* kimchi inoculated with *Lb. plantarum* strains incapable of producing BA contained lower level of tyramine (but not the other BA) than the uninoculated control. Therefore, utilizing LAB strains unable to produce (and/or able to degrade) BA as kimchi starter cultures may likely reduce the total BA content during kimchi fermentation.

Although the aforementioned studies have shown both direct and indirect methods of reducing BA content in kimchi, current commercial kimchi production processes do not appear to utilize the BA reduction techniques. This might be due to the application of BA reduction methods such as the use of additives, starter cultures, and adjusting the quantity of fermented seafood products have been reported to affect the flavor of kimchi products [69,72,73]. Consequently, inconsistent product quality is reflected in the wide range of BA content of kimchi products, including concentrations that exceed recommended limits for safe consumption. The high BA content reported for various kimchi products indicates that modern production methods require further preventative measures to ensure the safety of the fermented vegetable products, including practical application of research-based BA reduction techniques described above. Commercial kimchi production may greatly benefit from utilizing the aforementioned and novel strategies including control of fermentation conditions, utilizing starter cultures, alternative ingredients, and/or ingredients with low BA content. Furthermore, the establishment and expansion of regulations limiting BA content in fermented foods remain necessary to safeguard consumers against the potential BA intoxication.

6. Conclusions

The current study evaluated the BA content of kimchi, a term used to describe a group of Korean fermented vegetable products. Some kimchi samples have been reported to contain high concentrations of BA which exceeded recommended limits. Consumption of the fermented foods with high BA content may have detrimental effects on the body. Several factors contribute to the high BA concentrations in kimchi, which include BA production by microorganisms during fermentation and BA content of ingredients such as *Jeotgal* and *Aekjeot*. As variables such as ingredients, microorganisms, and initial BA content of *Jeotgal* that influence kimchi fermentation differed extensively, the reported BA concentrations of kimchi products also varied widely, even among the same varieties. Due to the large variations among kimchi products, standardization of kimchi production appears to be necessary to limit BA content. Furthermore, though several studies have described methods to indirectly reduce BA concentrations in kimchi by reducing the BA content of ingredients *Jeotgal* and *Aekjeot*, limited research has been conducted on the direct reduction of BA content in kimchi products.

To ensure the safe consumption of kimchi products, further research on methods to reduce the BA concentrations below recommended limits appears to be necessary. In conjunction with BA reduction studies, implementation of regulations such as continuous monitoring during production remains necessary to control BA content in kimchi and *Jeotgal* products.

Author Contributions: Conceptualization, Y.K.P. and J.-H.M.; Literature data collection, Y.K.P.; Writing—original draft, Y.K.P. and J.H.L.; Writing—review and editing, Y.K.P., J.H.L. and J.-H.M.; Supervision: J.-H.M.

Acknowledgments: The authors thank Young Hun Jin, Junsu Lee, and Alixander Mattay Pawluk of Department of Food and Biotechnology at Korea University for technical assistance.

References

1. Food Information Statistics System. Available online: https://www.atfis.or.kr/article/M001050000/view.do?articleId=2821&boardId=3&page=&searchKey=&searchString=&searchCategory= (accessed on 3 October 2019).
2. Lee, C.-H.; Ahn, B.-S. Literature review on Kimchi, Korean fermented vegetable foods I. History of Kimchi making. *Korean J. Food Cult.* **1995**, *10*, 311–319.
3. Cheigh, H.-S. *Kimchi Culture and Dietary Life in Korea*, 1st ed.; Hyoil Publishing Co.: Seoul, Korea, 2002; pp. 304–312.
4. Choi, S.-K.; Hwang, S.-Y.; Jo, J.-S. Standardization of kimchi and related products (3). *Korean J. Food Cult.* **1997**, *12*, 531–548.
5. Sim, S.G.; Shon, H.S.; Sim, C.H.; Yoon, W.H. *Fermented Foods*, 1st ed.; Jin Ro Publishing Co.: Seoul, Korea, 2001; pp. 233–286.
6. Codex Alimentarius Commission. *Codex Standard for Kimchi, Codex Stan 223–2001*; Food and Agriculture Organization of the United Nations: Rome, Italy, 2001.
7. Park, K.-Y. The nutritional evaluation, and antimutagenic and anticancer effects of Kimchi. *Korean J. Food Nutr.* **1995**, *24*, 169–182.
8. Park, K.-Y. Increased health functionality of fermented foods. *Korean J. Food Nutr.* **2012**, *17*, 1–8.
9. Jung, E.H.; Ryu, J.P.; Lee, S.-I. A study on foreigner preferences and sensory characteristics of kimchi fermented for different periods. *Korean J. Food Cult.* **2012**, *27*, 346–353. [CrossRef]
10. Jung, J.Y.; Lee, S.H.; Jeon, C.O. Kimchi microflora: History, current status, and perspectives for industrial kimchi production. *Appl. Microbiol. Biotechnol.* **2014**, *98*, 2385–2393. [CrossRef]
11. Cho, E.-J.; Rhee, S.-H.; Lee, S.-M.; Park, K.-Y. In vitro antimutagenic and anticancer effects of kimchi fractions. *J. Cancer Prev.* **1997**, *2*, 113–121.
12. Kim, J.-H.; Ryu, J.-D.; Song, Y.-O. The effect of *kimchi* intake on free radical production and the inhibition of oxidation in young adults and the elderly people. *Korean J. Community Nutr.* **2002**, *7*, 257–265.
13. Askar, A.; Treptow, H. *Biogene Amine in Lebensmitteln: Vorkommen, Bedeutung und Bestimmung*, 1st ed.; Verlag Eugen Ulmer: Stuttgart, Germany, 1986; pp. 21–74.
14. Silla Santos, M.H. Biogenic amines: Their importance in foods. *Int. J. Food Microbiol.* **1996**, *29*, 213–231. [CrossRef]
15. Ten Brink, B.; Damink, C.; Joosten, H.M.L.J.; Huis in't Veld, J.H.J. Occurrence and formation of biologically active amines in foods. *Int. J. Food Microbiol.* **1990**, *11*, 73–84. [CrossRef]
16. Gilbert, R.J.; Hobbs, G.; Murray, C.K.; Cruickshank, J.G.; Young, S.E.J. Scombrotoxic fish poisoning: Features of the first 50 incidents to be reported in Britain (1976–9). *Br. Med. J.* **1980**, *281*, 71–72. [PubMed]
17. Del Rio, B.; Redruello, B.; Linares, D.M.; Ladero, V.; Fernandez, M.; Martin, M.C.; Ruas-Madiedo, P.; Alvarez, M.A. The dietary biogenic amines tyramine and histamine show synergistic toxicity towards intestinal cells in culture. *Food Chem.* **2017**, *218*, 249–255. [CrossRef] [PubMed]
18. Del Rio, B.; Redruello, B.; Linares, D.M.; Ladero, V.; Ruas-Madiedo, P.; Fernandez, M.; Martin, M.C.; Alvarez, M.A. Spermine and spermidine are cytotoxic towards intestinal cell cultures, but are they a health hazard at concentrations found in foods? *Food Chem.* **2018**, *269*, 321–326. [CrossRef] [PubMed]
19. Del Rio, B.; Redruello, B.; Linares, D.M.; Ladero, V.; Ruas-Madiedo, P.; Fernandez, M.; Martin, M.C.; Alvarez, M.A. The biogenic amines putrescine and cadaverine show in vitro cytotoxicity at concentrations that can be found in foods. *Sci. Rep.* **2019**, *9*, 120. [CrossRef] [PubMed]

20. Ender, F.; Čeh, L. Conditions and chemical reaction mechanisms by which nitrosamines may be formed in biological products with reference to their possible occurrence in food products. *Z. Lebensm. Unters. Forsch.* **1971**, *145*, 133–142. [CrossRef]

21. Mah, J.-H.; Yoon, M.-Y.; Cha, G.-S.; Byun, M.-W.; Hwang, H.-J. Influence of curing and heating on formation of N-nitrosamines from biogenic amines in food model system using Korean traditional fermented fish product. *Food Sci. Biotechnol.* **2005**, *14*, 168–170.

22. Taylor, S.L. Histamine food poisoning: Toxicology and clinical aspects. *Crit. Rev. Toxicol.* **1986**, *17*, 91–128. [CrossRef]

23. Smith, T.A. Amines in food. *Food Chem.* **1981**, *6*, 169–200. [CrossRef]

24. Mah, J.-H.; Park, Y.K.; Jin, Y.H.; Lee, J.-H.; Hwang, H.J. Bacterial production and control of biogenic amines in Asian fermented soybean foods. *Foods* **2019**, *8*, 85. [CrossRef]

25. Cho, T.-Y.; Han, G.-H.; Bahn, K.-N.; Son, Y.-W.; Jang, M.-R.; Lee, C.-H.; Kim, S.-H.; Kim, D.-B.; Kim, S.-B. Evaluation of biogenic amines in Korean commercial fermented foods. *Korean J. Food Sci. Technol.* **2006**, *38*, 730–737.

26. Kang, K.H.; Kim, S.H.; Kim, S.-H.; Kim, J.G.; Sung, N.-J.; Lim, H.; Chung, M.J. Analysis and risk assessment of N-nitrosodimethylamine and its precursor concentrations in Korean commercial kimchi. *J. Korean Soc. Food Sci. Nutr.* **2017**, *46*, 244–250. [CrossRef]

27. Tsai, Y.-H.; Kung, H.-F.; Lin, Q.-L.; Hwang, J.-H.; Cheng, S.-H.; Wei, C.-I.; Hwang, D.-F. Occurrence of histamine and histamine-forming bacteria in kimchi products in Taiwan. *Food Chem.* **2005**, *90*, 635–641. [CrossRef]

28. Shin, S.-W.; Kim, Y.-S.; Kim, Y.-H.; Kim, H.-T.; Eum, K.-S.; Hong, S.-R.; Kang, H.-J.; Park, K.-H.; Yoon, M.-H. Biogenic-amine contents of Korean commercial salted fishes and cabbage kimchi. *Korean J. Fish. Aquat. Sci.* **2019**, *52*, 13–18.

29. Mah, J.-H.; Kim, Y.J.; No, H.-K.; Hwang, H.-J. Determination of biogenic amines in *kimchi*, Korean traditional fermented vegetable products. *Food Sci. Biotechnol.* **2004**, *13*, 826–829.

30. Jin, Y.H.; Lee, J.H.; Park, Y.K.; Lee, J.-H.; Mah, J.-H. The occurrence of biogenic amines and determination of biogenic amine-producing lactic acid bacteria in *Kkakdugi* and *Chonggak* kimchi. *Foods* **2019**, *8*, 73. [CrossRef]

31. Lee, J.-H.; Jin, Y.H.; Park, Y.K.; Yun, S.J.; Mah, J.-H. Formation of biogenic amines in *Pa* (green onion) kimchi and *Gat* (mustard leaf) kimchi. *Foods* **2019**, *8*, 109. [CrossRef]

32. Halász, A.; Baráth, Á.; Holzapfel, W.H. The influence of starter culture selection on sauerkraut fermentation. *Z. Lebensm. Unters. Forsch.* **1999**, *208*, 434–438. [CrossRef]

33. Kalač, P.; Špička, J.; Křížek, M.; Steidlová, Š.; Pelikánová, T. Concentrations of seven biogenic amines in sauerkraut. *Food Chem.* **1999**, *67*, 275–280. [CrossRef]

34. Lee, K.J. Westerner's view of Korean food in modern period-centering on analyzing Westerners' books. *Korean J. Food Cult.* **2013**, *28*, 356–370. [CrossRef]

35. Taylor, S.L.; Leatherwood, M.; Lieber, E.R. Histamine in sauerkraut. *J. Food Sci.* **1978**, *43*, 1030–1032. [CrossRef]

36. Mouritsen, O.G. *Tsukemono*—Crunchy pickled foods from Japan: A case study of food design by gastrophysics and nature. *Int. J. Food Des.* **2018**, *3*, 103–124. [CrossRef]

37. Handa, A.; Kawanabe, H.; Ibe, A. Content and origin of nonvolatile amines in various commercial pickles. *J. Food Hyg. Soc. Jpn.* **2018**, *59*, 36–44. [CrossRef] [PubMed]

38. Kung, H.-F.; Lee, Y.-H.; Teng, D.-F.; Hsieh, P.-C.; Wei, C.-I.; Tsai, Y.-H. Histamine formation by histamine-forming bacteria and yeast in mustard pickle products in Taiwan. *Food Chem.* **2006**, *99*, 579–585. [CrossRef]

39. Oh, S.-C. Influences of squid ink added to low-salted squid *Jeot-gal* on its proteolytic characteristics. *J. Korean Oil Chem. Soc.* **2013**, *30*, 348–355. [CrossRef]

40. Park, D.-C.; Kim, E.-M.; Kim, E.-J.; Kim, Y.-M.; Kim, S.-B. The contents of organic acids, nucleotides and their related compounds in *kimchi* prepared with salted-fermented fish products and their alternatives. *Korean J. Food Sci. Technol.* **2003**, *35*, 769–776.

41. Park, C.L.; Kwon, Y.M. A study on the kimchi recipe in the early Joseon Dynasty through [*Juchochimjeobang*]. *J. Korean Soc. Food Cult.* **2017**, *32*, 333–360.

42. Cha, Y.-J.; Lee, Y.-M.; Jung, Y.-J.; Jeong, E.-J.; Kim, S.-J.; Park, S.-Y.; Yoon, S.-S.; Kim, E.-J. A nationwide survey on the preference characteristics of minor ingredients for winter *kimchi*. *Korean J. Food Nutr.* **2003**, *32*, 555–561.

43. Lee, C.-H.; Lee, E.-H.; Lim, M.-H.; Kim, S.-H.; Chae, S.-K.; Lee, K.-W.; Koh, K.-H. Characteristics of Korean fish fermentation technology. *Korean J. Food Cult.* **1986**, *1*, 267–278.

44. Ha, S.-D.; Kim, A.-J. Technological trends in safety of jeotgal. *Korean J. Food Nutr.* **2005**, *38*, 46–64.

45. Kim, S.-M.; Lee, K.-T. The shelf-life extension of low-salted Myungran-Jeot 1. The effects of pH control on the shelf-life of low-salted Myungran-Jeot. *Korean J. Fish. Aquat. Sci.* **1997**, *30*, 459–465.

46. Um, I.-S.; Seo, J.-K.; Kim, H.-D.; Park, K.-S. The quality of commercial salted and fermented anchovy *Engraulis japonicas* sauces produced in Korea. *Korean J. Fish. Aquat. Sci.* **2018**, *51*, 667–672.

47. Mah, J.-H.; Han, H.-K.; Oh, Y.-J.; Kim, M.-G.; Hwang, H.-J. Biogenic amines in Jeotkals, Korean salted and fermented fish products. *Food Chem.* **2002**, *79*, 239–243. [CrossRef]

48. Moon, J.S.; Kim, Y.; Jang, K.I.; Cho, K.-J.; Yang, S.-J.; Yoon, G.-M.; Kim, S.-Y.; Han, N.S. Analysis of biogenic amines in fermented fish products consumed in Korea. *Food Sci. Biotechnol.* **2010**, *19*, 1689–1692. [CrossRef]

49. Cho, Y.-J.; Lee, H.-H.; Kim, B.-K.; Gye, H.-J.; Jung, W.-Y.; Shim, K.-B. Quality evaluation to determine the grading of commercial salt-fermented fish sauce in Korea. *J. Fish. Mar. Sci. Educ.* **2014**, *26*, 823–830.

50. Joung, B.C.; Min, J.G. Changes in postfermentation quality during the distribution process of anchovy (*Engraulis japonicus*) fish sauce. *J. Food Prot.* **2018**, *81*, 969–976. [CrossRef]

51. Saaid, M.; Saad, B.; Hashim, N.H.; Ali, A.S.M.; Saleh, M.I. Determination of biogenic amines in selected Malaysian food. *Food Chem.* **2009**, *113*, 1356–1362. [CrossRef]

52. Rosma, A.; Afiza, T.S.; Wan Nadiah, W.A.; Liong, M.T.; Gulam, R.R.A. Microbiological, histamine and 3-MCPD contents of Malaysian unprocessed 'budu'. *Int. Food Res. J.* **2009**, *16*, 589–594.

53. Kim, M.-J.; Kim, K.-S. Tyramine production among lactic acid bacteria and other species isolated from kimchi. *LWT-Food Sci. Technol.* **2014**, *56*, 406–413. [CrossRef]

54. Jeong, D.-W.; Lee, J.-H. Antibiotic resistance, hemolysis and biogenic amine production assessments of *Leuconostoc* and *Weissella* isolates for kimchi starter development. *LWT-Food Sci. Technol.* **2015**, *64*, 1078–1084. [CrossRef]

55. Guan, L.; Cho, K.H.; Lee, J.H. Analysis of the cultivable bacterial community in *jeotgal*, a Korean salted and fermented seafood, and identification of its dominant bacteria. *Food Microbiol.* **2011**, *28*, 101–113. [CrossRef]

56. Jeong, D.-W.; Han, S.; Lee, J.-H. Safety and technological characterization of *Staphylococcus equorum* isolates from jeotgal, a Korean high-salt-fermented seafood, for starter development. *Int. J. Food Microbiol.* **2014**, *188*, 108–115. [CrossRef] [PubMed]

57. Lim, E.-S. Inhibitory effect of bacteriocin-producing lactic acid bacteria against histamine-forming bacteria isolated from *Myeolchi-jeot*. *Fish. Aquat. Sci.* **2016**, *19*, 42. [CrossRef]

58. Mah, J.-H.; Ahn, J.-B.; Park, J.-H.; Sung, H.-C.; Hwang, H.-J. Characterization of biogenic amine-producing microorganisms isolated from Myeolchi-jeot, Korean salted and fermented anchovy. *J. Microbiol. Biotechnol.* **2003**, *13*, 692–699.

59. Bover-Cid, S.; Hugas, M.; Izquierdo-Pulido, M.; Vidal-Carou, M.C. Amino acid-decarboxylase activity of bacteria isolated from fermented pork sausages. *Int. J. Food Microbiol.* **2001**, *66*, 185–189. [CrossRef]

60. Coton, E.; Coton, M. Evidence of horizontal transfer as origin of strain to strain variation of the tyramine production trait in *Lactobacillus brevis*. *Food Microbiol.* **2009**, *26*, 52–57. [CrossRef]

61. Lucas, P.M.; Wolken, W.A.M.; Claisse, O.; Lolkema, J.S.; Lonvaud-Funel, A. Histamine-producing pathway encoded on an unstable plasmid in *Lactobacillus hilgardii* 0006. *Appl. Environ. Microbiol.* **2005**, *71*, 1417–1424. [CrossRef]

62. Marcobal, A.; de las Rivas, B.; Moreno-Arribas, M.V.; Munoz, R. Evidence for horizontal gene transfer as origin of putrescine production in *Oenococcus oeni* RM83. *Appl. Environ. Microbiol.* **2006**, *72*, 7954–7958. [CrossRef]

63. Wolken, W.A.M.; Lucas, P.M.; Lonvaud-Funel, A.; Lolkema, J.S. The mechanism of the tyrosine transporter TyrP supports a proton motive tyrosine decarboxylation pathway in *Lactobacillus brevis*. *J. Bacteriol.* **2006**, *188*, 2198–2206. [CrossRef]

64. Kim, S.-H.; Kang, K.H.; Kim, S.H.; Lee, S.; Lee, S.-H.; Ha, E.-S.; Sung, N.-J.; Kim, J.G.; Chung, M.J. Lactic acid bacteria directly degrade *N*-nitrosodimethylamine and increase the nitrite-scavenging ability in kimchi. *Food Control* **2017**, *71*, 101–109. [CrossRef]

65. Lee, G.-I.; Lee, H.-M.; Lee, C.-H. Food safety issues in industrialization of traditional Korean foods. *Food Control* **2012**, *24*, 1–5. [CrossRef]

66. Kang, H.-W. Characteristics of *kimchi* added with anchovy sauce from heat and non-heat treatments. *Culin. Sci. Hosp. Res.* **2013**, *19*, 49–58.
67. Becker, K.; Southwick, K.; Reardon, J.; Berg, R.; MacCormack, J.N. Histamine poisoning associated with eating tuna burgers. *JAMA* **2001**, *285*, 1327–1330. [CrossRef] [PubMed]
68. Mah, J.-H.; Kim, Y.J.; Hwang, H.-J. Inhibitory effects of garlic and other spices on biogenic amine production in *Myeolchi-jeot*, Korean salted and fermented anchovy product. *Food Control* **2009**, *20*, 449–454. [CrossRef]
69. Mah, J.-H.; Hwang, H.-J. Effects of food additives on biogenic amine formation in *Myeolchi-jeot*, a salted and fermented anchovy (*Engraulis japonicus*). *Food Chem.* **2009**, *114*, 168–173. [CrossRef]
70. Mah, J.-H.; Hwang, H.-J. Inhibition of biogenic amine formation in a salted and fermented anchovy by *Staphylococcus xylosus* as a protective culture. *Food Control* **2009**, *20*, 796–801. [CrossRef]
71. Kim, S.-H.; Kim, S.H.; Kang, K.H.; Lee, S.; Kim, S.J.; Kim, J.G.; Chung, M.J. Kimchi probiotic bacteria contribute to reduced amounts of N-nitrosodimethylamine in lactic acid bacteria-fortified kimchi. *LWT-Food Sci. Technol.* **2017**, *84*, 196–203. [CrossRef]
72. Ku, K.H.; Sunwoo, J.Y.; Park, W.S. Effects of ingredients on the its quality characteristics during kimchi fermentation. *J. Korean Soc. Food Sci. Nutr.* **2005**, *34*, 267–276.
73. Jin, H.S.; Kim, J.B.; Yun, Y.J.; Lee, K.J. Selection of kimchi starters based on the microbial composition of kimchi and their effects. *J. Korean Soc. Food Sci. Nutr.* **2008**, *37*, 671–675. [CrossRef]

Permissions

The contributors of this book come from diverse backgrounds, making this book a truly international effort. This book will bring forth new frontiers with its revolutionizing research information and detailed analysis of the nascent developments around the world.

We would like to thank all the contributing authors for lending their expertise to make the book truly unique. They have played a crucial role in the development of this book. Without their invaluable contributions this book wouldn't have been possible. They have made vital efforts to compile up to date information on the varied aspects of this subject to make this book a valuable addition to the collection of many professionals and students.

This book was conceptualized with the vision of imparting up-to-date information and advanced data in this field. To ensure the same, a matchless editorial board was set up. Every individual on the board went through rigorous rounds of assessment to prove their worth. After which they invested a large part of their time researching and compiling the most relevant data for our readers.

The editorial board has been involved in producing this book since its inception. They have spent rigorous hours researching and exploring the diverse topics which have resulted in the successful publishing of this book. They have passed on their knowledge of decades through this book. To expedite this challenging task, the publisher supported the team at every step. A small team of assistant editors was also appointed to further simplify the editing procedure and attain best results for the readers.

Apart from the editorial board, the designing team has also invested a significant amount of their time in understanding the subject and creating the most relevant covers. They scrutinized every image to scout for the most suitable representation of the subject and create an appropriate cover for the book.

The publishing team has been an ardent support to the editorial, designing and production team. Their endless efforts to recruit the best for this project, has resulted in the accomplishment of this book. They are a veteran in the field of academics and their pool of knowledge is as vast as their experience in printing. Their expertise and guidance has proved useful at every step. Their uncompromising quality standards have made this book an exceptional effort. Their encouragement from time to time has been an inspiration for everyone.

The publisher and the editorial board hope that this book will prove to be a valuable piece of knowledge for researchers, students, practitioners and scholars across the globe.

List of Contributors

Pierina Visciano, Maria Schirone and Antonello Paparella
Faculty of Bioscience and Technology for Food, Agriculture and Environment, University of Teramo, Via R. Balzarini, 1, 64100 Teramo, Italy

Jun-Hee Lee, Young Hun Jin, Young Kyoung Park, Se Jin Yun and Jae-Hyung Mah
Department of Food and Biotechnology, Korea University, 2511 Sejong-ro, Sejong 30019, Korea

Elvira S. Plakidi, Niki C. Maragou, Marilena E. Dasenaki, Nikolaos C. Megoulas, Michael A. Koupparis and Nikolaos S. Thomaidis
Laboratory of Analytical Chemistry, Department of Chemistry, National and Kapodistrian University of Athens, Panepistimioupolis Zografou, 15771 Athens, Greece

Han-Joon Hwang
Department of Food and Biotechnology, Korea University, 2511 Sejong-ro, Sejong 30019, Korea

Carla Daniela Di Mattia, Dino Mastrocola, Maria Martuscelli and Clemencia Chaves-Lopez
Faculty of Bioscience and Technology for Food, Agriculture and Environment, University of Teramo, Via R. Balzarini 1, 64100 Teramo, Italy

Johannes Delgado-Ospina
Faculty of Bioscience and Technology for Food, Agriculture and Environment, University of Teramo, Via R. Balzarini 1, 64100 Teramo, Italy
Grupo de Investigación Biotecnología, Facultad de Ingeniería, Universidad de San Buenaventura Cali, Carrera 122 # 6-65, Cali 76001, Colombia

Sònia Sánchez-Pérez, Oriol Comas-Basté, Judit Rabell-González, M. Teresa Veciana-Nogués, M. Luz Latorre-Moratalla and M. Carmen Vidal-Carou
Departament de Nutrició, Ciències de l'Alimentació i Gastronomia, Facultat de Farmàcia i Ciències de l'Alimentació, Universitat de Barcelona (UB), Av. Prat de la Riba 171, 08921 Santa Coloma de Gramenet, Spain

Institut de Recerca en Nutrició i Seguretat Alimentària (INSA•UB), Universitat de Barcelona (UB), Av. Prat de la Riba 171, 08921 Santa Coloma de Gramenet, Spain
Xarxa de Referència en Tecnologia dels Aliments de la Generalitat de Catalunya (XaRTA), C/ Baldiri Reixac 4, 08028 Barcelona, Spain

Sigrid Mayrhofer, Julia-Maria Schmidt, Ulrike Zitz and Konrad J. Domig
Institute of Food Science, Department of Food Science and Technology, BOKU - University of Natural Resources and Life Sciences Vienna, Muthgasse 18, A-1190 Vienna, Austria

Dalin Ly
Institute of Food Science, Department of Food Science and Technology, BOKU - University of Natural Resources and Life Sciences Vienna, Muthgasse 18, A-1190 Vienna, Austria
Faculty of Agro-Industry, Department of Food Biotechnology, RUA - Royal University of Agriculture, Dangkor District, Phnom Penh, Cambodia

Alberto Fernández-Reina
Departamento de Biología Molecular y Bioquímica, Facultad de Ciencias, Universidad de Málaga, 29071 Málaga, Spain

José Luis Urdiales and Francisca Sánchez-Jiménez
Departamento de Biología Molecular y Bioquímica, Facultad de Ciencias, Universidad de Málaga, 29071 Málaga, Spain
CIBER de Enfermedades Raras & IBIMA, Instituto de Salud Carlos III, 29010 Málaga, Spain

Bo Young Byun and Junsu Lee
Department of Food and Biotechnology, Korea University, 2511 Sejong-ro, Sejong 30019, Korea

Kwangcheol Casey Jeong
Department of Animal Sciences, University of Florida, Gainesville, FL 32611, USA
Emerging Pathogens Institute, University of Florida, Gainesville, FL 32611, USA

Rosanna Tofalo, Giorgia Perpetuini, Anna Chiara Manetta, Paola Di Gianvito, Fabrizia Tittarelli, Noemi Battistelli, Aldo Corsetti, Giovanna Suzzi and Giuseppe Martino
Faculty of Bioscience and Technology for Food, Agriculture and Environment, University of Teramo, Via R. Balzarini, 1, 64100 Teramo, Italy

Annvalisa Serio, Jessica Laika, Francesca Maggio, Giampiero Sacchetti, Flavio D'Alessandro, Chiara Rossi and Clemencia Chaves-López
Faculty of Bioscience and Technology for Food, Agriculture and Environment, University of Teramo, via Balzarini 1, 64100 Teramo, Italy

Jae Hoan Lee and Young Kyung Park
Department of Food and Biotechnology, Korea University, 2511 Sejong-ro, Sejong 30019, Korea

Index

Printed in the USA
CPSIA information can be obtained
at www.ICGtesting.com
JSHW060010020124
54623JS00006B/115

9 781641 168380